An Approach for Assessing U.S. Gulf Coast Ecosystem Restoration

A Gulf Research Program Environmental Monitoring Report

Committee on Long-Term Environmental Trends in the Gulf of Mexico

Gulf Research Program

A Consensus Study Report of

The National Academies of
SCIENCES · ENGINEERING · MEDICINE

THE NATIONAL ACADEMIES PRESS
Washington, DC
www.nap.edu

THE NATIONAL ACADEMIES PRESS 500 Fifth Street, NW Washington, DC 20001

This activity was supported by the Gulf Research Program of the National Academies of Sciences, Engineering, and Medicine. Any opinions, findings, conclusions, or recommendations expressed in this publication do not necessarily reflect the views of any organization or agency that provided support for the project.

International Standard Book Number-13: 978-0-309-26339-9
International Standard Book Number-10: 0-309-26339-5
Digital Object Identifier: https://doi.org/10.17226/26335

This publication is available from the National Academies Press, 500 Fifth Street, NW, Keck 360, Washington, DC 20001; (800) 624-6242 or (202) 334-3313; http://www.nap.edu.

Cover image: Researcher working in a marsh in Mobile Bay, Alabama, courtesy of Bethany Carl Kraft.

Printed in the United States of America

Suggested citation: National Academies of Sciences, Engineering, and Medicine. 2022. *An Approach for Assessing U.S. Gulf Coast Ecosystem Restoration: A Gulf Research Program Environmental Monitoring Report*. Washington, DC: The National Academies Press. https://doi.org/10.17226/26335.

The National Academies of
SCIENCES • ENGINEERING • MEDICINE

COMMITTEE ON LONG-TERM ENVIRONMENTAL TRENDS IN THE GULF OF MEXICO

HOLLY GREENING (*Chair*), CoastWise Partners
WALTER R. BOYNTON, Chesapeake Biological Laboratory, University of Maryland
HEIDA L. DIEFENDERFER, Coastal Sciences Division, Pacific Northwest National Laboratory
ALBERT A. GEORGE II, South Carolina Aquarium
KENNETH L. HECK, JR., Dauphin Island Sea Lab and University of South Alabama
BARBARA A. KLEISS, Department of River-Coastal Science and Engineering, Tulane University
CATHERINE L. KLING (NAS[1]), Dyson School of Applied Economics and Management, Cornell University
BETHANY A. CARL KRAFT, Volkert, Inc.
LARRY D. MCKINNEY, Harte Research Institute for Gulf of Mexico Studies, Texas A&M University
DEEPAK R. MISHRA, Department of Geography, University of Georgia
KODI GUILLORY, Sustainable Design Solutions[2]

National Academies of Sciences, Engineering, and Medicine Staff

DEBORAH GLICKSON, Director, Board on Earth Sciences and Resources and Water Science and Technology Board
LAURA WINDECKER, Program Officer, Gulf Research Program
MEGAN MAY, Associate Program Officer, Ocean Studies Board (until January 2022)
THELMA COX, Program Coordinator, Gulf Research Program

[1] National Academy of Sciences.
[2] Resigned from the committee in October 2020.

GULD RESEARCH PROGRAM

Division Committee

DAVID E. DANIEL (NAE[1]), Division Chair, University of Texas at Dallas
LAUREN ALEXANDER AUGUSTINE, Division Executive Director, Gulf Research Program
R. LYN ARSCOTT (NAE), International Association of Oil Gas Producers (Retired)
REGINA M. BENJAMIN (NAM[2]), Gulf States Health Policy Center
THOMAS P. BOSTICK (NAE), Bostick Global Strategies
VADM MANSON K. BROWN, U.S. Coast Guard (Retired)
BARBARA ENTWISLE, University of North Carolina
MARY L. LANDRIEU, Van Ness Feldman LLP
KERSTIN A. LEHNERT, Columbia University
MONICA MASON, Core Specialty Insurance
SARA N. ORTWEIN (NAE), Exxon Mobil Corporation (Retired)
AMELIE G. RAMIREZ (NAM), Salud America!, The University of Texas Health Science Center at San Antonio
FRANCIS K. WIESE, Stantec Consulting Services, Inc.
ROY E. WRIGHT, Insurance Institute for Business and Home Safety

Gulf Environmental Protection and Stewardship Board

Staff

DONALD BOESCH, Senior Scholar
LAURA WINDECKER, Program Officer
MEGHA KHADKA, Research Associate

---
[1] National Academy of Engineering.
[2] National Academy of Medicine.

Reviewers

This Consensus Study Report was reviewed in draft form by individuals chosen for their diverse perspectives and technical expertise. The purpose of this independent review is to provide candid and critical comments that will assist the National Academies of Sciences, Engineering, and Medicine in making each published report as sound as possible and to ensure that it meets the institutional standards for quality, objectivity, evidence, and responsiveness to the charge. The review comments and draft manuscript remain confidential to protect the integrity of the process.

We thank the following individuals for their review of this report:

Marcus Beck, Tampa Bay Estuary Program
Kim de Mutsert, University of Southern Mississippi
Judy Haner, The Nature Conservancy
Heidi Nepf, Massachusetts Institute of Technology
Stephen Polasky (NAS[1]), University of Minnesota
Gregory Steyer, U.S. Geological Survey
Paul Tschirky, APTIM Corp.
Robert Twilley, Louisiana State University

Although the reviewers listed above provided many constructive comments and suggestions, they were not asked to endorse the content of the report nor did they see the final draft before its release. The review of this report was overseen by **Barbara Schaal** (NAS), Washington University in St. Louis, and **John Boland**, Johns Hopkins University. They were responsible for making certain that an independent examination of this report was carried out in accordance with the standards of the National Academies and that all review comments were carefully considered. Responsibility for the final content rests entirely with the authoring committee and the National Academies.

[1] National Academy of Sciences.

Preface

On April 20, 2010, the *Deepwater Horizon* (DWH) drilling rig exploded in the Gulf of Mexico, resulting in the tragic loss of 11 lives and the discharge of at least 3 million barrels of oil. Oil reached over 2,100 km of coastline Gulf-wide and impacted all five U.S. Gulf states. The resulting civil and criminal litigation from the DWH oil spill led to over $16 billion in fines and penalties to be applied to economic recovery and environmental restoration-related activities in the Gulf region. Funds are administered through three major programs: the DWH Natural Resource Damage Assessment Trustees, the National Fish and Wildlife Foundation Gulf Environmental Benefit Fund, and the Gulf Coast Ecosystem Restoration Council (commonly known as the RESTORE Council).

As part of multiple legal settlements, the federal government requested that the National Academies of Sciences, Engineering, and Medicine (National Academies) establish the Gulf Research Program (GRP) to fund and conduct activities to advance three program areas: offshore energy safety; health and resilience; and environmental protection and stewardship.

The goals of the restoration activities are broader than recovery from the oil spill impacts alone, encompassing land acquisition, restoration of coastal and offshore habitats and the Gulf ecosystem, recovery of species, and water quality improvement. More than 10 years after the DWH explosion the Gulf Coast continues to recover from the impacts of the oil spill, and a multitude of academic studies, agency reports, and nongovernmental organizations assessments track ongoing impacts and the effects of recovery efforts. The tracking process is complicated by the presence of long-term background trends, such as those associated with climate change and land use changes, which can obscure the effects of the spill and subsequent restoration efforts.

In early 2020, GRP initiated this study, which is intended to be the first study in a series on environmental protection and stewardship. The study series is focused on long-term environmental trends in the Gulf and is intended to advance GRP's strategic approach to monitor progress and change and document how environmental conditions in the Gulf evolve over time. This report addresses monitoring and assessment of the cumulative effects of restoration projects beyond the project scale within the context of long-term environmental change. The study scope builds on research results and publications from many sources, including the previous National Academies reports Effective Monitoring to Evaluate Ecological Restoration in the Gulf of Mexico; Understanding the Long-Term Evolution of the Coupled Natural-Human System: The Future of the U.S. Gulf Coast; and the Progress toward Restoring the Everglades report series. The committee's 10 members brought to the study expertise in a variety of fields, coastal ecosystem restoration and cumulative effects; natural resource management and policy; coastal ecosystem ecology, including wetlands, benthic, and fisheries ecology; water quality; ecosystem modeling; remote sensing and emerging technologies; river science and engineering; environmental economics; and data management and synthesis.

The study scope was broadly defined, with an option of focusing on restoration projects within a geographic region to keep the scope manageable. After deliberation, the committee decided that the geographic scope should include all five U.S. Gulf states, with a focus on assessing the cumulative effects of landscape-scale efforts encompassing multiple restoration projects in coastal areas. The committee hopes to reach a broad audience, including regional program managers, state resource managers, federal agencies, and other Gulf-wide entities with an interest in restoration. Entities particularly positioned to consider the recommendations of this report include those entities funded by the DWH settlements and agreements.

The committee held four information-gathering meetings in 2020 and multiple meetings in closed session in 2021 to develop this report. All meetings and discussions were held virtually due to the COVID-19 pandemic. The committee members heard presentations from representatives from state and federal agencies, nonprofit organizations, and academia. Speakers shared their knowledge and expertise in Gulf environmental trends, restoration, ecology, and DWH settlement-funded restoration programs.

Throughout 18 months of deliberations, the committee developed an increasing sense of urgency to encourage the synthesis of data and information already collected and use that information to inform future DWH settlement–funded restoration efforts. The DWH Project Tracker website reports that more than 570 environmental restoration projects have been completed or are under way, including 152 focused on habitat restoration and enhancement, 82 on species restoration, and 47 focused on water quality restoration and maintenance (see https://dwhprojecttracker.org/). With committed and expended DWH settlement recovery funds approaching half of the total amount available, and data and information from completed restoration projects becoming available, synthesis and analysis of successful and (especially) less successful efforts is both timely and essential to ensure effective restoration efforts and wise use of the remaining restoration funds. The committee members recognize the challenges facing the Gulf Coast environmental restoration community (several from personal experience), not just recovery from the DWH oil spill, but also from multiple hurricanes and other climatic events. I continue to be amazed at the Gulf Coast communities' hard work and resiliency in making progress on the recovery efforts in the face of these difficult conditions. Our conclusions and recommendations are provided to assist in supporting successful restoration efforts now and in the future.

This report is the result of the collective expertise and experience of some of the nation's leading experts in environmental restoration theory and application. I want to express my deep appreciation to every member of the committee for their insight and expertise, as well as their humor, collegiality, and commitment to our collective effort. Consensus is not easy, and all have contributed to ensuring a strong and consistent message. Thank you.

I also want to acknowledge the many individuals who assisted the committee and the National Academies staff in their task to create this report. Over the course of our four virtual information-gathering meetings, the committee engaged with interested colleagues in academia, nongovernmental organizations, and government. The committee greatly appreciated the chance to learn from the participants at these meetings.

On behalf of the entire committee, kudos to our outstanding National Academies staff for their excellent support, guidance, and contributions to the report. Study directors Laura Windecker (Gulf Research Program), Deborah Glickson (Board on Earth Sciences and Resources/Water Science and Technology Board), and Megan May (Ocean Studies Board), as well as program coordinator Thelma L. Cox (Gulf Research Program) were instrumental in coordinating and guiding the committee's progress throughout and in producing the final report. I have thoroughly enjoyed working with such a professional and dedicated team.

Holly Greening, Chair
Committee on Long-Term Environmental Trends in the Gulf of Mexico

Contents

Summary

The Gulf of Mexico (GoM) has a combination of ecological richness, economic value, and physical location that makes it unique among seas. Its habitats and ecosystems include oyster reefs, salt marshes, seagrass beds, mangrove forests, estuaries, barrier islands, coral reefs, and sandy beaches, as well as the water column and bottom habitats of the Gulf itself. The 21st-century GoM is also vulnerable to ecosystem pressures and stressors, such as altered freshwater flows, coastal development, pollution, overfishing, and landscape alterations. It is further susceptible to the effects of fundamental natural and anthropogenic driving forces, such as climate change.

When the *Deepwater Horizon* (DWH) platform explosion and resulting oil spill occurred in April 2010, it resulted in yet another substantial impact to Gulf ecosystems from Texas to Florida, which were already experiencing many anthropogenic stressors. The resulting civil and criminal claims, fines, and penalties from the spill included approximately $16.7 billion for economic and environmental restoration activities.

Now, nearly a decade into these restoration efforts, there is a need to assess the impacts of these activities and to lay a foundation for restoration efforts that will continue beyond the allocation of DWH funds. The need for this effort was identified through discussions with stakeholders and representatives from several DWH funding entities,[1] and from information in publications from Gulf-based entities, nongovernmental organizations, and the National Academies of Sciences, Engineering, and Medicine.

The Gulf Research Program (GRP) at the National Academies initiated this report, with a committee of volunteer experts, to assess the cumulative effects of multiple restoration projects along the U.S. Gulf of Mexico coast in the context of long-term environmental trends; to consider effects of acute events and long-term environmental changes; to discuss synergistic and antagonistic effects of multidecadal restoration activities; and to recommend adaptive management strategies to address these factors. In addition, the committee was asked to assess the relevant existing resources, including the available data, for informing decision making and to consider what additional efforts are needed (see Box 1.1 in Chapter 1 for the full Statement of Task). The geographic scope of restoration efforts considered by this report looks beyond the project scale to the estuary/watershed scales (i.e., coastal areas connected by characteristics of their hydrology) and even larger scales (e.g., regional or Gulf-wide scale).

[1] When used in this report, the "DWH funding entities" are defined as the following: the RESTORE (Resources and Ecosystems Sustainability, Tourist Opportunities, and Revived Economies) Council state and federal members (states of Alabama, Florida, Louisiana, Mississippi, and Texas; U.S. Department of Commerce; U.S. Department of the Interior (DOI); U.S. Environmental Protection Agency (EPA); U.S. Department of Agriculture (USDA); U.S. Coast Guard; and U.S. Department of the Army); the DWH NRDA (Natural Resource Damage Assessment) Trustees (National Oceanic and Atmospheric Administration [NOAA], EPA, DOI, USDA, and the five Gulf states); the National Fish and Wildlife Foundation Gulf Environmental Benefit Fund; the U.S. Fish and Wildlife Service North American Wetlands Conservation Fund; the Centers of Excellence; the NOAA RESTORE Science Program; and the Gulf Research Program.

LONG-TERM ENVIRONMENTAL TRENDS AFFECTING RESTORATION EFFORTS

Large-scale and long-term changes in hydrology, geomorphology, biogeochemistry, trophic status, and species range distributions have been occurring within the GoM in response to both natural and anthropogenic forces. Further, these forces are themselves impacted by system-wide stressors, such as those associated with the DWH oil spill, and their rate of change is accelerating as a result of climate change. Assessment of long-term environmental trends that describe the changing conditions of an ecosystem are essential for providing decision makers and restoration practitioners with useful insight in their assessment of cumulative effects of multiple restoration actions.

The changing climate is a major stressor in the GoM, and some of its associated stressors, such as relative sea level rise, can reflect both natural and anthropogenic forces. Overall, long-term trends show that relative sea level is rising, that intense hurricanes are increasing in frequency, that sea surface temperature is gradually increasing, that subsurface waters are acidifying at a rate that is higher than the global surface ocean rate, and that more tropical species (including mangroves and fish species) have been observed year-round in the northern GoM. Inputs of freshwater, dissolved and particulate nutrients, and suspended sediments act to influence salinity regimes, water residence times and other features of physical circulation, trophic status and keystone habitats, and water clarity and sediment accretion rates. And over previous decades, changes in land uses and human activities have modified the annual patterns and magnitude of these inputs. Current and projected climate change effects will further influence these trends.

Better understanding and assessment of long-term environmental trends is critical from a restoration planning perspective. Although some parameters needed to assess environmental trends in the GoM are well monitored, data collection and analysis for many aspects are limited. The Gulf Coast of today is unlikely to look like the Gulf Coast of the future; these changes are likely to influence the success or failure of ongoing and future GoM restoration projects.

ASSESSING THE CUMULATIVE EFFECTS OF RESTORATION: CURRENT AND EMERGING APPROACHES

Estuary- and watershed-scale restorations in the GoM are involved undertakings. Measurable changes as a result of restoration actions are often confounded by the effects of multiple interacting stressors, including long-term environmental trends. Several approaches can be used to consider and address these challenges.

Defining Cumulative Effects of Restoration The cumulative effects of restoration refer to the collective additive, synergistic, and antagonistic effects of all restoration activities that occur within a setting defined by common or connected characteristics of hydrology, geomorphology, ecology, ecological function, and/or biodiversity. Assessment of the cumulative effects of restoration may occur at various geographic landscape scales, such as a marsh complex, bay, estuary, watershed, or the Gulf Coast itself. The scale of assessment may also be defined by specific interests in the outcomes, such as ecosystem processes (e.g., sedimentation), biodiversity or specific organisms (e.g., oysters), performance targets (e.g., water quality), type of restored system (e.g., wetland restoration), political boundaries (e.g., state boundaries), or type of restoration method (e.g., living shorelines).

Antagonism and Synergism in Restoration Efforts Diverse pressures in estuarine and coastal waters, both natural and anthropogenic, generate multiple stresses on ecosystem structure and function. The effect of those multiple stressors can be additive (equal to the sum of their individual effects), synergistic (greater than the sum of their individual effects), or antagonistic (less than the sum of their individual effects); they may also be judged either beneficial or detrimental relative to program goals and objectives. The cumulative effects of restoration efforts in a given estuary or watershed may also be additive, synergistic, or antagonistic and similarly judged beneficial or detrimental. Exploring how to make use of ecological synergies and avoid antagonistic interactions could improve benefits of multi-project restoration efforts.

Conceptual Models and Hypothesis Development Conceptual models—graphical representations of interrelationships between drivers, pressures and stressors, restoration actions, and ecosystem response, based on one or more hypotheses—are often used to represent understanding of the current and future states of the ecosystem. They are crucial for determining restoration project priorities and assessing future projections. Preparing a conceptual model can enhance understanding of the current state of an ecosystem and raise informative questions about underlying assumptions. Their applicability to the evaluation of cumulative effects of large-scale restoration bears emphasis. For example, a conceptual model can capture the potential synergistic and antagonistic effects that occur as a result of interactions among habitats being restored, knowledge that may be tested by future studies and assessment of the effectiveness of actions over time.

Multiple Lines of Evidence—An Approach for Assessing Cumulative Effects Evaluating the effects of a restoration effort often involves an individual body of water or watershed and is therefore unreplicable. This means that the usual experimental design with which most ecologists and environmental scientists are familiar, with randomly allocated treatments and replication, is not possible. However, this does not mean that rigorous analysis of system-wide restoration projects cannot be done, only that strict assignment of cause and effect cannot be made via standard methods of statistical analysis alone. An evidence-based evaluation methodology that utilizes multiple lines of evidence and causal criteria can compensate for the inability to use traditional experimental designs, the lack of reference conditions, the lack of replication, the difficulties in establishing causality, and the likely shortage of appropriate data.

Tools for gathering multiple lines of evidence include:

- **research on critical ecological uncertainties**, which is important for avoiding unexpected outcomes;
- **evidence-based review of the literature**, which is a systematic approach to assessing environmental cause and effect via information synthesized from multiple publications;
- **physics-based and ecosystem models**, which encompass both the flow of materials and energy in addition to capturing complex interactions among ecosystem components, processes, and services;
- **meta-analysis of restoration action effectiveness**, which includes the assessment of interim reports and data in lieu of published scientific literature when the timing of publication in scientific literature is inadequate for decision making;
- **analysis of data and modeling of target species**—for example, population models that predict or simulate population dynamics of species within an ecosystem due to changes in habitat characteristics from a set of pressures and stressors;
- **models for cumulative net ecosystem improvement**, which are methods for estimating whether and to what degree an ecosystem may or may not have improved as a result of an intervention such as restoration; and
- **change analysis on the landscape setting**, using a suite of tools that rely on data-driven models and that can be quite effective in teasing apart complex relationships among ecosystem stressors and responses.

APPLICATIONS OF SYNTHESIS AND CUMULATIVE EFFECTS ASSESSMENT IN THE GULF OF MEXICO

The projects supported by the legal settlements of the 2010 DWH oil spill represent an opportunity and a challenge for assessment and learning because of the unprecedented number and diversity of projects and organizations involved and the 5–30-year time frame of payments from the settlements. Individual project monitoring can help ensure that what is learned from initial projects benefits successive projects. The ongoing, large-scale, long-term restoration is also an opportunity to assess cumulative impacts of multiple projects on estuary/watershed and larger scales.

Prior Assessments Quantitative analyses of the effectiveness of different types of restoration, such as the effects of seagrass or oyster reef restoration on a local-bay or statewide basis, remain uncommon in the GoM. The committee was unsuccessful in obtaining enough information to quantitatively demonstrate the feasibility of synthesizing DWH-funded project-level monitoring data and information for the assessment of cumulative effects and achievement of restoration objectives. Despite numerous calls from various committees and organizations for consistent and transparent monitoring and assessment since the DWH oil spill, barriers exist for the collation of monitoring and assessment of successes or failures of restoration projects. These barriers include varied or nonexistent monitoring requirements, depending on funding stream, and a lack of publicly available final reports and/or monitoring data. In addition, examples of syntheses in the GoM that have attempted to detect beneficial cumulative effects of multiple restoration efforts at the estuary/watershed- and Gulf-wide scales are very limited.

Several long-term, science-based resource management programs located in GoM estuaries offer examples of the application of multiple lines of evidence to assess cumulative effects of the many restoration actions implemented in each area over the last three decades. Although not designed to assess cumulative effects, the Tampa Bay Estuary Program and its partners, the Galveston Bay National Estuary Program in partnership with the Galveston Bay Foundation, and the Comprehensive Everglades Restoration Plan each utilized tools and techniques similar to the theoretical approaches outlined above. In addition, each program developed and now applies program-specific approaches to assessing cumulative effects of restoration efforts within their study areas.

MOVING FORWARD

Because GoM ecosystems cross many political boundaries, coordination across geographic and jurisdictional lines is also needed. There are several actions to consider that would contribute to supporting the scientific efforts (e.g., monitoring, modeling, and research) needed for effective adaptive management and, in turn, assessment of cumulative effects of restoration efforts Gulf-wide.

Data Resources Long-term environmental trends for some parameters and species are now available at local and regional scales, but data collection, analysis, and reporting are often inconsistent, and existing efforts are not adequate to detect all important Gulf-wide trends. Possessing baseline and trend data for important environmental variables when evaluating restoration efforts provides fundamental support for the synthesis activities needed to inform cumulative effects assessment and adaptive management actions. Building on existing monitoring efforts, filling in known data gaps, and moving toward a Gulf-wide ecosystem monitoring network could help practitioners move forward. State and federal resource agencies also collect long-term environmental monitoring data specific to their agency and state missions, and this information can be useful for assessing the cumulative impact of multiple restoration projects at estuarine or larger scales.

However, such efforts are hampered by the lack of a unifying GoM analysis and synthesis activity for many key stressors. No one entity has the resources and the explicit responsibility to accomplish this objective. Thus, this type of analysis and synthesis activity, so essential for accurate assessments of cumulative effects of large-scale restoration activities, remains to be undertaken.

Finally, it is important that all data and information regarding restoration be available to all users, regardless of where data are deposited. FAIR principles (Findable, Accessible, Interoperable, Reusable; e.g., Wilkinson et al., 2016) are widely accepted and can provide guidance for data management and stewardship.

Emerging Technologies Recent advancements in data-driven techniques, such as artificial intelligence (AI), machine learning, deep learning, and cloud and edge computing, are expected to fundamentally transform many domains of human endeavor, including post-restoration monitoring. Traditional remote sensing, combined with new sensing technologies and AI-driven techniques, can generate high-quality, long-term monitoring data across terrestrial and coastal ecosystems. Although some large-scale remote sensing studies to monitor GoM-wide water quality and wetland habitats have been conducted in the past, the committee could not identify any integrated remote sensing and emerging-technology-driven monitoring studies adopted for Gulf restoration projects. These newer data-driven frameworks are not expected to replace traditional restoration science–driven ecosystem monitoring, but they can complement and strengthen them.

Program-Level Adaptive Management Strategies Environmental background trends, especially those associated with climate change, are exhibiting higher variability over time; in turn, restoration practices that have been successful in the past may no longer be adequate to compensate for the effects of anticipated changes in background trends. Adaptive management techniques can provide restoration program managers with the ability to revisit and update large-scale restoration strategies, based on periodic review of monitoring data and progress toward programmatic goals. However, successful implementation of adaptive management in Gulf restoration has been limited; this lack of success is not isolated to the region, but there are examples of successful implementation elsewhere.

The Importance of Data Synthesis Synthesis efforts are needed to determine how much the many localized restoration efforts, collectively, have resulted in measurably improved coastal and estuarine ecosystems across the GoM region. In addition, such analyses provide a mechanism for adjusting efforts to produce better restoration outcomes. A synthesis framework makes it possible to address difficult and exceedingly complex environmental questions and provide answers that lead to increased understanding of coastal and estuarine system dynamics and, ultimately, better management decisions. Synthesis at scales relevant to management groups is also particularly needed in the Gulf because of strong and concerning trends in both chronic (e.g., sea level rise, tropicalization) and acute (e.g., hurricanes, floods) stressors that can, directly and indirectly, strongly influence the success of restoration projects at all scales.

A comprehensive monitoring database is needed for any synthesis activities, yet current data collection in the GoM is often inadequate for regional or Gulf-wide synthesis, and there are considerable inconsistencies in the types of variables and techniques used in monitoring programs across the Gulf. There is a lack of centralized data management, storage, and access across the many restoration projects. Funding for long-term monitoring efforts is difficult to obtain. Without serious synthesis efforts, it seems unlikely that it can be quantitatively determined whether and to what degree GoM habitats and ecosystems have improved.

Steps Needed to Assess Cumulative Effects Figure S.1 summarizes the key components needed for assessing cumulative effects of multiple restoration efforts. It includes the development of conceptual models, assessment of data needs and acquisition of additional data (if needed), implementation of the multiple lines of evidence approach, and data synthesis. The cycle can be repeated as part of adaptive management and related decision-making efforts.

The assessment of the largest ecological restoration investment in history is an unprecedented challenge and opportunity. Significant funds have been expended or committed to date on DWH-funded projects, with valuable progress having been made in advancing monitoring and modeling capabilities. The restoration community has made much progress on recovery and restoration efforts related to the DWH oil spill and from the impacts of hurricanes and other climatic events. Learning achieved through the remainder of

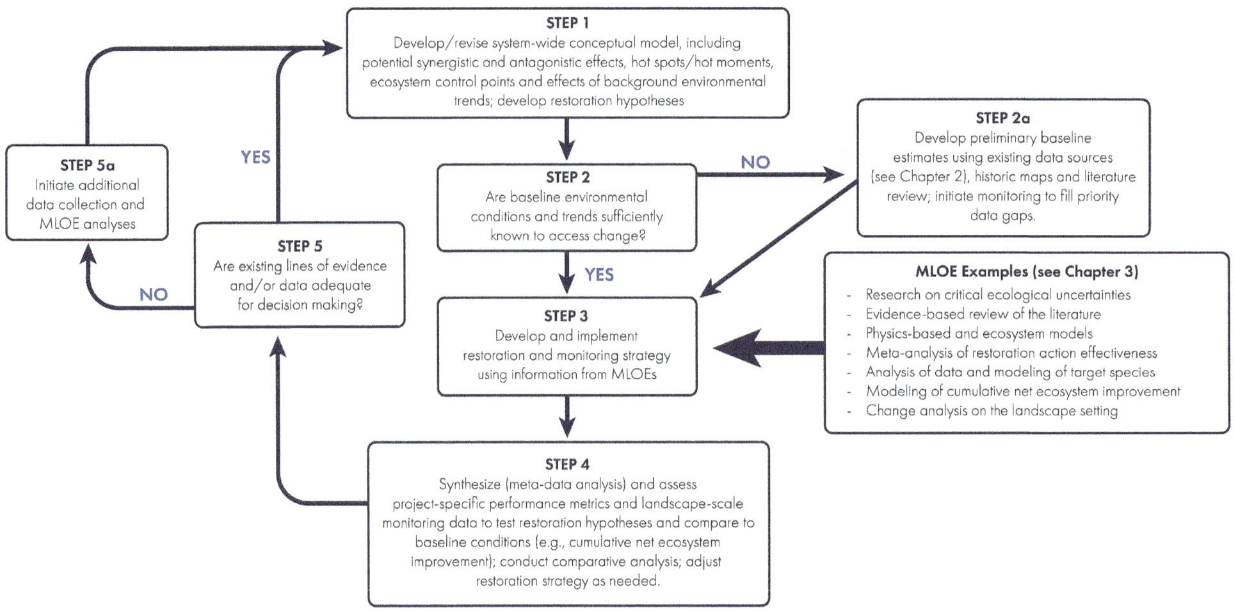

FIGURE S.1. A flow chart approach for environmental restoration to assess cumulative effects of multiple restoration projects, using multiple lines of evidence (MLOEs) and an iterative adaptive management approach. The large arrow between MLOE examples and Step 3 indicates that one or more lines of evidence can be used to develop the restoration and monitoring strategy. The steps in this approach may not be applicable for all existing and planned large-scale or multi-project restoration in the GoM, and other information and lines of evidence could be included at each step.

the settlement period will be the foundation for the next generation of managers, who inherit the responsibility for GoM ecosystems and communities. An underlying theme of this report is the need for integration of science and management of restoration activities. It is envisioned that each DWH funding entity, within its programmatic authority, can work cooperatively with others to realize this integration.

CONCLUSIONS AND RECOMMENDATIONS

Conclusion 1: Adequate scientific evidence needed to evaluate cumulative effects of restoration on a regional scale in the Gulf of Mexico is currently not available and, to date, no entity has been tasked to develop and implement a strategy to assess cumulative effects of environmental restoration efforts. Environmental benefits associated with multiple restoration projects have been observed within some Gulf of Mexico estuaries and watersheds, although not at larger scales. Without a focused effort and strategy, rigorous synthesis of the effects of multiple restoration projects at a regional or Gulf-wide scale cannot be conducted.

Conclusion 2: Because environmental changes can influence the success or failure of restoration efforts and can hinder the ability to detect potential cumulative effects of multiple restoration efforts, a thorough understanding of long-term environmental trends is essential for decision makers and restoration practitioners. Advanced monitoring techniques and approaches, including satellite remote sensing, connected sensor networks, and automation, can greatly assist in determining long-term environmental trends and assessing acute events. In addition, long-term environmental trends derived from targeted monitoring efforts can inform a range of analytical tools. The output from these tools can support the development of adaptive management actions that will subsequently improve restoration success and protect existing investments. Significant spatial and temporal gaps in monitoring Gulf of Mexico-wide environmental indicators and data collection and dissemination efforts limit development of this important and valuable capability. Specifically:

- Long-term environmental trends across the Gulf Coast states are monitored by a patchwork of government agencies, nonprofit organizations, and industries for a variety of reasons (e.g., regulatory, environmental tracking, performance evaluation). Study designs, data collection methods, analyses, and data availability vary and are often not comparable, making synthesis very difficult.
- One-time Gulf-wide monitoring studies are useful, but without periodic updates, do not generate enough information to determine long-term background trends needed for cumulative effects assessments.
- Key metrics to assess landscape-scale changes and support adaptive management include those necessary to estimate environmental trends associated with climate change; freshwater, nutrient, and sediment loading to coastal waters; land use/land cover; ambient water quality; status and trends of finfish and shellfish species, marine mammals, turtles, and birds; and primary and secondary production. For example, enhanced efforts and standardization of methods are needed for:
  - ambient water quality, the measurement of which can be enhanced by using high-spatiotemporal-resolution satellite data on chlorophyll *a*, suspended sediments, colored organic matter, and harmful algae;
  - tide gage data and subsidence measurements to estimate local, relative sea level rise;
  - estimating the extent and effects of ocean and coastal acidification, information essential for successful restoration and maintenance of commercially important shellfish and in the selection of materials for restoration efforts (e.g., oyster shells, limestone);
  - tracking, targeting needed research, and managing the effects of tropicalization on fishery species, other species, and habitats; and
  - estimating the ecological functioning of restored habitats, something not often measured even though the spatial extent of restored habitats is usually monitored and reported.

Recommendation A: Enhanced, consistent, and sustained long-term monitoring, analysis, synthesis, and reporting of environmental trends and indicators are urgently needed to enable the detection and tracking of cumulative effects of multiple restoration projects. Monitoring efforts should focus on developing the lines of evidence to support the assessment of cumulative effects at estuarine, regional, and larger scales. The *Deepwater Horizon* funding entities should immediately evaluate methods, identify funding mechanisms, and charge an entity to lead efforts to coordinate and enhance long-term priority monitoring efforts and promote consistent data collection, analysis, synthesis, and reporting among programs; support periodic assessments of collected data; assess the use of advanced techniques; and ensure data availability, with the goal of implementing these changes within 3–5 years.

Conclusion 3: The Gulf Coast environmental restoration community (federal agencies, states, nongovernmental organizations, and local public and private entities) has an opportunity to incorporate what has been learned from past and ongoing ecosystem restoration to inform future projects and programs supported by the remaining DWH funds. However, unless data and information from existing projects are made accessible and identification of information needed to assess cumulative effects of restoration efforts is undertaken more expeditiously, opportunities to improve the likelihood of success in the many projects remaining to be implemented will be greatly reduced or even permanently lost. Although it may be too early to fully assess cumulative effects of DWH-funded restoration efforts because of lag times between implementation and detection of effects, applying "lessons learned" from existing restoration efforts can help mitigate future risks of failure and ensure that DWH funds are invested wisely to increase the likelihood of meaningful and long-term Gulf of Mexico recovery and resilience.

Recommendation B: Restoration funding entities should adopt guidance to ensure that, as soon as they are available, all data, reports, and other project-specific information are deposited into freely accessible repositories that follow FAIR (Findable, Accessible, Interoperable, Reusable) principles. The *Deepwater Horizon* funding entities should identify and allocate resources to ensure that these data repositories remain functional throughout the life of each program, and additional support (as needed) should be sought to maintain data access in the future.

Recommendation C: The *Deepwater Horizon* funding entities should expedite the issuance of guidance for adaptive management and cumulative effects assessment at the programmatic scale for *Deepwater Horizon*-funded large-scale and multiple-restoration efforts. Guidance should include consistency in monitoring criteria that facilitate cumulative effects assessments.

Recommendation D: The *Deepwater Horizon* funding entities should immediately initiate a synthesis of available information from *Deepwater Horizon*-funded projects to assess characteristics of successful and unsuccessful restoration efforts. Results should be utilized in designing and implementing effective large-scale restoration projects within geographic areas of concern, and/or adjusting restoration approaches and techniques with the remaining funds from the *Deepwater Horizon* settlement.

Conclusion 4: Natural and anthropogenic drivers create multiple ecosystem pressures and stressors that act on restoration efforts over broad spatial scales, ranging from individual projects to entire ecosystems. The cumulative impacts of these pressures and stressors are often complex, resulting in synergistic and antagonistic effects of ecological significance. However, synergistic and antagonistic effects of large-scale restoration efforts in the Gulf of Mexico have not been assessed to date, and results from a limited number of assessments are mixed.

Recommendation E: The *Deepwater Horizon* funding entities should evaluate mechanisms that support cross-state and Gulf-wide collaboration among researchers, resource managers, and practitioners, with an objective to design and implement restoration efforts that allow assessment of antagonistic and synergistic effects.

Conclusion 5: The use of multiple lines of evidence to develop a framework to help assess cumulative effects for large-scale restoration efforts in the Gulf of Mexico has been proposed and, in some cases, applied. Assessment of cumulative effects of large-scale restoration is a recent research area and work on applying this research to restoration implementation is needed.

Conclusion 6: Opportunities exist now to prepare for the assessment of cumulative effects and restoration success from existing regional or large-scale restoration efforts in the Gulf of Mexico. These include:

- applying methods to assess functional equivalency between restored and natural sites;
- assessing the degree of environmental stress from natural and anthropogenic sources;
- applying a multiple-lines-of-evidence approach to assess cumulative effects at the estuary or watershed scale in preparation for Gulf-wide efforts;
- undertaking comparative analysis of estuaries or watersheds across the Gulf of Mexico to develop a greater understanding of similarities and differences among these systems; and
- evaluating expected benefits of a restoration effort as compared to a future condition without the effort.

These opportunities will involve consideration of changing environmental trends and a commitment to monitor, analyze, synthesize, and report results.

Recommendation F: To take advantage of the unprecedented opportunity to assess cumulative effects and inform restoration efforts ongoing and planned in the Gulf of Mexico, the *Deepwater Horizon* funding entities should evaluate and implement mechanisms necessary to address priority research needs and support efforts assess cumulative effects within the next 3–5 years. Mechanisms could include providing explicit responsibility to and support for existing Gulf-wide entities; development of an independent, regional, multidisciplinary, multiagency team; or a distribution of effort between existing entities.

Recommendation G: As additional monitoring data and scientific evidence become available, *Deepwater Horizon* program managers should continue to collaboratively develop and implement an adaptive management strategy for the Gulf of Mexico restoration effort, including the development of ecosystem conceptual models. Evaluation of priority issues should use the best available tools and methods, focus on progress of cumulative effects assessments and restoration objectives, and identify necessary changes to restoration approaches if needed. Mechanisms to continue these efforts beyond the eventual sunset of *Deepwater Horizon* restoration programs should be identified and implemented.

Chapter 1
Introduction

THE GULF OF MEXICO

The physical location, ecological richness, and economic value of the Gulf of Mexico (GoM) makes it unique among America's coastal seas. At 582,103 square miles (1,507,639 km2), it is the world's largest gulf and ninth-largest waterbody, effectively separating North and South America both physically and ecologically (Turner and Rabalais, 2019). The GoM's bowl-like shape is created in part by the Florida and Yucatan peninsulas that, along with Cuba, segregate it from the Atlantic Ocean and Caribbean Sea. Its average depth is 5,299 feet (1,615 m), or nearly a mile, which is contrasted by an expansive and shallow coastal zone that makes up 38 percent of its total area (Davis, 2017).

The GoM exists at the interface of tropical and temperate climates, creating a complex and diverse marine ecosystem. This is reflected in its high biological diversity, with a recent inventory noting over 15,000 species (Felder et al., 2009), and in its productivity, generating 633.7 million metric tons of commercial seafood landings in 2019 (NMFS, 2021). This productivity is sustained by extensive habitats along coastal margins, but also via its offshore oceanic ecosystems (Davis, 2017).

The GoM has helped shape the culture and economy of not only the five Gulf states but also the United States as a whole.[1] However, these transformations have come with costs, including habitat loss from coastal development, pollution and ecological alteration from oil and gas extraction, ecological alteration from overfishing, and coastal erosion from channelization and dredging. These ecosystem stressors are exacerbated by climate change (Brown et al., 2011; Twilley et al., 2001). For example, oyster reefs that were once dominant across the northern GoM have seen losses from 50 to 99 percent per bay compared with historical abundances (Beck et al., 2011). Over half of GoM wetlands have been lost (Brown et al., 2011), and the remainder are vulnerable to continued relative sea level rise (Osland et al., 2017b). Hypoxia and harmful algal blooms also routinely affect the GoM (e.g., Rabalais and Turner, 2019; Tominack et al., 2020).

THE *DEEPWATER HORIZON* OIL SPILL AND RESULTING SETTLEMENTS AND AGREEMENTS

On April 20, 2010, the *Deepwater Horizon* (DWH) drilling rig exploded about 80 km southeast of the Mississippi River Delta, resulting in the tragic loss of 11 lives and culminating in the largest marine oil spill in history. While the actual amount of hydrocarbons released is unknown, the resulting settlement under the

[1] See https://www.epa.gov/gulfofmexico/why-it-important-protect-gulf-mexico.

Clean Water Act recognized a net discharge of 3.19 million barrels of oil.[2] A significant amount of the hydrocarbons released ended up in the deep waters of the GoM, along with all of the released methane (Joye, 2015; McNutt et al., 2012). Oiling occurred in all five Gulf states, impacting over 2,100 km of coastline and at least 3,200 km^2 of deep-sea sediments (Nixon et al., 2016; Valentine et al., 2014). The civil litigation resulting from the DWH oil spill led to approximately $20.8 billion[3] in civil claims and $4 billion in criminal fines and penalties[4]—see Appendix A. The settlement and agreement funds specifically related to restoration activities were allocated to three primary funding sources and designated for specific purposes (Diamond et al., 2014; ELI, 2020):

- The Natural Resource Damage Assessment (NRDA) under the Oil Pollution Act (1990) led to the allocation of approximately $8.8 billion, to be overseen by the DWH NRDA Trustee Council.[5] The Trustee Council funds evaluation of impacts; restoration planning and implementation; and related activities, including monitoring and adaptive management.
- The Resources and Ecosystems Sustainability, Tourist Opportunities, and Revived Economies of the Gulf Coast States Act (RESTORE Act), passed by Congress in 2012, dedicated 80 percent of all Clean Water Act civil penalties for both environmental and economic recovery projects, including scientific research (more detail below).[6] This funding is divided into five areas or "buckets" with different authorized uses and decision makers for each bucket, totaling approximately $5.33 billion.
- Approximately $2.54 billion from the criminal agreements with responsible parties led to the establishment of the Gulf Environmental Benefit Fund of the National Fish and Wildlife Foundation (NFWF). These projects must "remedy harm and eliminate or reduce the risk of future harm to Gulf Coast natural resources"[7] in the five Gulf states. Half of its funding is allocated to Louisiana for barrier island and river diversion projects.

Through the criminal agreements, the North American Wetlands Conservation Fund, administered through the U.S. Fish and Wildlife Service, also received $100 million for bird habitat and populations.[8] In addition to funding restoration, several science programs were also established as part of the DWH settlements and agreements. The RESTORE Act included the establishment of the National Oceanic and Atmospheric Administration (NOAA) RESTORE Science Program (approximately $133 million), which aims to support science for a sustainable GoM ecosystem, with a focus on fisheries.[9] The RESTORE Act also allocated funds (approximately $133 million) for establishing Centers of Excellence in each of the five Gulf states, with the goal of supporting science, technology, and monitoring (Diamond et al., 2014). The Gulf Research Program (GRP)[10] of the National Academies of Sciences, Engineering, and Medicine (National Academies), which sponsored this study, was also established directly through the criminal agreements. This $500 million program, with funding that sunsets in 2043, supports activities related to environmental stewardship and protection, offshore energy safety, and human health and resilience.

When used in this report, the "DWH funding entities" are defined as the following:

- the RESTORE Council state and federal members (states of Alabama, Florida, Louisiana, Mississippi, and Texas; U.S. Department of Commerce; U.S. Department of the Interior; U.S. Environmental Protection Agency; U.S. Department of Agriculture; U.S. Coast Guard; and U.S. Department of the Army);

[2] See https://www.science.org/content/article/after-geoscientists-joust-judge-rules-bp-gulf-spill-totaled-319-million-barrels-oil.
[3] See https://www.justice.gov/opa/pr/us-and-five-gulf-states-reach-historic-settlement-bp-resolve-civil-lawsuit-over-deepwater.
[4] See https://www.justice.gov/opa/pr/bp-exploration-and-production-inc-agrees-plead-guilty-felony-manslaughter-environmental.
[5] See https://www.gulfspillrestoration.noaa.gov/co-trustees.
[6] See https://www.restorethegulf.gov/sites/default/files/RESTORE%20ACT%20July2012.pdf.
[7] See https://www.nfwf.org/gulf-environmental-benefit-fund/faqs.
[8] See https://www.fws.gov/southeast/infographic/north-american-wetlands-conservation-fund-deepwater-horizon-oil-spill.
[9] See https://restoreactscienceprogram.noaa.gov/about/faqs.
[10] See https://www.nationalacademies.org/gulf/about.

- the DWH NRDA Trustees (National Oceanic and Atmospheric Administration, U.S. Environmental Protection Agency, U.S. Department of the Interior, U.S. Department of Agriculture, and the five Gulf states);
- the NFWF Gulf Environmental Benefit Fund;
- the U.S. Fish and Wildlife Service North American Wetlands Conservation Fund;
- the Centers of Excellence;
- the NOAA RESTORE Science Program; and
- the Gulf Research Program.

DWH RESTORATION ACTIVITIES

As of March 2020, of the approximately $16.7 billion set aside for economic and environmental restoration-related activities, just under 30 percent ($4.65 billion) had been allocated to projects across the three funding sources (ELI, 2020)—see Appendix A. The three primary restoration entities share similar goals and objectives for restoration across a broad swath of natural resources and make efforts to coordinate their work; this is especially true of the RESTORE Council and DWH NRDA Trustees, which have significant overlap in membership. An example of their coordinated efforts is the DWH NRDA Trustees' guidance provided in the *Monitoring and Adaptive Management Manual V.1* (DWH NRDA Trustees, 2017a), which indicates that the trustees intend to share lessons learned in restoration projects with other DWH-funded programs (section 2.6.2). Another example is the RESTORE Council's efforts in coordination, collaboration, and connection among GoM restoration activities, illustrated by activities such as facilitating the Council Monitoring and Assessment Workgroup (RESTORE Council, 2016). Additionally, the RESTORE Council Monitoring and Assessment Program funded the Gulf-wide Monitoring Community of Practice, which is coordinated by the Gulf of Mexico Alliance (GOMA) and formed with the aim of improving coordination, accessibility, and comparability of monitoring information; this effort ended in 2021 but the work will be continued by GOMA's new Gulf-wide Monitoring effort in 2022 in accordance with its new Governors' Action Plan IV.[11]

Both the DWH NRDA Trustee Council and the RESTORE Council acknowledge the interconnected nature of the GoM's natural resources and the need to think comprehensively about restoration at scales beyond the project level (DWH NRDA Trustees, 2016; RESTORE Council, 2016). For example, the *DWH Final Programmatic Damage Assessment and Restoration Plan and Final Programmatic Environmental Impact Statement* (PDARP/PEIS) (DWH NRDA Trustees, 2016) contains the results of the injury assessments conducted and lays out a plan for recovery of injured resources. The DWH NRDA Trustees note that, because of the links among resources and habitats across GoM ecosystems, the injuries incurred as a result of the spill cannot be fully articulated for any single species or habitat and instead constitute "an ecosystem-level injury"[12] and that the DWH NRDA Trustees' approach to restoration needs to be integrated to best address injuries (DWH NRDA Trustees, 2016). Currently, monitoring efforts exist primarily at the project scale, although there is recognition that data collected at larger scales is needed (Brown et al., 2011; DWH NRDA Trustees, 2016; RESTORE Council, 2019).

STUDY ORIGIN AND RELATED ACTIVITIES

In 2019, GRP initiated a study series to document key challenges and monitor progress toward achieving a safe, healthy, and resilient Gulf of Mexico. A main goal of this series is to help GRP advance its strategic approaches as outlined in its 2020–2024 Strategic Plan: monitoring for progress and change, advancing scientific understanding, bridging knowledge to action, and building partnerships and engaging networks.[13]

[11] Personal communication from Laura Bowie, Gulf of Mexico Alliance (March 25, 2022).

[12] See http://www.gulfspillrestoration.noaa.gov/restoration-planning/gulf-plan.

[13] See https://www.nationalacademies.org/gulf/about.

These studies focus on topics associated with three GRP primary program areas, respectively: offshore energy safety, health and resilience, and environmental protection and stewardship (this study). It is intended that this first set of studies will comprise the initial volumes of a study series with three reports to be completed approximately every 3 years for the duration of the GRP.

ABOUT THIS REPORT

This report focuses on monitoring and assessment of GoM restoration projects beyond the project scale within the context of long-term environmental change. The need for this study was identified primarily via discussions with stakeholders and representatives from the DWH funding entities during a GRP meeting held in New Orleans in 2019. The study scope was developed based on these conversations, as well as on information provided in publications from Gulf-based entities, nongovernmental organizations, and the National Academies—see Box 1.1 for the formal Statement of Task. Relevant National Academies reports include *Effective Monitoring to Evaluate Ecological Restoration in the Gulf of Mexico* (NASEM, 2017), *Understanding the Long-Term Evolution of the Coupled Natural-Human System: The Future of the U.S. Gulf Coast* (NASEM, 2018a), and the *Progress Toward Restoring the Everglades* report series (NASEM, 2018b).

AUDIENCE

The audience for this report is broad and includes the DWH funding entities and their working groups (such as the RESTORE Council Monitoring and Assessment Workgroup, the DWH NRDA Trustees' associated Trustee Implementation Groups [TIGs], and the Cross-TIG Monitoring and Adaptive Management Working Group); restoration practitioners, resource managers, and regional program managers affiliated with federal, state, local, and nongovernmental efforts; and representatives of federal agencies, academic institutions, and other entities with an interest in restoration (whether in the Gulf or elsewhere).

BOX 1.1
STATEMENT OF TASK

The National Academies of Sciences, Engineering, and Medicine will convene an ad hoc committee to:

1. Examine the cumulative effects of multiple restoration projects (e.g., regional-scale benefits or improvement in ecosystem function) within the context of long-term environmental trends. To keep the scope manageable, the panel may focus on restoration projects within a geographic region and provide justification for the geographic boundaries selected.

2. Discuss acute events (e.g., hurricanes and storm surge, flooding associated with extreme precipitation events, large oil spills) and long-term environmental changes (e.g., relative sea level rise, changing land use/coastal development, chronic environmental disasters) that could have profound effects on the individual impact and cumulative effects of restoration projects in the Gulf of Mexico.

3. Consider the potential synergistic and antagonistic effects across multi-decadal restoration activities, and recommend adaptive management strategies that address the acute and long-term environmental changes that could affect restoration projects.

4. Identify existing resources (e.g., environmental monitoring data and modeling efforts) that could be used to address the three bullets above. Are these data and efforts sufficient to inform decision making on environmental stewardship and protection? What additional observations may be needed?

STUDY SCOPE AND APPROACH

The Committee on Long-Term Environmental Trends in the Gulf of Mexico was convened to address the tasks outlined in Box 1.1. The committee's 10 members brought to the study expertise in a variety of fields: coastal ecosystem restoration, including cumulative effects; natural resource management and policy; coastal ecosystem ecology, including wetlands, benthic, and fisheries ecology; water quality; ecosystem modeling; remote sensing and emerging technologies; river science and engineering; environmental economics; and data management and synthesis—see Appendix B for committee member and staff biographies.

In its deliberations, the committee decided that the geographic scope would include restoration efforts in all five Gulf states. Further, the spatial scale considered would focus on restoration efforts at the watershed/landscape scale and larger, and would include restoration efforts initiated before the DWH disaster, as well as those funded by DWH settlement funds, because of their integrated relationship on the landscape. Although coastal ecosystems exist as part of a continuum from the upland watersheds to the deep ocean, the committee focused the study scope on U.S. GoM coastal areas, including estuaries and bays. GoM rivers, particularly the Mississippi River system, are discussed as important driving forces in coastal ecosystems through their delivery of sediment, nutrients, and freshwater. However, noncoastal landscapes and ecosystems, including GoM waters beyond the coast, are not the primary focus of the report.

Definitions of the scales of restoration or background environmental effects vary widely, as noted in NASEM (2017). The committee determined that the geographic scope of the study would include restoration efforts at the estuary/watershed scale and larger and could include not only DWH-funded restoration efforts but also those funded by other sources. Table 1.1 summarizes terms and definitions regarding geographic, resource, and programmatic scales used in this report.

The committee held four information-gathering meetings in 2020 (August 11–13; September 29–30; November 9–10 and 16; December 14–16) and additional meetings in closed session in 2020–2022 to develop this report. All open- and closed-session meetings were held virtually because of the COVID-19 pandemic. The committee members heard presentations from 24 invited speakers, including representatives

TABLE 1.1. Terms and Definitions of Scale as Used in This Report

Spatial Scale	Definition
Project Scale	Activity, or set of dependent activities, within a specifically defined geographic area.
Estuary/Watershed Scale	Defined by connected characteristics of hydrology (e.g., hydrolic unit codes [HUCs]).
Regional Scale	Defined by common and/or connected characteristics of geomorphology, ecology, ecological function, and biodiversity. Regional scale can also be defined by political boundaries, such as a state, county, tribal nation, or parish boundary. This scale may include more than one estuary.
Gulf of Mexico Scale	Across all five Gulf states in the continental United States (referred to as "Gulf scale" or "Gulf-wide" in this report). This is consistent with the availability of DWH funds for restoration of coastal regions, estuaries, watersheds, and resources.
Resource Scale	Managed populations or habitats (e.g., oysters, birds, marshes).
Program Scale	Projects supported by a specific funding process, such as DWH NRDA or the RESTORE Council.

SOURCE: Committee on Long-Term Environmental Trends in the Gulf of Mexico.

from state and federal agencies, nonprofit organizations, and academia. Speakers shared their knowledge and expertise in GoM environmental trends, ecological restoration, and natural resource management. For longer-term context, the committee also heard from an expert in restoration efforts following the *Exxon Valdez* oil spill. Invited speakers are listed in Appendix C.

The report is divided into five chapters. Following this introductory chapter, Chapter 2 summarizes long-term background environmental trends and associated indicators affecting GoM coastal ecosystems and restoration projects (Task 2). Chapter 3 provides an overview of the theory, approaches, and considerations for assessing cumulative effects of multiple restoration projects, including synergistic and antagonistic effects (Tasks 1 and 3). Chapter 4 considers the current scientific progress toward assessing cumulative effects of multiple restoration projects and adaptive management in the GoM (Task 1). Chapter 5 discusses data resources and new observation methods to enable assessment of cumulative effects beyond the project scale; new applications of adaptive management, including program-level strategies; and the need for and approaches to conduct synthesis activities (Task 3). Identification and assessment of existing resources to assess cumulative effects (Task 4) are included throughout the report.

Addressing GoM restoration needs extends far beyond the available funding associated with the three primary DWH restoration entities (NRDA, RESTORE Council, and NFWF)—it is a "multi-generational undertaking" (RESTORE Council, 2016, p. 5). The recommendations in this report are intended to be a starting point for a long-term approach to addressing restoration needs in the context of changing conditions in the GoM, with applicability for and beyond DWH-related expenditures.

Chapter 2

Environmental Trends and Indicators

INTRODUCTION

Large-scale, long-term changes in hydrology, geomorphology, biogeochemistry, trophic status, and species range distributions have been occurring in the U.S. Gulf of Mexico (GoM) in response to both natural and anthropogenic forces. These forces are themselves impacted by system-wide stressors, such as the Deepwater Horizon (DWH) oil spill (see, e.g., Carl Kraft and Crandall, 2019; NRC, 2013), and the rate of change is accelerating due to climate change (see, e.g., Comeaux et al., 2012; Dee et al., 2019; Fujiwara et al., 2019; IPCC, 2021; Kemp et al., 2016; Moser et al., 2014).

Predominant natural influences in the GoM include temperate and tropical influences, ocean dynamics created by the Loop Current, hurricanes, subsidence, and freshwater influences that come primarily from the Mississippi River. These influences are all associated with and affected by a changing climate. The combined effects of these influences have created an ecologically varied and productive Large Marine Ecosystem (UNIDO, 2014). At 1.5 million km^2 (Turner and Rabalais, 2019), the subtropical GoM is large enough to include a high diversity of habitats, from coral and oyster reefs and seagrass to sawgrass. GoM estuaries make up nearly 42 percent of continental U.S. estuarine areas (excluding Alaska) (EPA, 1999) and are nurseries to vast numbers of commercially and recreationally important fish, crab, and shrimp, with functional food webs that support them.

In its Statement of Task (see Box 1.1 in Chapter 1), the committee was asked to "examine the cumulative effects of multiple restoration projects (e.g., regional-scale benefits or improvement in ecosystem function) within the context of long-term environmental trends." In this chapter, the committee focuses on the latter part of this charge. The chapter first defines several terms used throughout the report (drivers, pressures, stressors, and indicators) (see Box 2.1). It then summarizes the current understanding of indicators that would be needed to assess cumulative effects of multiple restoration efforts relative to background trends, identifies spatial and temporal gaps in existing data collection and interpretation, and discusses the need for robust environmental data to assess critical background environmental trends that impact restoration efforts.

The committee identified long-term trends and indicators that are grounded in science and can provide decision makers and restoration practitioners with useful insight in assessing opportunities for positive cumulative effects of multiple restoration actions, recognizing that these trends differ across the Gulf region—in terms of which trends are driving change and to what extent. The committee also developed a conceptual diagram to help visualize the effects of these long-term trends, as well as acute and chronic pressures and stressors, on restoration efforts in the GoM (see Figure 2.1). These trends and indicators provide important information about ecological processes that occur across multiple disciplines, at regional or Gulf-wide scales, and across political boundaries. Collecting, analyzing, synthesizing, and reporting robust long-term environmental data are critical components necessary for assessing cumulative effects of multiple restoration projects (Beck et al., 2019; Carl Kraft and Crandall, 2019; Schiff et al., 2016).

BOX 2.1
Ecosystem Drivers, Pressures, Stressors, and Indicators

The committee considered several conceptual approaches to ecological assessment in forming the framework for this report (Borja et al., 2006; Bradley and Yee, 2015; Harwell et al., 2019; Maxim et al., 2009; Oesterwind et al., 2016). Definitions for terms have changed, and continue to change over time, with evolving understanding of the ecological assessment process. The definitions below, a simplification of those provided by Harwell et al. (2019), are used throughout the report.

Drivers are fundamental natural or anthropogenic forces that act on a system. They tend to be large-scale, long-term forces that are not easily controlled or diverted. The Earth's rotation is an example of a natural driver, while population growth and agriculture are examples of anthropogenic drivers.

Pressures are human activities or natural processes that generate environmental stressors. They also tend to be large scale and long term, but often can be highly variable over space and time. Natural pressures include sediment dynamics, such as sedimentation, and episodic events, such as hurricanes. Examples of anthropogenic pressures include coastal development and recreational or commercial fishing.

Stressors are what the ecosystem experiences directly as a result of the impacts of drivers or pressures, and where direct action on the part of resource managers and restoration practitioners may have the most impact. Biological stressors in the GoM include overfishing and invasive species. Notable chemical stressors include nutrient inputs, endocrine disruptors, and oil and chemical spills. Physical stressors include altered freshwater inflows, changes in water clarity, habitat alteration, and hypoxia.

Environmental **indicators** of drivers, pressures, and stressors describe the status and changing conditions of an ecosystem. Ideally, indicators represent complex aspects of environmental quality or ecological integrity through simple measurements (Freedman, 2015), indices, or other forms easily understood and adaptable to modeling or the development of frameworks (Harwell et al., 2019). The value of a particular indicator may vary depending on spatial and temporal parameters. For example, an indicator for the population growth driver could be a change in the areal extent of a metropolitan area, and a stressor indicator could be a measurement of nutrient loading in a particular waterbody.

Indicators are organized into two broadly defined categories: those that reflect the status of background drivers, pressures, or stressors (e.g., climate change, riverine inflows, land cover, oil spills), and those that are indicators of specific value to the assessment of progress toward regional-scale environmental benefits and improved ecosystem function (e.g., coastal and estuarine habitats; fisheries, birds, turtles, and mammals; water quality).

CLIMATE CHANGE INFLUENCES ON LONG-TERM ENVIRONMENTAL CHANGE

Changing climate is a major factor driving change in the GoM. The committee focused on both natural and human-induced climate change effects on GoM coastal environments, including relative sea level rise, tropical storms and hurricanes, sea surface temperature, ocean acidification, and tropicalization. Some of these drivers or pressures, like relative sea level rise, can reflect both natural and anthropogenic forces. Considering the impacts of climate change at the restoration program scale can be particularly challenging due to nonstationarity—the concept that what was once considered a normal time series is not normal or stable any longer. Taking nonstationarity into account when planning restoration projects and programs can improve the likelihood of successful outcomes, even as ecological processes and conditions become less stable over time (Rollinson et al., 2021).

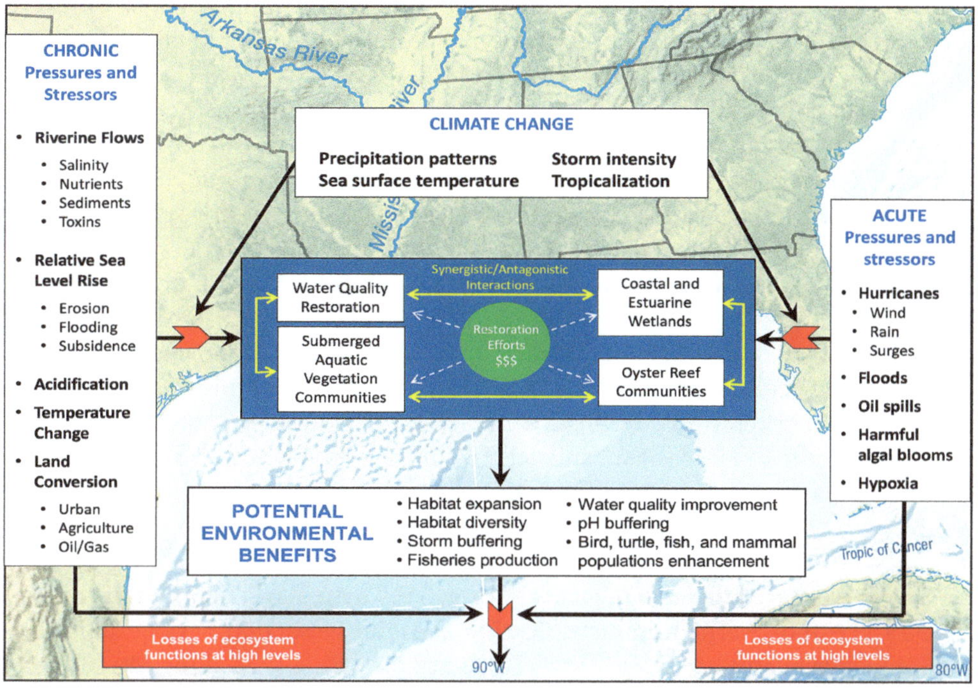

FIGURE 2.1. Conceptual diagram of the effects of long-term environmental trends associated with the climate change driver and both acute and chronic pressures and stressors, including climate change interactions, on restoration efforts in the Gulf of Mexico. **Climate change** indicates longer time scales of changing influence, as well as more explicitly linking the land to coastal waters. **Terrestrial influence** is shown by several large rivers delivering freshwater, nutrients, and sediments to the coast, as well as decadal-scale land use changes that can result in increased pollutant loads to coastal waters. In addition to climate change, two types of system-scale influences are shown: **chronic** (continuous) and **acute** (short-lived or episodic), on the left and right sides of diagram. Four major restoration types are shown in the center of the diagram and include water quality restoration (mainly wastewater treatment plant upgrades), coastal and estuarine wetlands, submerged aquatic vegetation (SAV) communities, and oyster reef communities. Dashed white lines indicate funds expended in restoration efforts, while yellow solid lines connecting the restoration types indicate possible synergistic or antagonistic interactions. Potential environmental benefits resulting from the cumulative effects of restoration efforts are shown as a rectangle below the restoration panel. Finally, the red interaction arrow exiting the environmental benefits box indicates that chronic and acute pressures and stressors can have negative effects on restored communities. SOURCE: Committee.

Relative Sea Level Rise

Relative sea level rise (RSLR) is the combination of global sea level rise (GSLR) (e.g., from reduced glacial ice volume and the thermal expansion of water) and the changes in an area's land surface elevation, such as subsidence (see, e.g., Sweet et al., 2018). These two drivers function differently in different parts of the GoM—GSLR is generally the dominant factor in Florida, Alabama, and Mississippi, while subsidence is a more important factor in Louisiana and much of Texas.[1]

As a result of increases in oceanic thermal expansion, along with ice loss from the Antarctic and Greenland glaciers and ice sheets, the rate of GSLR increased from 1.7 mm/yr in the 20th century to 3.1 mm/yr in the early 21st century (Argus et al., 2018). Concurrently, the World Climate Research Programme produced an averaged global mean sea level rise estimate of 3.35 mm/yr (WCRP Global Sea Level Budget Group, 2018). More recently, the Intergovernmental Panel on Climate Change (IPCC) found that GSLR averaged 3.7 [3.2 to 4.2] mm/yr between 2006 and 2018 (IPCC, 2021). While the current rates of GSLR are somewhat lower in the GoM, it cannot be assumed that this is a permanent trend.[2]

---

[1] Information from Torbjörn Törnqvist, Tulane University (September 29, 2020).

[2] Information from Torbjörn Törnqvist, Tulane University (September 29, 2020).

In comparison with GSLR, subsidence has significant spatial and temporal variation, presents challenges for accuracy and precision of measurement, and will therefore need additional monitoring to assess and predict regional and local impacts (Cahoon et al., 2020; Lane et al., 2006; NASEM, 2018a; Nolte et al., 2013; Russell et al., 2022). This is due to the multiple factors that can result in subsidence, as well the temporal and spatial scales on which these factors operate. Subsidence is caused by the interactions of at least six factors: tectonics, Holocene sediment compaction, sediment loading, glacial isostatic adjustment, anthropogenic fluid adjustment, and surface water drainage and management. Estimates for subsidence rates from these factors range from 0.6 to 2.0 mm/yr for glacial isostatic adjustment, which occurs across the Gulf, to as high as 23 mm/yr for anthropogenic fluid withdrawal, which is focused in areas of hydrocarbon production (Yuill et al., 2009).

The relative significance of these factors depends partially on the general characteristics of GoM geomorphology, geology, and ecology (summarized in NASEM, 2018a). For example, Florida is mostly built on a geologically inactive substrate, which results in hard-bottom reef and seagrass habitats with sandy barriers and spits and less than 1 mm/yr of subsidence. In contrast, Louisiana and northeastern Texas have a mud substrate with a variable sand veneer. There are extensive areas of coastal wetlands and marsh, many of which are subsiding and converting to open water, and subsidence can be greater than 20 mm/yr in various locations (NASEM, 2018a).

RSLR is generally estimated through the use of tide gages. Particularly common is the use of a subset of National Oceanic and Atmospheric Administration (NOAA) tide gages that have periods of record of more than 30 years and in some cases up to 50 years.[3] There are only nine sites along the GoM coast that meet these criteria.[4] While there are many other tide gages on the GoM coast, most have short or incomplete records that make them currently unsuitable for measuring RSLR (NASEM, 2018a).

Tide gage measurements from different parts of the GoM can provide insight into regional or local water level trends. For example, in addition to the use of NOAA gages, there have been efforts by the U.S. Army Corps of Engineers (USACE) to document, provide quality control, and publish records from over 30 local tide gages in Louisiana (Veatch, 2017). The USACE analysis of these gages suggests that while the long-term Louisiana NOAA gage at Grand Isle records one of the highest rates of RSLR in the United States (9.16 mm/yr)—see Figure 2.2—the limited NOAA tide gage coverage may fail to identify some of the most problematic areas. Two USACE gages at the mouth of the Mississippi River (Southwest Pass at East Jetty and South Pass at Port Eads) have 50-year subsidence estimates of greater than 22 mm/year, a measure that has implications for marsh persistence, the long-term viability of towns near the mouth of the river, and even the navigation of the lower Mississippi River. In spite of these extreme values, research suggests that tide gages in Louisiana systematically underestimate subsidence (Keogh and Törnqvist, 2019), especially shallow subsidence in low-elevation coastal areas.

By using an alternative approach that directly measures shallow subsidence using rod surface-elevation table-marker horizons, deep subsidence using the global navigation satellite system stations, and sea level rise from satellite altimetry, Nienhuis et al. (2017) created a new subsidence map of Louisiana (see Figure 2.3). Rates of subsidence using this methodology are typically substantially greater than those inferred from tide gage data. The high rates of subsidence along the Louisiana and northeast Texas coasts are expected to continue to constitute most of the RSLR in these areas over the next 50 years (Boesch, 2020).

Elsewhere in the GoM, GSLR will play a somewhat less drastic but still significant role in future decades (Argus et al., 2018). As noted by Törnqvist et al. (2020), long-term trends in RSLR have to be factored into all present and future wetland restoration projects in the GoM. Maintenance of coastal marshes depends on an equilibrium between RSLR and the vertical accretion of organic material through primary productivity of marsh plants and organic matter from elsewhere in the system (e.g., Bianchette et al., 2016; Glick et al., 2013; Nyman et al., 2006; Van de Broek et al., 2018) and sediment deposition from settling processes, including marsh plants trapping sediment (e.g., Fagherazzi et al., 2013). Marsh progradation, erosion, and

[3] See https://tidesandcurrents.noaa.gov/sltrends/sltrends.html.

[4] See https://tidesandcurrents.noaa.gov/sltrends/regionalcomparison.html?region=USTG.

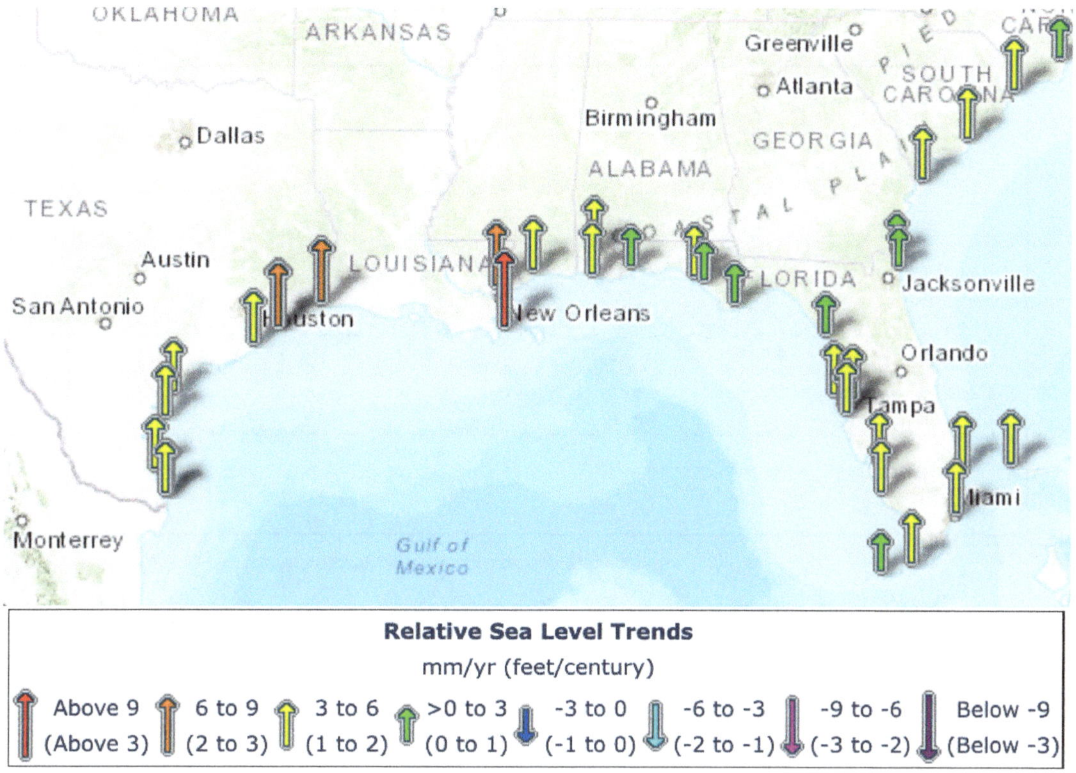

FIGURE 2.2a. Relative sea level trends, in mm/yr. The two red arrows correspond with Eugene Island and Grand Isle, Louisiana. SOURCE: https://tidesandcurrents.noaa.gov/sltrends/sltrends.html.

collapse are also important processes in microtidal environments (Ganju et al., 2017). The 2017 Louisiana State Master Plan utilized a "marsh collapse value" of 7 mm of RSLR/year and showed that for marshes in Louisiana to survive at rates beyond this, sediment replenishment would be necessary (White et al., 2019b). Others posit that if RSLR is greater than 3 mm/yr, marsh drowning occurs within a few centuries, and if relative sea level rise exceeds 6–9 mm/yr, Louisiana marsh conversion into open water could happen within a 50-year timeframe (Jankowski et al., 2017; Törnqvist et al., 2020). Many locations within Louisiana and Texas are at the 6–9 mm/yr[5] rate (Figure 2.2), which has serious implications for the viability of restoration projects within the western GoM when considering the future of wetland restoration projects. Although subsidence is not as prevalent in states like Florida, Alabama, and Mississippi, the 3 mm/yr threshold has also been reached in many locations in those states.[6]

Subsidence is the principal stressor causing RSLR in Louisiana and the northern Texas coast. RSLR rates in this region are among the highest in the nation. Long-term trends in other areas in the GoM also show an increase in RSLR, a trend that is expected to continue.

The current rate of RSLR is already considered a major stressor on the decline of marsh area and resilience. There are areas in the GoM where marsh plant growth and organic matter accumulation cannot keep up with RSLR rates without additional inputs of sediment, resulting in marsh collapse. This is a critical issue in environmental restoration planning.

[5] See https://tidesandcurrents.noaa.gov/sltrends/regionalcomparison.html?region=USTG.

[6] See https://tidesandcurrents.noaa.gov/sltrends/regionalcomparison.html?region=USTG.

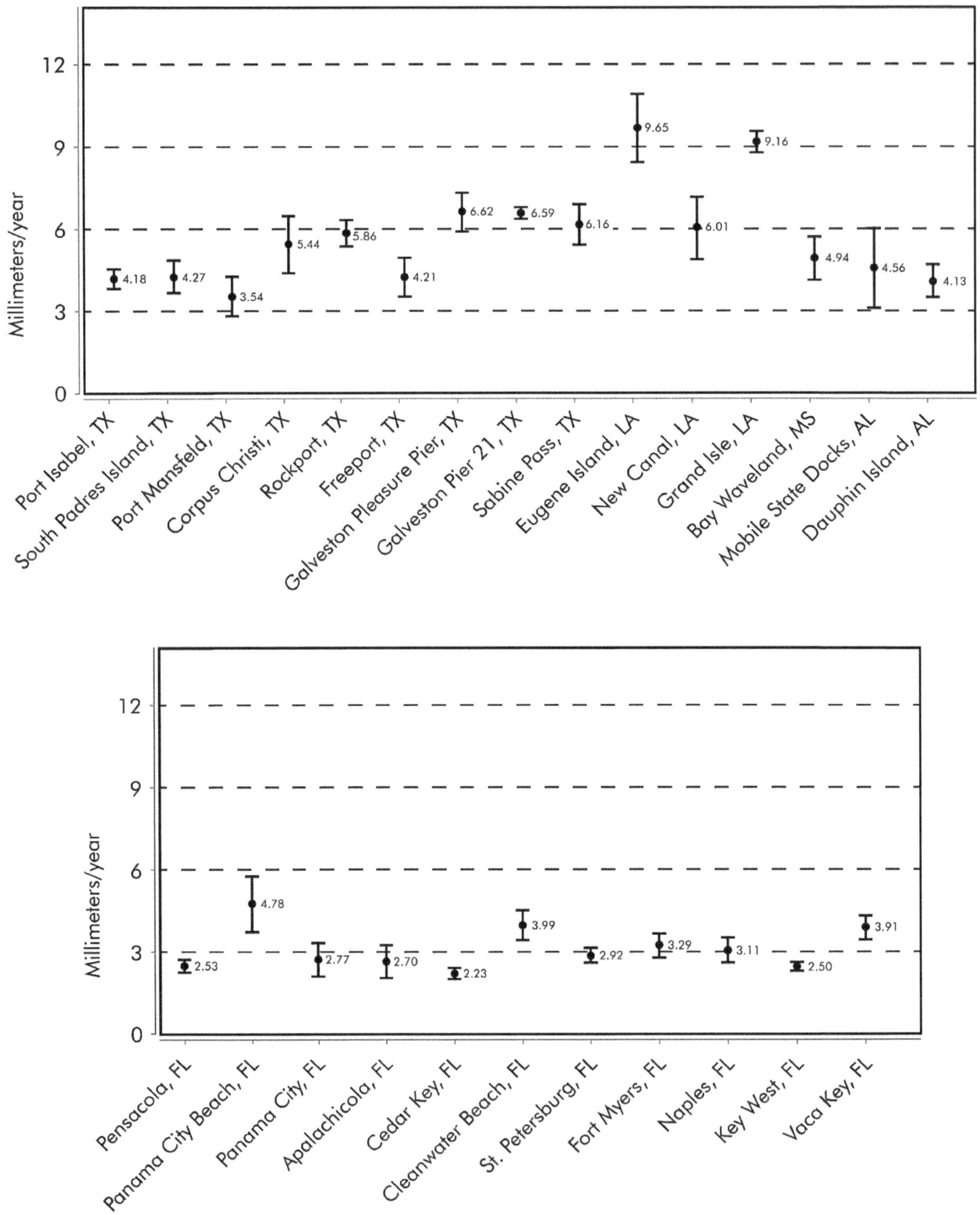

FIGURE 2.2b. Relative sea level trends (mm/yr) in the Gulf of Mexico. Sites from Texas, Louisiana, Mississippi, and Alabama are shown in the top panel, while Florida sites are shown in the bottom panel. Trends with the narrowest confidence intervals are based on the longest data sets. Trends with the widest confidence intervals are based on only 30-40 years of data. SOURCE: Adapted from NOAA: https://tidesandcurrents.noaa.gov/sltrends/regionalcomparison.html?region=USTG.

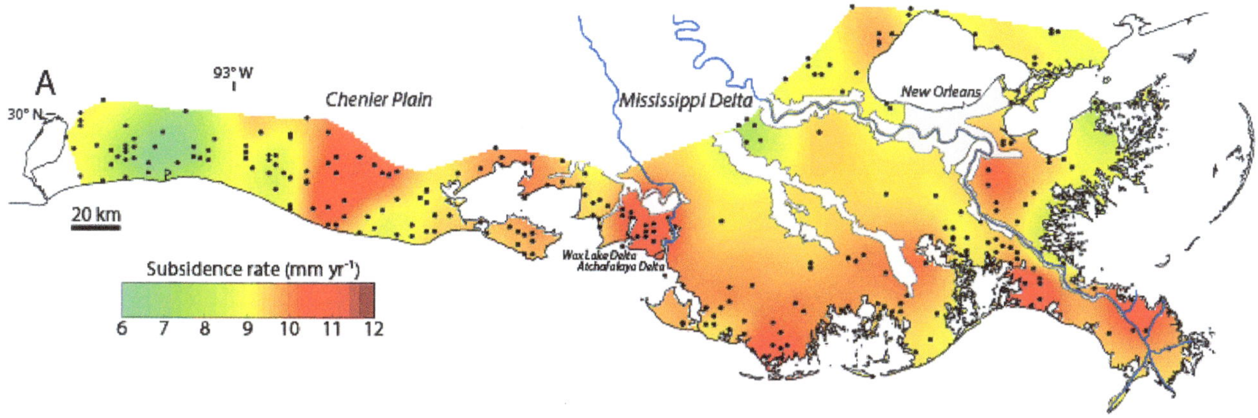

FIGURE 2.3. Estimates of subsidence in coastal Louisiana from the evaluation of data collected through Louisiana's Coastwide Reference Monitoring System program, based on geostatistical interpolation of observations (black dots) of land surface subsidence. SOURCE: Nienhuis et al., 2017.

Tropical Storms and Hurricanes

Hurricanes and tropical storms not only endanger human life and infrastructure, but also dramatically reshape or change physical landscapes and coastal ecosystems, creating hot moments of substantial change (discussed in Chapter 3). Thus, hurricanes are critical to consider when assessing restoration success and determining potential cumulative effects. Prior studies have developed empirical relationships between sea surface temperatures (SST) and the maximum potential intensification rate of tropical cyclones (DeMaria and Kaplan, 1994; Whitney and Hobgood, 1997; Xu et al., 2016; Zeng et al., 2007), leading to heightened concern of the effect of future hurricanes on the GoM coast. However, the current understanding of the situation is somewhat more complicated (Berardelli, 2019). One of the difficulties is that satellite-derived intensity data has only been collected since the early 1980s and therefore only span about four decades; this relatively short period of record makes it difficult to distinguish clearly between trends because of anthropogenic climate change and natural trends that occur on decadal to multidecadal time scales (Ramsay, 2017).

Since 1985, hurricane frequency has remained consistent, with an average of about 80 tropical cyclones forming globally each year (Ramsay, 2017). Holland and Bruyère (2014) found "no anthropogenic signal in annual global tropical cyclone or hurricane frequencies" (p. 617). However, they did find a strong signal in the proportion of both weaker and stronger hurricanes. Category 4 and 5 hurricanes have increased at a rate of about 25–30 percent per °C of global warming, and the proportion of Category 1 and 2 storms have decreased a similar amount (Holland and Bruyère, 2014). The increased frequency of stronger hurricanes means that the intensity of the "common hurricane" has become stronger. Tropical storms and hurricane strikes also vary over time and space. Temporally, periods of hyperactivity were seen in south Florida from the 1920s through the 1950s and then again in the 2000s (Figure 2.4) (Keim et al., 2007). According to Landsea et al. (2010), century-scale analysis of SST and hurricane frequency is needed to establish the decade-to-decade shift in hurricane frequency and intensification trends due to climate change.

The effects of hurricanes and tropical storms do not uniformly affect the GoM; the southern part of Florida and the north-central Gulf are disproportionately affected by both (Figure 2.5). When planning restoration projects, practitioners need to be aware of the timespan for developing resilience of their projects to hurricane conditions, a critical consideration in areas where the return period is between 5 and 7 years for a hurricane and 14 and 22 years for a major hurricane landfall.[7]

[7] See https://www.nhc.noaa.gov/climo/.

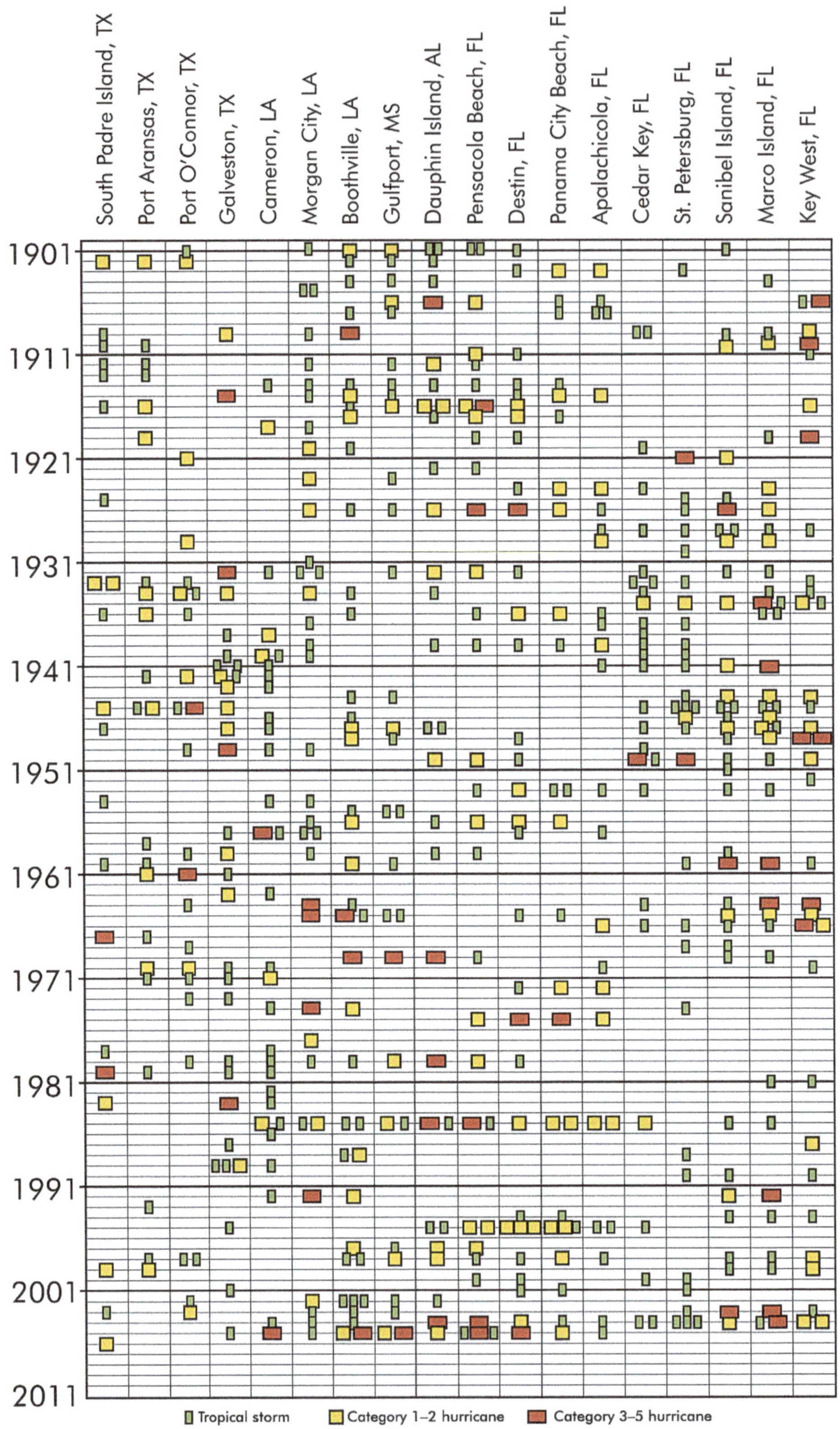

FIGURE 2.4. Time series from 1901 to 2005 of Gulf of Mexico tropical storm (green boxes), hurricane (Category 1–2; yellow boxes), and severe hurricane (Category 3–5; red boxes) strikes at 18 locations across the five U.S. Gulf Coast states. SOURCE: Redrawn from Keim et al., 2007.

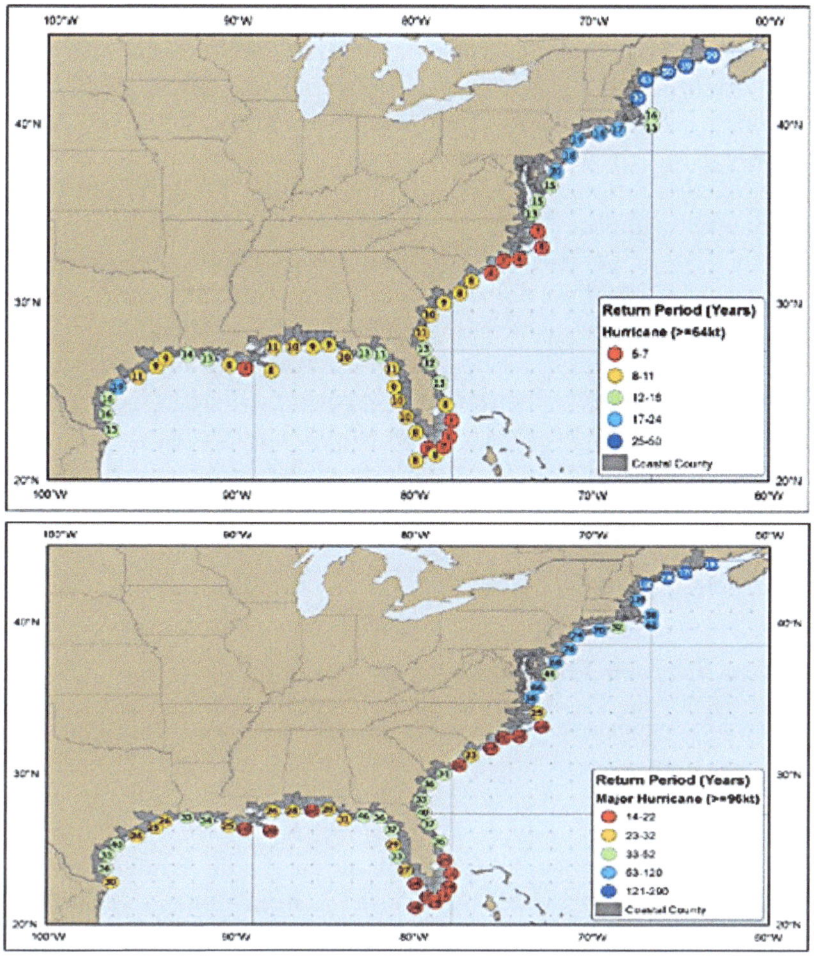

A

B

FIGURE 2.5. Estimated return period in years for (a) hurricanes and (b) major hurricanes passing within 50 nautical miles of various locations on the U.S. coast. SOURCE: National Hurricane Center and Central Pacific Hurricane Center, 2020.

Hurricanes can also destroy monitoring equipment, creating data gaps and hampering the ability to measure the effects of restoration efforts. For example, Hurricanes Katrina and Rita hit in 2005, the year that the initial Coastwide Restoration Monitoring System in Louisiana was being constructed. Many stations were destroyed, causing the system to have to be reestablished in 2006 and 2007.[8]

Needham and Keim (2012) provide a comprehensive overview of storm surges from tropical storms and hurricanes in the GoM. Storm surges have dramatic effects on coastal environments, damaging habitat, destroying forests and crops, inundating the coastline with saltwater, and moving and cutting channels through barrier islands. These effects are also not uniformly distributed across the GoM. In addition to hurricane strikes, coastal geomorphology, water depth, and man-made features such as levees can all influence storm surge. Texas, Louisiana, and Mississippi coasts tend to have the greatest depths of storm surge, with both the first and second greatest storm surge depth ever measured (8.47 and 7.5 m) occurring in Pass Christian, Mississippi (Needham and Keim, 2012). Knowledge of the depth and probability of storm surge is critical to ecosystem restoration planning. Some restoration projects are also designed with the idea that the presence of wetland and coastal features will serve to minimize the effects of storm surge. A meta-analysis of coastal marshes conducted by Shepard et al. (2011) concluded that marshes beneficially attenuate waves from storm surges and provide shoreline stabilization across a range of different environments. According to Wamsley et al. (2010), the magnitude of the attenuation is dependent on the surrounding coastal landscape and the strength and duration of the storm forcing, with some coastal/wetland configurations actually increasing surge.

[8] Information from Gregory Steyer, U.S. Geological Survey, personal communication, October 11, 2021.

Hurricanes can have far reaching impacts on restoration efforts, particularly wetland restoration projects. For example, at a site in Barataria Bay, Louisiana, Sapkota and White (2019) documented average marsh erosion of 0.35 cm/day between 2015 and 2019. Based on measurements taken immediately following Hurricane Laura in 2020, these same marshes eroded at 22 cm/day during the storm, a rate 63 times faster than the 4-year period immediately prior (Gibbens, 2020). The effect of acute events like hurricanes on coastal ecosystems can cause large-scale disruption of the system (Conner et al., 1989; D'Sa et al., 2011; Mo et al., 2020; Reja et al., 2017) and potentially impact restoration efforts.

Morton and Barras (2011) suggest that 25–35 percent of all wetland loss in Louisiana since the 1940s was due to direct and indirect storm-induced losses. These damages range from winds and waves physically stripping the vegetation from the marsh, to storm surges bringing saltwater into freshwater wetland areas, resulting in salt-induced diebacks. Conversely, there have been several accounts of hurricanes bringing sediment and nutrients into wetland systems through freshwater influx, providing benefits to the surviving wetlands (Castañeda-Moya et al., 2020; Nyman et al., 1995b; Turner et al., 2006, Box 2.2).

Hurricanes are becoming more intense and destructive in the GoM, a trend that is expected to continue. Wind and wave energy; storm surge; and freshwater, sediment, and nutrient inflows associated with storm-related rainfall have potential to significantly disrupt coastal ecosystems and restoration efforts.

Sea Surface Temperature

Like much of the global ocean, the GoM is experiencing rising sea surface temperatures (Figure 2.6), although rates of increase in the GoM are less than in many other regions (Huang et al., 2015). Using monthly offshore SST calculated from satellite data, Karnauskas et al. (2017) found that SST has gradually increased in the GoM since the early 1980s and that the western GoM has warmed more rapidly than the eastern Gulf. Changes in GoM SST are generally driven by anthropogenic climate change (e.g., Gil-Agudelo et al., 2020) as well as climate variations. The Atlantic Multidecadal Oscillation state has been shown to affect several elements of the GoM physical system, including ocean currents, freshwater inflows, and SST (Karnauskas et al., 2017). SST can also be affected by episodic factors, including weather fronts (e.g., Wang et al., 2020b). Rising SST can increase coastal hypoxia (areas of low dissolved oxygen) by decreasing oxygen solubility in the water column as well as by contributing to stratification (Altieri and Gedan, 2014; discussed further in a later section).

Sea surface temperatures in the GoM have gradually increased since the 1980s. Annual average SST is as much as 2° F higher than the historical mean SST. Increased water temperature may affect fisheries restoration efforts, and hypoxia and stratification in coastal waters, resulting from increasing temperatures, could impact restoration efforts for living resources.

Temperature and Precipitation Patterns

The IPCC *Sixth Assessment Report* estimated that global surface temperatures were 1.09 °C higher during the last decade (2011–2020) compared with the historical past (IPCC, 2021). The temperature increases were even higher over land than over oceans and primarily driven by further warming of the atmosphere. These warming trends are significant everywhere, including in the GoM. There are many physical manifestations of these increasing temperature trends, from observed temperature extremes to the duration of heatwave days during summer, from longer summers to warmer nights and frequent flash droughts, and from record freezing temperatures in winter to sudden cold snaps (Biasutti et al., 2012). The mostly human-induced warming trends will result in extreme variability in precipitation patterns (IPCC, 2021), which would result in fluctuations in freshwater flow to GoM coastal ecosystems.

BOX 2.2
Acute and Chronic Environmental Events in the GoM: A Hurricane Case Study

Climate change controls both the magnitude and frequency of acute environmental events, and in some cases, they are turning into chronic events (USGCRP, 2017). As an example, hurricanes in the GoM are considered acute environmental disturbances that adversely impact coastal habitats. However, increases in the number of high-intensity hurricanes in the last few decades due to climate change has turned them into a chronic series of events, causing irreversible change to some ecosystems (USGCRP, 2017). Further, as concluded in the Intergovernmental Panel on Climate Change's (IPCC's) recently released Sixth Assessment Report, frequency of tropical cyclones and hurricanes will increase in future El Ni o years (IPCC, 2021).

A similar argument can be made for droughts, which alter estuarine water quality and wetland productivity. One of the positive effects of the high frequency of these acute events is that it can reverse the course of existing chronic trends at a site by temporarily alleviating them. For example, a wetland site suffering from a chronic lack of freshwater can be temporarily altered due to the impact of a hurricane producing massive freshwater runoff to the wetlands. However, high frequencies of these acute events can create irreversible damage, hampering the resiliency of the ecosystem. The decline in ecosystem functions and resiliency can result in either a state change (a complete change in the ecosystem type), or an alternate stable state (the ecosystem will continue to oscillate between a number of possible locally stable configurations) (Beisner et al., 2003). Ryo et al. (2019) demonstrated the concept in the following figure.

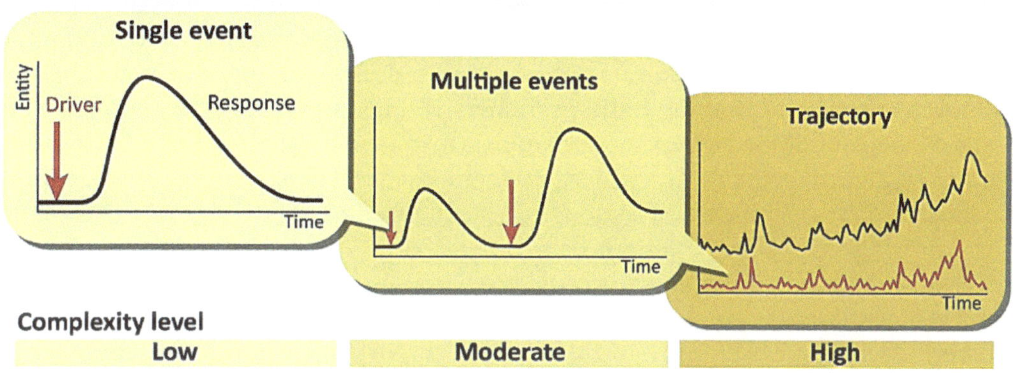

FIGURE 2.2.1. Impact of acute or disturbance events on ecosystem functions and resiliency. The return of the ecosystem function back to the status quo depends on the frequency and magnitude of the events. SOURCE: Ryo et al., 2019.

As an example, because of climate change, hurricanes are becoming stronger in the U.S. Gulf Coast (Holland and Bruyère, 2014). The 2020 hurricane season was one of the worst on record, with several severe storms that hit the southeast Atlantic coast. The wind and wave energy associated with these storms can produce severe damage, su√Ωch as defoliation in the mangrove ecosystem. However, a study by Casta eda-Moya et al. (2020) showed that hurricanes can be natural fertilizing events that contribute a significant proportion of phosphorus to the nutrient-limited mangrove soil. Their study revealed that Hurricane Irma supplied as much as 49–98 percent of the annual phosphorus to the Everglades mangrove's soil nutrient pool. Vertical soil accretion in that event alone was 6.7 to 14.4 times more than the past 100-year annual accretion rate. It shows that one hurricane can episodically supply nutrients and freshwater that compensates for the structural damage to the vegetation canopy and, in the long term, will enhance gross primary production of the ecosystem. But if the annual frequency of these hurricanes increases (as seen in the last decade), the compounding effect may no longer allow the ecosystem to bounce back from its structural damage, ultimately causing an irreversible loss in net and gross primary production and habitat shrinkage due to erosion.

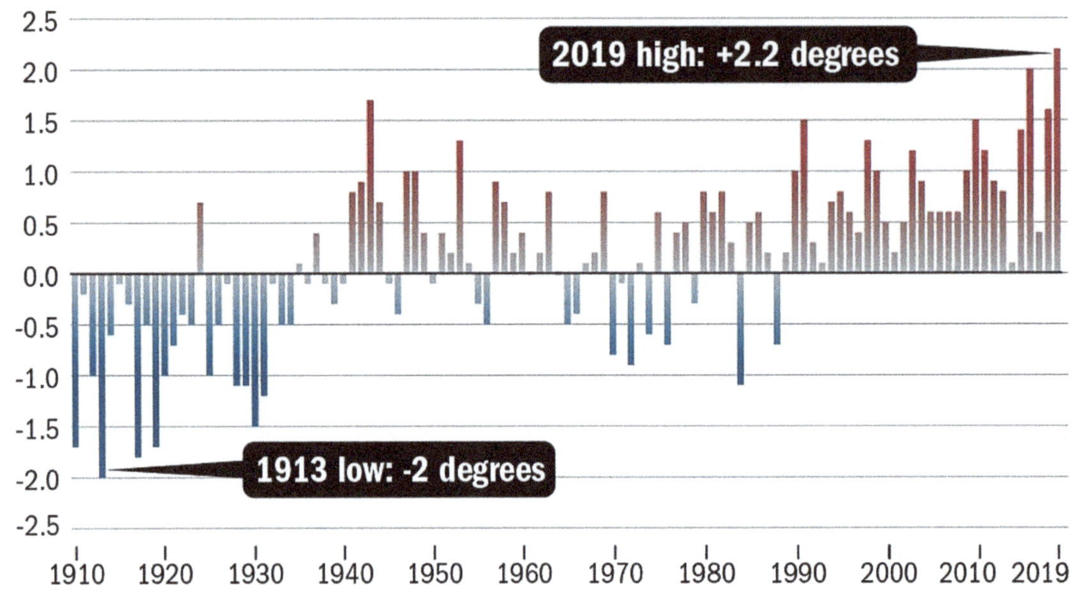

FIGURE 2.6. SST in the GoM from 1910 to 2019, shown as °F above or below the mean SST. SOURCE: NOLA.com.

Biasutti et al. (2012) provided a comprehensive summary of GoM climate trends and projections. According to their study, "In the Gulf, a typical winter in the last decades of this century will be as warm as the warmest winter ever recorded, and the coolest summers will be as hot or hotter than any summer in the last century; in 95 percent of the years, summer temperatures will be unprecedented" (Biasutti et al., 2012, p. 827-828). Seasonal mean precipitation will experience a modest change in magnitude, but not in the frequency distribution. However, the driest seasons will be extraordinarily dry, leading to flash droughts and extreme variability in soil moisture conditions.

These extreme fluctuations in temperature and precipitation patterns will have a significant impact on GoM wetland ecological functions and biophysical status (Comeaux et al., 2012; Osland et al., 2017a). Alterations in precipitation-induced freshwater flow will alter nutrient and sediment supply to the Gulf Coast estuaries (Kemp et al., 2016; McCarthy et al., 2018a). Temperature fluctuations and longer summer seasons will affect salt marsh biomass and productivity patterns, since both are strongly coupled (Feher et al., 2017).

Air temperatures have increased and precipitation patterns have shown extreme fluctuations in recent decades, and both are projected to become more extreme over time, with many associated effects. Restoration efforts may be impacted, as increased air temperature will affect marsh plant productivity and biomass, and changes in freshwater flows will alter nutrient and sediment supply delivery patterns to marshes and estuaries.

Coastal Ocean Acidification

Rising atmospheric carbon dioxide (CO_2) levels due to anthropogenic activities and the ability of the ocean to partially absorb this CO_2 increase has led to alterations in the chemistry of seawater, including decreasing pH (Doney et al., 2020). In coastal systems, the amplitude of variability in the carbonate system (including pH) is generally much greater than in the open ocean (Carstensen and Duarte, 2019). This phenomenon is driven by several interacting factors, most notably land-based inputs of nutrients, organic matter, and freshwater. Nutrient input leads to increased primary production and in turn, pH, while the decomposition of this newly produced or land-derived organic matter by heterotrophs leads to increased acidification (e.g., Cai et al., 2011; Gilbert et al., 2010). Freshwater contributes to decreasing pH because it lowers the alkalinity (buffering capacity) of seawater. Each of these factors is in varying degrees controllable by management strategies. In the northern GoM in particular, eutrophication-induced acidification due to decomposition processes is likely widespread, given the common occurrence of seasonal hypoxia both on the GoM shelf and within its estuaries (e.g., Laurent et al., 2017; Wang et al., 2020a).

Ecological effects of ocean acidification are many and include impacts from species level (e.g., metabolic and developmental effects) to community level (e.g., phytoplankton species composition) to ecosystem scale (e.g., coral reef health, pelagic food webs). Understanding the nuances of these effects is an area of active research, especially in the context of multiple stressors (e.g., acidification plus warming) (Doney et al., 2020, and references therein). Acidifying seawater has been shown to affect the formation of calcium carbonate shells and skeletons in a range of organisms, including reef-building corals and commercially important mollusks such as oysters and mussels (Doney et al., 2020). In addition, ocean acidification can affect fish metabolism (e.g., Heuer and Grosell, 2014) and behavior (e.g., Leduc et al., 2013), with potential implications for fisheries. Some organisms, however, have demonstrated the capacity to adapt to a more acidified environment (e.g., Engström-Öst et al., 2019). Some ecosystems have also shown the potential for adaptation; for example, seagrass ecosystems can modify pH within their canopy, potentially offsetting acidification effects on their associated communities (e.g., Bergstrom et al., 2019; Hendriks et al., 2014).

Monitoring of coastal acidification is also increasing. NOAA's Ocean Acidification Program has funded deployments of pCO_2 (partial pressure of carbon dioxide) and pH sensors at a limited number of locations in the northern GoM since 2009. And since 2002, NOAA has also collected underway sea surface pCO_2 data on some ships of opportunity, including cruise liners (Gledhill et al., 2008) and NOAA fishery survey vessels, which have been incorporated into the global Surface Ocean CO_2 Atlas database (Bakker et al., 2016). However, especially in the southern GoM, many coastal and estuarine areas and subsurface waters currently have few or no measurements.

A recent study based on the Surface Ocean CO_2 Atlas revealed that sea surface pCO_2 trends are higher in the northwestern GoM and the West Florida Shelf but lower in the central GoM compared with the extent of pCO_2 increase due to air CO_2 uptake (Kealoha et al., 2020). Subsurface waters in the northwestern GoM are acidifying at a rate greater than the global surface ocean rate, with the majority (59–70 percent) of the acidification being respiration driven (Hu et al., 2018). Multidecadal acidification has also been reported for a majority of the Texas and Florida estuaries, based on state agency–collected data (Hu et al., 2015; Robbins and Lisle, 2017).

Although data are scarce or lacking for many coastal and estuarine areas, recent observations indicate that acidification has been occurring in many Texas and Florida estuaries and that subsurface waters in the northwestern GoM are acidifying at a rate greater than the global rate. Increased acidification affects the formation of calcium carbonate shells and skeletons in corals and mollusks, potentially impacting the long-term success of oyster reef restoration and coral reef recovery.

Tropicalization

Climate-driven changes in species composition and the novel ecological interactions that follow them are occurring throughout the world. As pointed out by Hyndes et al. (2016), a rapidly expanding literature has documented poleward range shifts in terrestrial and marine plants and animals; changes in their body sizes; and altered behavioral and ecological interactions, such as competition, herbivory, and predation (e.g., Pecl et al., 2017; Poloczanska et al., 2013; Vergés et al., 2014, 2018; Wernberg et al., 2013). Collectively, such changes have been termed *tropicalization*, defined as an increase in the ratio of tropical to temperate taxa in a given region (cf. Wernberg et al., 2013) or as the entire suite of changes in species composition, abundances, and interactions (Vergés et al., 2014). The latter, more comprehensive definition is used here.

Tropicalization has repeatedly been documented to produce rapid shifts in marine ecosystem structure and functioning (Hyndes et al., 2016; Scott et al., 2018; Vergés et al., 2014). Examples of recent GoM tropicalization include the replacement of temperate salt marsh species by tropically associated mangroves in many eastern and northwestern Gulf locations (Jackson et al., 2021; Osland et al., 2017b, 2020, 2021) and the occurrence of tropically associated fish species that now occur consistently in the northern Gulf (Fodrie et al., 2010; Marshak and Heck, 2017; Purtlebaugh et al., 2020). For other coastal fish species like Southern flounder, tropicalization is generating range-wide declines due to increased climate vulnerability (Erickson et al., 2021; Montalvo et al., 2012).

A direct pathway from the tropical Caribbean to the GoM is provided by the Loop Current, a mass of warm water that carries larval, juvenile, and adult organisms from the tropics northward. Many tropical organisms transported into the GoM have until recently been unable to survive its winter temperatures. However, the decreasing frequency of lethal temperatures has allowed many formerly rare tropical species to become increasingly common in coastal waters of all five Gulf states (Osland et al., 2021) (Figure 2.7). As global climate change continues to alter climate conditions, such as temperature regimes, it is very likely that increasing numbers of tropical species will become established throughout the GoM, and this will produce changes in the relative abundance and species composition of, for example, wetland plants and seagrasses (Osland et al., 2021; Rodriguez and Heck, 2020), corals (Aronson and Precht, 2016), and fishes (Fodrie et al., 2010; Purtlebaugh et al., 2020).

These changes can produce large and cascading effects in the abundance and composition of the many species of organisms that occupy these habitats, with large shifts in their structure and function (Heck et al., 2015; Scheffel et al., 2018). In their review article, Vergés et al. (2014) note that climate-driven interactions can profoundly alter ecological communities, particularly when they affect foundation species. They cite overgrazing of temperate microalgae by tropical herbivorous fish as an example. Hyndes et al. (2016), in their overview assessment of the poleward movement of tropical seagrasses and herbivores, conclude that novel ecosystem configurations are likely to appear. All have implications for the long-term sustainability of habitat restoration in the GoM.

Data from fisheries surveys and habitat monitoring programs will be of critical importance for tracking tropicalization within the GoM. These data can provide information needed to model expected changes in future finfish and shellfish harvests, as well as the value of ecosystem services provided by historically productive wetlands, seagrass meadows, and reefs.

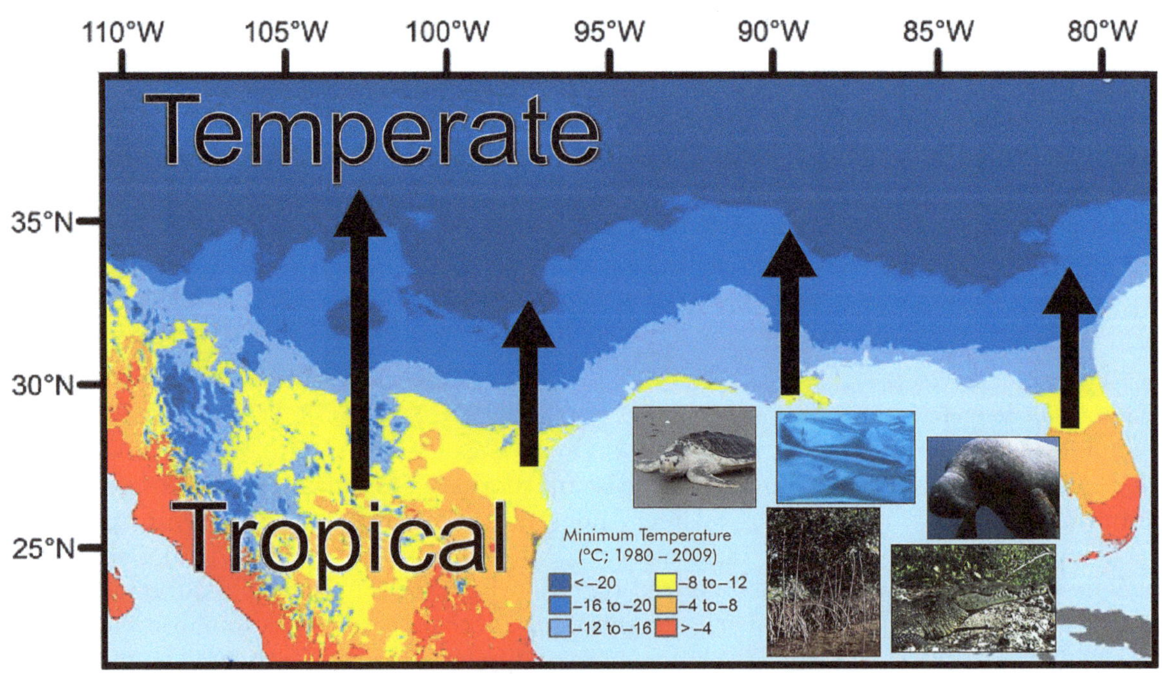

FIGURE 2.7. Tropical–temperate climate and ecological transition zones in North America. Tropical zones are shown with warm colors. The color scale shows the minimum temperature in °C from 1980 to 2009. Photos show examples of tropical cold-sensitive organisms whose ranges are governed by winter cold temperature extremes, including the red mangrove, American crocodile, cobia, manatee, and Kemp's ridley sea turtle. SOURCE: After Osland et al., 2021.

Evidence of GoM tropicalization includes replacement of temperate salt marsh species by tropically associated mangroves and the occurrence of tropical fish species in the northern GoM throughout the year. Increasing tropicalization effects can impact restoration objectives, approaches, and outcomes.

WATER, NUTRIENT, AND SUSPENDED SEDIMENT INFLOW TRENDS

Inputs of freshwater, dissolved and particulate nutrients (primarily nitrogen and phosphorus), and suspended sediments are important drivers of estuarine and coastal marine ecosystems (Hobbie, 2000). These inputs influence salinity regimes, water residence times, and other features of physical circulation, trophic status and keystone habitats, and water clarity and sediment accretion rates. Changes in land uses and human activities have modified and will continue to modify the annual patterns and magnitude of these inputs. Current and projected climate change effects will further influence these trends.

Anthropogenic alteration of rivers and streams has affected streamflow in more than 86 percent of streams in the United States (Carlisle et al., 2011). These alterations may cause ecological impairment in river and stream ecosystems, but also affect water quality, salinity, sediment transport, and ecology of the receiving estuaries. Historic alterations in streamflow, along with future modifications, are likely to influence the success or failure of ongoing and future GoM restoration projects.

Freshwater inflows control salinities in GoM estuaries and shallow coastal waters, thus defining, at least in part, environments and habitats. Riverine inputs largely dominate nutrient loading to coastal waters (NRC, 2000; Oelsner and Stets, 2019). Additionally, suspended sediments and bed materials carried by rivers can influence GoM restoration efforts (Allison et al., 2012). Clays and fine suspended materials may affect turbidity and light penetration and can negatively impact algal primary productivity, seagrass beds, and filter feeders. Conversely, efforts to compensate for land loss by diversions or dredging and barrier island restoration are highly dependent on sediments transported by rivers.

Freshwater Inflows

Rodgers et al. (2018, 2020) assembled daily streamflow discharge data from 139 stream gages located on tributaries and streams flowing into the GoM with continuous daily flow data from 1950 to 2015. A model was developed to estimate streamflow from ungaged streams. Since 1950, the sites showed a downward trend in streamflow (Rodgers et al., 2020), which appears to strongly contrast with the steadily increasing streamflow trends for the Mississippi River (Zhang and Schilling, 2006). The Rodgers et al. (2020) study was funded through the RESTORE Council;[9] however, this valuable synthesis is currently not scheduled to be updated with additional data. The Mississippi and Atchafalaya River system is by far the greatest source of freshwater to the GoM, contributing an estimated range of 79–90.5 percent of all freshwater streamflow between 1990 and 2019 (USGS, report in preparation, 2021). The outsized impact of the Mississippi and Atchafalaya River system on freshwater flows is illustrated graphically in Orlando et al. (1993): see Figure 2.8.

Increasing flows of the Mississippi River seem to be associated with increased runoff from the upper Mississippi and Ohio River basins and have been attributed to changes in climate and land use (Schilling et al., 2010; Tao et al., 2014). One stark indication of these increased discharges is the increase in the operation of the Bonnet Carré Spillway. The spillway is designed to divert about 20 percent of the floodwaters from the Mississippi River through Lake Pontchartrain into Mississippi Sound and the Gulf of Mexico, relieving stress on the river levees and reducing flood risk for the City of New Orleans.[10] Although the spillway was opened only eight times from 1931 (when it was completed) until 2008, it has been opened seven times since then (2008, 2011, 2016, and consecutively in 2018, 2019 [twice], and 2020). Gledhill et al. (2020) detail the

[9] See https://www.usgs.gov/centers/lmg-water/science/streamflow-alteration-assessments-support-bay-and-estuary-restoration-gulf.

[10] See https://www.mvn.usace.army.mil/Missions/Mississippi-River-Flood-Control/Bonnet-Carre-Spillway-Overview.

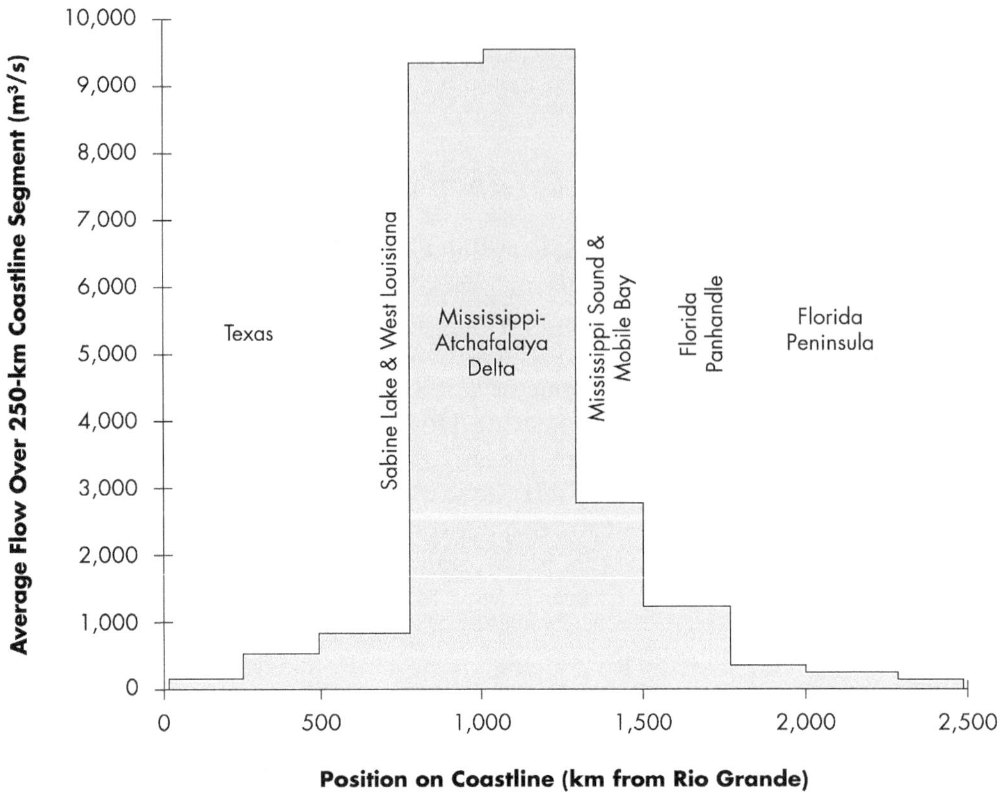

FIGURE 2.8. General variation of river flow around the Gulf of Mexico. The authors note that inflow is not a smooth function of coastline position and that these data are smoothed over 250 km segments. SOURCE: Redrawn from Orlando et al., 1993.

deleterious effects of the two 2019 openings of the spillway—an unusual mortality event for dolphins, a harmful algal bloom, and the decimation of the oyster population in western Mississippi Sound. A freshwater plume extended as far away as 180 km from the spillway.

Daily streamflow discharge data (excluding the Mississippi River) generally document a downward trend in freshwater inflow to the GoM since 1950. Conversely, there is evidence that Mississippi River discharge has increased during this period. Freshwater inflow controls salinity in downstream coastal waters and nutrient and sediment input, which can affect the restoration success of marshes, submerged aquatic vegetation, and in particular, oyster reefs.

Nutrient Inputs to the GoM

Nutrient inputs or loading (nitrogen and, to some degree, phosphorous) from terrestrial sources as well as from atmospheric deposition are critical to the function of coastal and estuarine systems. However, excessive nutrient inputs, especially in more recent decades, have led to eutrophication via increased organic matter production (Nixon, 1995; NRC, 2000). This excessive nutrient enrichment has contributed to widespread effects on coastal habitats, including decreased water clarity, increased acidification due to organic matter decomposition, hypoxia, and shifts in species composition (e.g., phytoplankton and fish), among other effects (Diaz and Rosenberg, 2008; Dorgham, 2014). In addition, Montagna et al. (2018) found that the likely effects of climate change on regional GoM hydrology will influence nutrient input rates and subsequent effects on estuarine and coastal habitats and restoration projects.

Hypoxia, in particular at the mouth of the Mississippi River, is one of the most visible effects of excessive nutrient inputs in the GoM (e.g., Rabalais and Turner, 2019) and the source of much study. Several time series have been developed using flow and concentration data (e.g., Whitall, 2008 [1974–2008]; Murphy et al., 2013 [1980–2010]). These data have also been used for statistical modeling of loads, which emphasize interannual variation over decadal time periods. Modeling has also been used to link nitrogen loads to hypoxia and to better understand dominant nitrogen and phosphorus sources in the basin (e.g., Robertson and Saad, 2021). Kleiss et al. (2021) demonstrated statistical tools for use in Mississippi River water quality trends, including loads during the past 40 years, at numerous sites throughout the basin.

Long-term temporal estimates of nitrogen and phosphorous loads to other parts of the Gulf system are not as comprehensive as those for the Mississippi River, but several studies with more limited spatial and temporal scope have been completed. For example, Dunn (1996) developed nutrient inflows to the GoM for the time period from 1972 to 1993 and found that median yields of total nitrogen and phosphorous were significantly lower in the 13 streams west of the Sabine River (in Texas) than in the 24 streams to the east. Turner and Rabalais (1999) measured or modeled load data for 32 of 39 GoM systems located between Florida Bay and the Yucatan. They found annual nitrogen loads ranging from 1.4 to 70 g N m^{-2} yr^{-1} and a Gulfwide average of 22 g N m^{-2} yr^{-1}, within the ranges reported by Boynton and Kemp (2008) based on a global sampling of nutrient loading rates. Rebich et al. (2011) modeled annual nutrient loads for nine Texas systems, with highest yields from the Trinity/Galveston Bay systems. Harned et al. (2004) developed a model of nutrient loads from 18 sub-basins of Mobile Bay as part of the U.S. Geological Survey (USGS) National Water Quality Assessment program and reported a decline in nitrogen loads between 1975 and 1997, from both agriculture and urban areas, and found a clear link between stream water quality and adjacent land uses. Montagna et al. (2018) developed modeled nutrient yields to four Texas estuaries ranging from mesohaline to hypersaline systems and used this spatial gradient as a substitute for expected hydrologic changes due to climate change. USGS researchers have used the SPARROW (SPAtially Referenced Regression On Watershed attributes) model to estimate sources of nutrients and sediments in rivers and streams throughout the United States (e.g., Hoos and Roland, 2019; Robertson and Saad, 2019; Wise et al., 2019). USGS has also assembled a dataset of regional patterns of anthropogenic influences on U.S. rivers and streams from the 1970s through 2012 (Falcone et al., 2019). Additional applications of this dataset to GoM issues could prove useful. Despite these efforts, a need still exists for better coverage of nutrient inputs and analysis and interpretation of their effects both in space and time in GoM ecosystems.[11]

When nutrient inflow data is collected over decades, it can be very useful for assessing management actions related to restoration. For example, in Tampa Bay, Florida, Greening et al. (2014) used long-term (1976–2010) nutrient loading rates and other water quality data to evaluate restoration progress and found that despite lag times on the order of a decade, submerged aquatic vegetation (SAV) expansion did occur following a 50 percent reduction in nitrogen loading. Similar results occurred in Sarasota Bay Florida (Tomasko et al., 2005), as well as for Chesapeake Bay in more recent years (Lefcheck et al., 2018). More recently, Beck et al. (2019) were able to evaluate broad regional outcomes of restoration in Tampa Bay with a focus on best management actions and found that nutrient point source reductions were most effective in restoring water clarity and SAV.

Gulfwide nutrient loading to the GoM has been modeled for specific points in time, but trends are not available. From the limited number of estuaries and the Mississippi River system where long-term loading estimates exist, trends are variable. Nutrient load reduction has been found to be the most effective restoration technique for water quality and SAV in several Florida estuaries.

[11] Kirk Rogers, presentation to committee, September 29, 2020.

Sediment Inflows

Sediments transported by rivers play a critical role in coastal regions by building land, supporting healthy wetlands, and bringing nutrients into coastal ecosystems (Anderson et al., 2016; Elsey-Quirk et al., 2019; Keogh et al., 2019). River-transported sediment supports building and replenishing barrier islands, which are important for storm surge mitigation, maintenance of estuarine salinity gradients (Montagna et al., 2018), and habitat for beach-nesting birds.[12] However, sediments suspended in the water column may have detrimental effects on some restoration or species recovery projects. Suspended sediments, particularly clays from fine alluvial soils, can interfere with filter feeders, such as oysters, and bury the hard substrate needed for their reproduction. SAV communities may also be negatively affected due to decreased light penetration into the water column (Adams et al., 2016; Carr et al., 2010; Choice et al., 2014).

Trends in sediment inflow estimates are less readily available than nutrient estimates, both in long-term measurements and in analysis of the data to produce loads and trends. Efforts were made to include sediment in the USGS SPARROW modeling efforts. Hoos and Roland (2019) discuss the variability and uncertainty of the sediment data. Some data and trends exist for major rivers, such as the Rio Grande, the Brazos, and the Apalachicola,[13] although long-term efforts to establish suspended sediment budgets and trends have been largely restricted to the Mississippi River (Allison et al., 2012, 2017; Keown et al., 1986; Meade and Moody, 2010; Mize et al., 2018). The Mississippi River is the principal contributor of suspended sediment to the GoM; however, annual riverine inflows of freshwater and sediments have varied over the past several decades (Allison et al., 2017) and are anticipated to be further affected by factors such as climate change, which may impact storm frequency and intensity as well as regional to continental-scale rainfall patterns.

Declines in suspended sediment loads in the Mississippi River, decreased connectivity between the river and its floodplain, sea level rise, and subsidence have contributed to the loss of Louisiana coastal wetlands (Kesel, 1988; Tweel and Turner, 2012). These declines are particularly evident in the lower part of the river, where floodplain trapping, loss of stream power, and existing natural and man-made diversions contribute to channel aggradation (Allison et al., 2012). A principal concern for Louisiana is whether enough river-transported sediment is available to support restoration plans (Allison et al., 2012, 2017; Blum and Roberts, 2009), especially plans to construct large-scale diversions of river water to augment wetland acreage in coastal marshes.

Long-term trends in sediment inflow are limited to the few estuarine rivers where long-term sediment data have been collected; these trends show annual variations over time. Declines in Mississippi River sediment loads have contributed to marsh loss and may affect restoration plans by reducing the expected amount of sediment transported from the river.

AMBIENT WATER QUALITY OF ESTUARINE AND COASTAL RECEIVING WATERS

Changes in ambient water quality[14] in estuaries are largely driven by the inputs discussed above, namely, hypoxia and harmful algae blooms (HABs) (denoted as acute pressures and stressors in Figure 2.1). These can directly affect the success of restoration projects, individually and collectively. These projects include seagrass restoration, which depends on such factors as water transparency (summarized in Larkum et al., 2006; Choice et al., 2014) and oyster reef creation, which is sensitive to salinity, dissolved oxygen, and water temperature (Coen and Luckenbach, 2000; NASEM, 2017). Trend analyses of water quality measurements in GoM coastal and estuarine waters can provide a means to assess change in ecosystem services and can serve as an indicator of cumulative effects of multiple habitat restoration and water quality improvement projects (see Figure 2.1) (Lester and Fairweather, 2011; Sherwood et al., 2016).

---

[12] See https://la.audubon.org/news/barrier-islands-critical-restoration-project-people-and-birds.

[13] See https://nrtwq.usgs.gov/nwqn/#/.

[14] As defined by the U.S. Environmental Protection Agency (EPA): "[A]mbient refers to open waters such as rivers, lakes, and streams, as opposed to closed water supply systems that distribute treated water or wastewater" (https://www.epa.gov/wqs-tech/supplemental-module-human-health-ambient-water-quality-criteria, para. 1).

Assessing cumulative effects of multiple restoration projects using water quality indicators has been successfully applied in regional GoM examples, including Tampa Bay (Greening et al., 2014; Tomasko et al., 2018) and Galveston Bay (HARC, 2020). Water quality monitoring can provide information to assess the effects of acute events on habitats; for example, tracking sediment plumes after hurricanes or heavy rainfall (Moreno-Madrinan et al., 2010), or oil spills (summarized in NRC, 2013). Monitoring and tracking water quality parameters can also show effects of long-term environmental trends, such as changing land use patterns, improvements in wastewater treatment, or shifts in the use of onsite septic systems (Lusk et al., 2017; McCarthy et al., 2018b; Withers et al., 2014. However, Baldera and colleagues (2018) found that ongoing and planned restoration programs will likely not be sufficient to fully restore water quality in areas larger than individual estuaries and waterbodies.

NASEM (2017) recommended that restoration programs "would greatly benefit from working together to identify strategic opportunities in the Gulf of Mexico to maximize the effectiveness and utility of monitoring while also reducing the overall cost of long-term monitoring across the Gulf region" (p. 4). Toward that end, NOAA and USGS formed the collaborative Council Monitoring and Assessment Program (CMAP) to prepare an inventory of existing habitat and water quality monitoring and mapping programs for GoM programs, to provide "essential information to support the development, selection, and application of effective management and restoration alternatives, and inform adaptive management decisions at the local, state, and regional levels" (NOAA and USGS, 2019, p. 1). To date, the CMAP inventory assessment of water column monitoring programs has found that 68 percent (247 programs) are conducted at the local (rather than state or federal) level. The majority of the water column monitoring is done in estuarine, nearshore marine, and riverine settings (NOAA and USGS, 2019).

Salinity

Although salinity in the open Gulf is generally constant, GoM estuaries reflect the integrated effects of highly variable mixing processes, including freshwater inflows; connectivity between estuaries and the open ocean; physical forcing from wind, currents, and tides; and precipitation and evaporation (Solis and Powell, 1999). Changes and trends in salinity are a critical aspect of understanding the ecological structure of faunal communities (Christensen et al., 1997).

Orlando et al. (1993) provided a synthesis of salinity information for 26 principal GoM estuaries. At the time of their study, the authors noted the inadequacy of standardized salinity data for GoM estuaries and said that "the characterization of salinity variability at certain time scales is limited or impossible" (Orlando et al., 1993, p. *v*). Their report supplies typical seasonal distributions of salinity within the estuaries under normal hydrologic conditions at that time. They also proposed a classification of estuaries by salinity behavior, which included magnitude of variability and average annual salinity.

The Naval Research Laboratory processed MODIS-Aqua satellite imagery covering the Gulf of Mexico for a 5-year period (January 2005 to December 2009). The resulting maps provided a broad overview of GoM spatial salinity patterns and indicated seasonal fluctuations in the extent of salinity ranges associated with rainfall and runoff patterns: see Figure 2.9. Similar to those efforts, the National Aeronautics and Space Administration (NASA) conducted a series of successful efforts to measure sea surface salinity and temperature, capturing events such as the 2011 Mississippi River flood and the effects of Hurricane Katia.[15] Additionally, Vazquez-Cuervo et al. (2018) compared in situ salinity measurements to four remotely sensed salinity datasets to help describe seasonal variability.

As part of an analysis of the effects of drought, Petkewich and colleagues (2019) developed a Coastal Salinity Index that included some GoM stations. More recently, in situ monitoring data compiled by Rodgers and Swarzinski (2019, revised 2021) included available salinity measurements from Upper Laguna Madre, Texas, to Rookery Bay, Florida, over the period 1990–2019. They found that while the overall station coverage was sufficient to assess regional trends, several spatial gaps remain, including the absence of monitoring

[15] See https://salinity.oceansciences.org/science-results.htm.

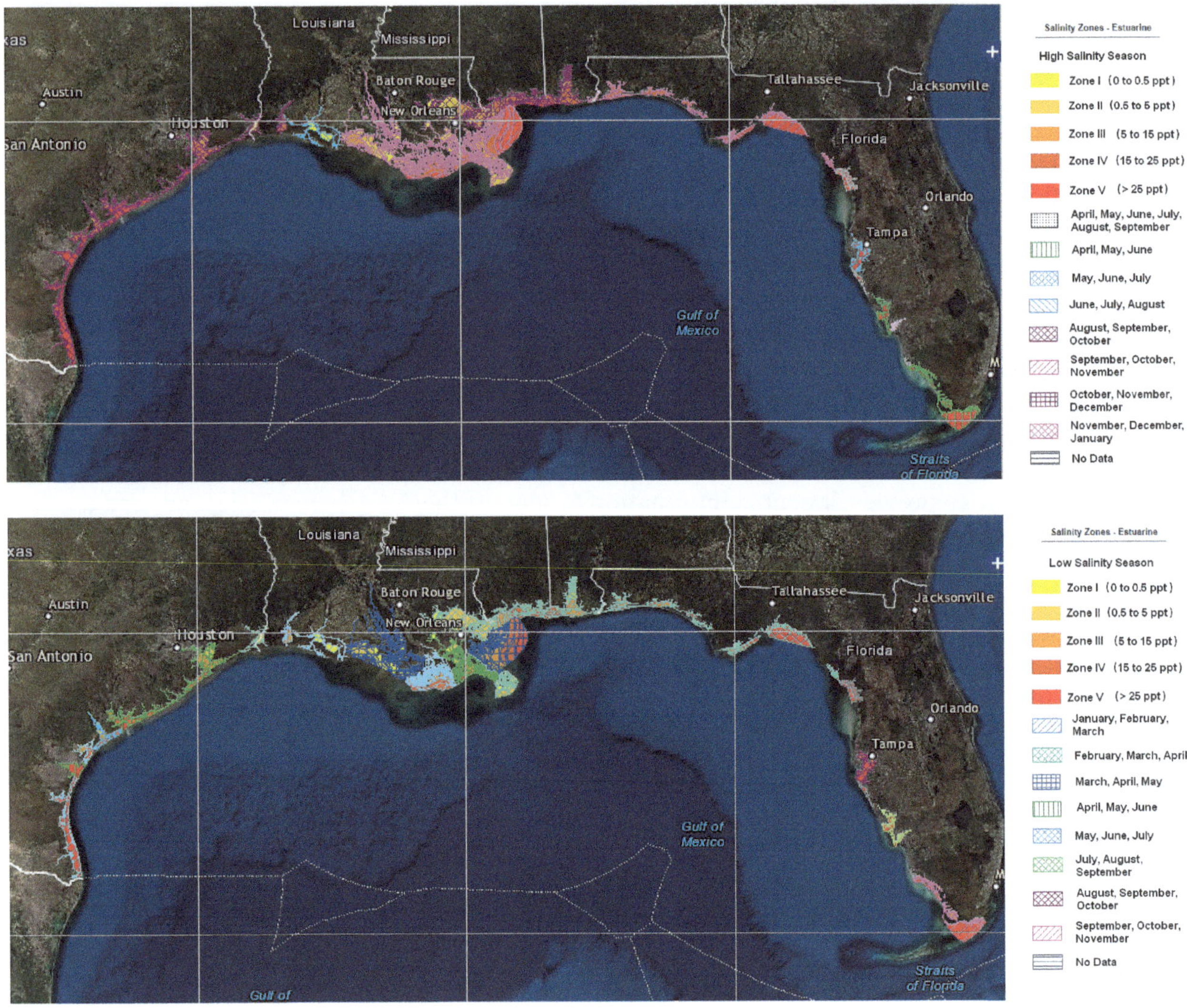

FIGURE 2.9. A five-zone scheme for high and low salinity seasons in the Gulf of Mexico. (A) The "high" salinity season is typically the time of year with the lowest freshwater inflow to estuaries. (B) The "low" salinity season is typically the time of year with the highest freshwater inflow to estuaries, which differs among estuaries. The low-salinity season in Aransas Bay, Texas, occurs from January to March, but in Sabine Lake (Texas and Louisiana), the low-salinity season occurs from March to May. SOURCE: Nelson, 2015; accessed at https://www.ncei.noaa.gov/maps/gulf-data-atlas/atlas.htm?plate=Salinity-Zones.

stations in the Florida Panhandle (USGS, report in preparation, 2021). In addition, spatial distribution within the estuaries is often incomplete and does not cover the range of salinity zones from freshwater to marine waters.

Climate change has the potential to cause changes in streamflow, sea level, oceanic salinity, wind stress, and hurricane intensity, all of which can affect coastal salinity (Ross et al., 2015). Severe tropical storm events have the potential to decrease the salinity of waters over oyster reefs due to heavy rainfall (ADCNR et al., 2021), while also pushing saline water far inland into brackish and freshwater wetlands (Middleton, 2009). The IPCC found that recent observations support that sea level rise (SLR) increases seawater intrusion and raises estuarine salinity, and noted that salinization in estuaries is projected to continue in response to SLR, warming, and droughts. This will cause stress to ecosystem functions and biodiversity (IPCC, 2019). GoM estuaries can be expected to exhibit a range of conditions—those with increased freshwater inflows may experience decreases in salinity, while estuaries primarily driven by SLR and oceanic salinity are likely to increase in salinity.

While salinity data are available, gaps remain in their collection and analysis. This limits time-series analyses of salinity for the major estuaries. Changes in salinity due to climate change are likely to affect the extent of saltwater intrusion in GoM estuaries, which will affect plant and animal distribution and thus impact restoration and recovery efforts.

Nutrients, Eutrophication, Hypoxia, and HABs

Although a number of regional and statewide water quality datasets have been used to track trends in ambient water quality both before and after DWH, an assessment of GoM-wide trends in water quality is not readily available. EPA's National Coastal Condition Assessment (NCCA) reports are currently available for four periods (1999–2001, 2005–2006, 2010, and 2015). They provide a snapshot of conditions at locations across the GoM and allow statistical changes to be estimated, but are not adequate to provide trends (EPA, 2015, 2021). The GoM estuaries show steady increases in the percent area rated "poor" for the estuarine eutrophication index from 2005 (14 percent) to 2010 (25 percent) to 2015 (28 percent) (EPA, 2021). Nutrients, chlorophyll *a*, water clarity, and dissolved oxygen are included in the estuarine eutrophication index (EPA, 2021).

Kennicutt (2017) summarized water quality results from local, state, and federal sources for the 1990s through the mid-2000s (prior to DWH). This includes two NOAA National Eutrophication surveys (Bricker et al., 1999, 2008), NCCA reports (EPA, 2015), and regional and local water quality programs. The NOAA 1999 National Eutrophication Survey collated water quality information collected by local and regional programs for 38 estuarine areas around the GoM (Bricker et al., 1999). Their assessment concluded that "the expression of high eutrophic conditions is extensive, and human influence is substantial, in the Gulf of Mexico region" (Bricker et al., 1999, p. 32). The most significant issues were low dissolved oxygen and loss of submerged aquatic vegetation. The 2007 assessment (Bricker et al., 2008) found similar results, with few areas exhibiting improved eutrophic conditions since the 1999 assessment.

Large areas of seasonal bottom-water oxygen deficiency and hypoxic conditions on the northern GoM continental shelf adjacent to the Mississippi River have been documented since the 1950s, shown in Figure 2.10, with an acceleration of worsening severity during the 1970s (Rabalais and Turner, 2019). The area of hypoxic bottom waters in the GoM west of the Mississippi River can reach up to 23,000 km^2 in midsummer[16]

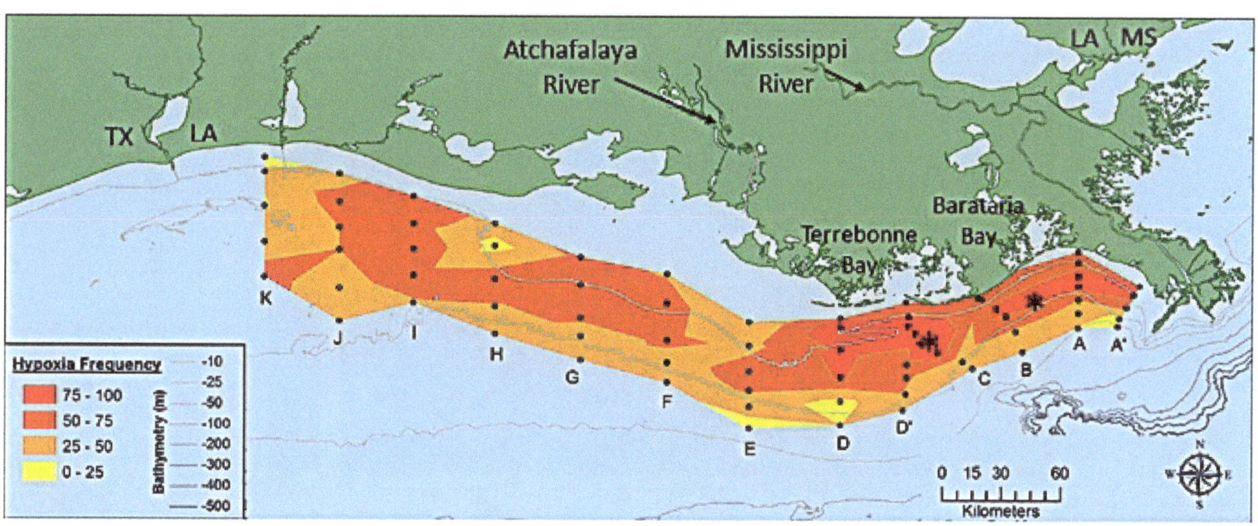

FIGURE 2.10. Frequency of bottom-water hypoxia from shelf-wide mapping, 1985–2014. Frequency is determined from stations for which there are data for at least half all cruises. Asterisks indicate locations of near-bottom oxygen meters.
SOUCRE: Rabalais and Turner, 2019.

[16] See http://www.gulfhypoxia.net.

and correlates best with the nitrate-nitrogen load of the Mississippi River in the previous May (Turner et al., 2012). Openings of the Bonnet Carré Spillway have also been associated with bottom-water hypoxia in the Mississippi Sound (Gledhill et al., 2020): see Figure 2.11. Hypoxia also occurs east of the Mississippi River, where it is becoming more frequent (Dzwonkowski et al., 2018), and in nearshore GoM waters where it can impact nearshore and estuarine restoration efforts (NASEM, 2017).

Rising SST can increase coastal hypoxia by decreasing oxygen solubility in the water column as well as by contributing to stratification (Altieri and Gedan, 2014). Stratification isolates waters of differing physical characteristics, such as temperature and salinity, which can allow biological processes to deplete oxygen (DiMarco et al., 2012; Rabalais and Turner, 2019) and result in the production of hypoxic waters. The GoM's northern shoreline prevents aquatic species sensitive to hypoxia from migrating further north in response to warming temperatures and hypoxia events. Their options are to adapt, to spend more time deeper in the water column, or to die off (Roman et al., 2019). Coastal zooplankton and fish do not appear to have adapted to the increasing frequency of low dissolved oxygen and seasonal hypoxia (McBryan et al., 2013). Current research suggests that warming SST and hypoxia will affect the entire coastal food web, from primary producers to zooplankton to commercially important shrimp, oysters, and fish species (Karnauskas et al., 2017; Rybovich et al., 2016; Trifonova et al., 2019).

HABs can also be fueled by eutrophication (summarized in Anderson et al., 2021). HABs occur when unusually large numbers of specific phytoplankton come to dominate nearshore, estuarine, bay, or lagoonal waters, displacing the normal seasonal association of algae and causing harm to humans or other species (Anderson et al., 2012; Backer et al., 2013). Red tides caused by *Karenia brevis* are the most well-known and can result in fish kills, respiratory irritation in humans and mammals, and neurotoxic shellfish poisoning (NRC, 2000). The GoM also experiences a variety of other HABs, including several that can directly affect human health and result in shellfish harvesting closures and fish species consumption bans.[17] HABs have increased in coastal waters (Anderson et al., 2021; Glibert et al., 2014) and will likely continue to increase (Schmale et al., 2019). HABs are also occurring more frequently with climate change (Griffith and Gobler, 2020).

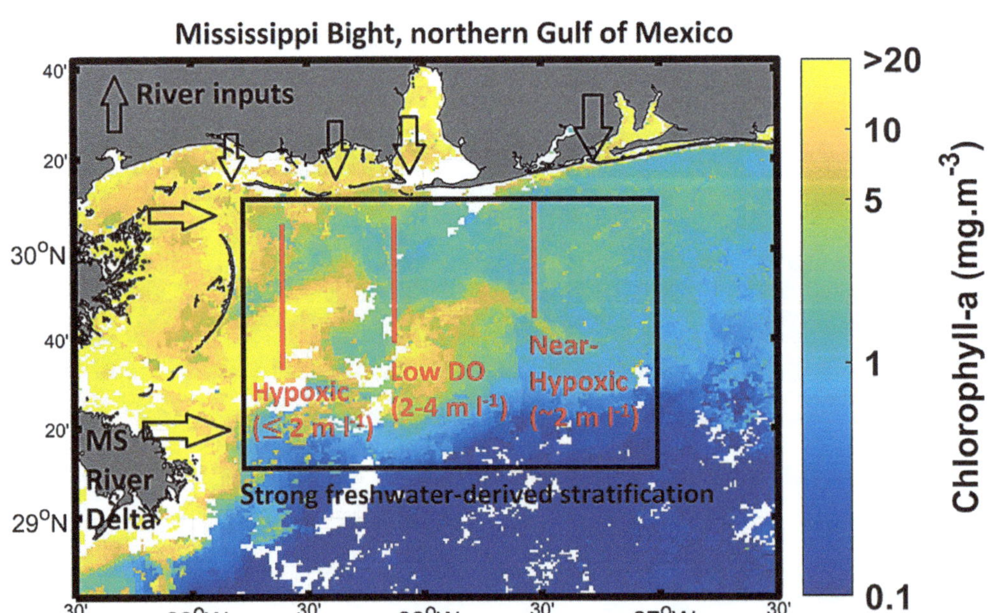

FIGURE 2.11. Hypoxia and low dissolved oxygen in Mississippi Bight. Multiyear observations (2010–2016). SOURCE: Dzwonkowski et al., 2018.

[17] See https://oceanservice.noaa.gov/hazards/hab/gulf-mexico.html.

In addition to having significant human health effects, high concentrations of chlorophyll *a* in HAB blooms can directly affect coastal restoration efforts via reduced light availability for submerged aquatic vegetation restoration (summarized in Anderson et al., 2021). HAB and macroalgae blooms in Tampa Bay and Sarasota Bay have been associated with a reduction in mapped seagrass acreage between 2016 and 2020 (Beck et al., 2021). Increased oxygen demand as HAB cells die and decompose can result in low dissolved oxygen, impacting restored and created oyster reefs and other living resources (Coen and Luckenbach, 2000).

NOAA routinely monitors[18] and forecasts[19] red tide blooms in the GoM, more frequently during periods of likely occurrence and when a bloom is reported; this is the only extant Gulf-wide HAB monitoring effort. All Gulf states have shellfish and beach monitoring programs that can detect HABs, but they are not designed to directly monitor HAB occurrence. A number of efforts have been undertaken to develop monitoring programs, most notably the Gulf of Mexico Coastal Ocean Observing System (GCOOS)'s Harmful Algal Bloom Integrated Observing System, but none have yet moved beyond planning proposals (e.g., Nowlin et al., 2015). Regional monitoring and assessment products developed by NOAA's National Centers for Coastal Ocean Science Algal Bloom Monitoring System[20] deliver near real-time products for use in locating, monitoring, and quantifying algal blooms. Products are currently available for selected regions, including Lake Pontchartrain in Louisiana and the Southwest Florida coast.

Statistically significant decreases in the area percentage rated "good" by the eutrophication index developed by the National Coastal Condition Assessment were observed between two periods, from 1999–2001 to 2005–2006 (16 percent) and from 2005–2006 to 2010 (10 percent), although these data are not adequate to detect long-term trends. Annual assessment of the extent of hypoxia in the nearshore northern GoM show variability associated with spring rainfall in the Mississippi River watershed.

Factors contributing to difficulties in synthesizing water quality monitoring in GoM coastal and estuarine areas include lack of consistent data collection and analysis methods, intermittent funding resulting in spatial and temporal gaps in data collection efforts, and lack of coordinated monitoring objectives that do not include current priority issues such as assessing the cumulative effects of multiple restoration projects. Excess nutrient inputs can result in reduced water clarity, hypoxia, and increases in HABs, which affect restoration efforts for submerged aquatic vegetation.

OTHER SIGNIFICANT GULF OF MEXICO ENVIRONMENTAL TRENDS

Shifting environmental trends driven by climate change are significant in the GoM, as in many parts of the world. However, other anthropogenic drivers, pressures, and stressors also affect coastal resilience in the region, including changing land use patterns, habitat loss, and impaired water quality. In addition to indicators associated with long-term environmental trends, other indicators that illuminate the status and trends of important GoM foundation species are considered below.

Coastal and Estuarine Habitats

A significant percentage of the RESTORE, Natural Resource Damage Assessment (NRDA), and National Fish and Wildlife Foundation (NFWF) funds expended for ecological purposes thus far have been focused on habitat restoration activities (see Chapter 1), including land acquisition, restoration, and enhancements of wetlands and barrier islands; projects to improve water quality; and projects to restore and replenish living coastal and marine resources, including oysters and birds (ELI, 2020). Changes in the occurrence and distribution of GoM salt marshes, seagrass meadows, mangroves, oyster reefs, and barrier islands are useful

[18] The NOAA Harmful Algal BloomS Observing System contains information for the Gulf of Mexico: https://habsos.noaa.gov/.
[19] See https://coastalscience.noaa.gov/research/stressor-impacts-mitigation/hab-forecasts/gulf-of-mexico/.
[20] See https://coastalscience.noaa.gov/research/stressor-impacts-mitigation/hab-monitoring-system/.

indicators of long-term environmental trends because of their broad distribution, their well-understood relationships with economically important finfish and shellfish (recently summarized by Hollweg et al., 2020), and the availability of historical information about their areal extent (Dahl and Stedman, 2013).

Emergent wetlands of the GoM make up around 55 percent of total U.S. salt marshes (e.g., Duke and Kruczynski, 1992; Mendelssohn and McKee, 2000) and are most common in the northern Gulf. Seagrass meadows are broadly distributed across the GoM (Handley and Lockwood, 2020) but are abundant along the Florida coasts and, to a lesser degree, those of Texas. Mangroves now occur across the entire GoM but are patchily distributed across the northern Gulf and usually only represented there by *Avicennia germinans* (the black mangrove), the most cold-tolerant mangrove species (Osland et al., 2013, 2020), while multispecies, mature mangrove forests are currently found only in south Florida. Oyster reefs occur across the entire GoM (Beck et al., 2011), but they are especially widespread in the northern Gulf.

Emergent Wetlands

The predominant emergent wetland type along the northern GoM is the salt marsh, which is dominated by erect and rooted herbaceous hydrophytes (Cowardin et al., 1979). Wetlands make up ~20 percent of the total land cover in the GoM coastal area, even though approximately 1,000 square miles of wetland were lost between 1996 and 2010, primarily due to open water conversion and, to a lesser extent, from development and urbanization (Karnauskas et al., 2017): see Figure 2.12.

Wetland areal extent and change vary geographically, with the most wetlands and wetland loss acreage occurring in coastal Louisiana. Wetland loss is primarily attributed to subsidence, erosion (due to multiple factors), saltwater intrusion, the lack of sediment replenishment needed to sustain marshes, and impacts from storms (Karnauskas et al., 2017). Human activity has altered hydrology in Barataria Bay by decreasing riverine flows and increasing runoff and channelization of wetlands, resulting in significant emergent wetland loss (Day et al., 2021). Industrial activities can also contribute to wetland loss and subsidence. For example, oil and gas development contributed almost 26 percent of net wetland loss in Louisiana from 1955 to 1978 (Baumann and Turner, 1990), and there are an estimated 8,000–10,000 miles of canals that support oil and gas extraction in the GoM and contribute to wetland loss (Turner, 1997; Turner and Cahoon, 1988). Dredging of canals converts some wetlands to open water and others to spoil banks, and canals can change sediment sources and distribution as well as salinity and water levels (Turner and Cahoon, 1987).

The U.S. Fish and Wildlife Service's (USFWS's) National Wetlands Inventory[21] provides access to reports that summarize more than 50 years of monitoring data for the conterminous United States. The most

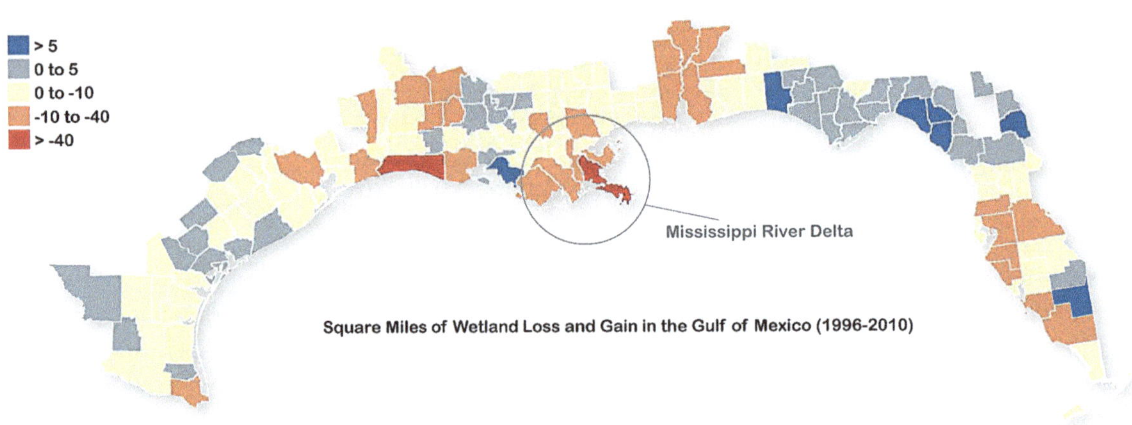

FIGURE 2.12. Square miles of wetland loss and gain by county from 1996 to 2010. SOURCE: Adapted from Karnauskas, 2017.

[21] See https://www.fws.gov/wetlands/status-and-trends/Status-and-Trends-2004-2009.html.

recent assessment (Dahl, 2011) provides information about the Gulf of Mexico, but does not include data on specific bays or estuaries. NOAA's Digital Coast,[22] through the Coastal Change Analysis Program's Regional Land Cover and Change reports, provides inventories of land cover use for the GoM and other regions. This is a resource for sustaining change analyses and indicator development; an independent assessment of its capabilities by McCombs et al. (2016) provides confidence in its accuracy. In addition, the USGS Wetlands and Aquatic Research Center maintains an ongoing program assessing impacts of coastal and watershed changes (Couvillion et al., 2017).

A GoM-wide coastal wetland dataset (Couvillion, 2021; Couvillion et al., 2021) was recently published, and it could be used to create a baseline and change analysis of coastal wetlands utilizing standardized methods. Most National Estuary Program–designated sites in the GoM maintain regular monitoring programs that include emergent wetlands assessments. For example, Galveston Bay National Estuary Program prepares an annual report card[23] that includes the status of emergent wetlands. The Tampa Bay Estuary Program regularly updates a Habitat Master Plan that includes status and trends analysis.[24] The Mobile Bay National Estuary Program addresses status and trends updates in annual work plans.[25] At the state level, the Louisiana Coastal Protection and Restoration Authority maintains a Coastal Information Management System that can be used to assess land change, including emergent wetlands, over time.[26] Gulf-wide assessments of emergent wetlands include the Handley et al. (2012) status and trends report, which updated the Duke and Kruczynski (1992) report on emergent and submerged vegetated habitats. *The Coastal Wetlands Initiative: Gulf of Mexico Review*[27] also provides a comprehensive discussion, including recommendations for undertaking emergent wetlands assessments. Handley et al. (2012) assessed emergent wetlands trends in the northern GoM from 1950 to 2010, focusing on eight systems: Corpus Christi/Nueces/Aransas Bays, Galveston Bay, Barateria/Terrebonne Bay, the Mississippi Delta, Mississippi Sound, Mobile Bay, Florida Panhandle, and Tampa Bay.

Wetlands make up ~20 percent of the total land cover in the GoM coastal area, although significant amounts of wetlands have been lost due to open water conversion and, to a lesser extent, due to development and urbanization. Emergent wetland restoration is a key component of the DWH restoration strategy. Several agencies collect, map, and interpret emergent wetland information, but interpretation and reporting generally lag data collection by several years.

Seagrass

The GoM accounts for approximately 50 percent of all seagrass in the continental United States (Congdon et al., 2018; Handley et al., 2007). Seagrass meadows can be found from Texas to Florida to depths of 30–40m in shallow bays and lagoons offshore on the western shelf of Florida (Zieman and Zieman, 1989). Handley and Lockwood (2020) identified two primary data sources for seagrass occurrence and distribution: the Bureau of Ocean Energy Management (BOEM) and NOAA Marine Cadastre national database and the Florida Seagrass Integrated Mapping and Monitoring Program. Additionally, Handley et al. (2018) provided details of a monitoring approach for Gulf of Mexico seagrasses to sustain ongoing efforts. Meiman and Segura (2019) described in detail a seagrass monitoring program for GoM national parks. On a state level, the Texas Coastal Report Card uses seagrass as an indicator for both coastwide assessment and subregional analyses (McKinney et al., 2019), and the Florida Fish and Wildlife Conservation Commission provides periodic seagrass mapping and monitoring information throughout Florida's coastal waters (Yarbro and Carlson, 2016).

[22] See https://coast.noaa.gov/digitalcoast/.
[23] See https://www.galvbaygrade.org/wp-content/uploads/2015/07/Galveston_Bay_Full_Report_updweb.pdf.
[24] See https://drive.google.com/file/d/1Hp0l_qtbxp1JxKJoGatdyuANSzQrpL0I/view.
[25] See http://www.mobilebaynep.com/images/uploads/library/MBNEP_EPAworkplan2020_2021_ Final.pdf.
[26] See https://cims.coastal.louisiana.gov/Viewer/Map.aspx?guid=f8ec2690-bbb1-4879-ac30-aa44f5878b7f.
[27] See https://www.epa.gov/sites/default/files/2015-04/documents/gulf-of-mexico-review.pdf.

Three seagrass status and trends assessments (Duke and Kruczynski, 1992; Handley et al., 2007, 2020) provide a foundation on which future status and trend analyses can be based. Handley and Lockwood, 2020) noted that gains in seagrass areal extent can be linked to improvements in water quality, while losses are often associated with degradation of water quality, although losses can also be attributed to the effects of hurricanes (Tomasko et al., 2020). Handley and Lockwood (2020) compared gains and losses between the most recent areal extents available in 2007 (as collated by Handley et al. [2007] from information collected between 1987 and 2002) and maps developed from information collected between 2004 and 2017, collated from a variety of monitoring programs. The authors found that, for the areas where seagrass maps are available, the Gulf experienced an overall gain in seagrass extent of almost 24 percent (an increase of 127,910 ha) between these two time periods. Texas, Alabama, and Florida exhibited seagrass gains between the 1990s and the latest statewide assessment; Louisiana and Mississippi both showed seagrass losses between the earlier periods and 2017 reporting periods (Table 2.1). These data on seagrass areal extent in each state can provide useful standards by which the cumulative impacts of seagrass restoration projects can be assessed, with consideration of other pressures and stressors, but only if those restoration sites were within the originally mapped polygons and if the same areas were mapped again.

Gulf-wide estimates of seagrass extent are not available. For the limited areas for which seagrass maps are available, an overall gain in seagrass extent of almost 24 percent between two time periods (1987–2002 and 2004–2017) was observed. Texas, Alabama, and Florida exhibited gains between the 1990s and the latest assessment; Louisiana and Mississippi showed losses. Gains are associated, in part, with improved water quality. The ability to map changes in seagrass extent over time is an important component to measure restoration progress.

Mangrove Forests

The recent expansion of mangroves across the northern GoM and the displacement of salt marshes by the black mangrove (Armitage et al., 2015) has implications for future salt marsh restoration projects (as discussed under "Tropicalization"). Black mangrove is expanding northward into *Spartina*-dominated salt marshes, one of the coastal habitats often targeted for restoration (Comeaux et al., 2012; Spies et al., 2016; Osland et al., 2021). Thus, the effects of the ongoing mangrove colonization of GoM salt marshes will likely be an important consideration in planning for the restoration of salt marshes (Feller et al., 2017).

Although there is currently no known assessment of mangrove areal extent in the GoM, national programs, such as the USFWS National Wetlands Status and Trends program (Dahl, 2011), map different categories of wetlands, including an estuarine intertidal forested/shrub category that is dominated by mangrove shrub in the northern Gulf and mangrove trees in south Florida. The most recent National Wetlands Status and Trends report estimated an areal extent loss of 0.2 percent for this category between 1998 and 2004. Day et al. (2018) provide guidance to developing resilience indicators for mangrove ecosystems that could be of interest to restoration planners. Lewis (2005) summarized the status of mangrove restoration up to that time, providing guidance for future restoration. Although mangrove restoration efforts have been hindered by poor restoration strategies and the decline of mangroves across the tropics (Feller et al., 2017), large-scale mangrove restoration has been successful in south Florida (Lewis, 2005; Rey et al., 2012; Begam et al., 2017).

Although long-term mapping results of mangrove areal extent at the estuary scale are available for some locations, trends in Gulf-wide mangrove areal extent are not known. The last assessment to include GoM mangrove estimates was based on information collected between 1998 and 2004. The effects of ongoing colonization of salt marshes in the northern GoM by mangroves will be an important consideration in salt marsh restoration projects.

TABLE 2.1. Acreage Change by State Since the 2007 Seagrass Status and Trends Report (Handley et al., 2007)

State	Status & Trends 2007 (1987/88 through 2002)		Seagrass Update 2020 (2004 through 2017)		Gain/Loss +/-
	Date	Hectares/ Acres	Date	Hectares/(Acres)	
Texas	1987/88; 1992/1994; 1998	88,393 (218,418)	2004/07/12	92,475 (228,506)	+4,083 (+10,088)
Louisiana	1995	4,512 (11,149)	2011	1,058 (2,614)	-3,454 (-8,535)
Mississippi	2002	1,194 (2,950)	2007	816 (2,017)	-378 (-933)
Alabama	2002	2,690 (6,646)	2015	4,440 (10,570)	+1,588 (+3,924)
Florida	1992-2002	443,412 (1,095,675)	2007-2017	569,484 (1,407,196)	+126,072 (+311,512)

SOURCE: Handley and Lockwood, 2020.

Oyster Reefs

Oyster reefs are widely distributed throughout GoM estuaries (Shepard et al., 2018) and near river mouths, as shown in Figure 2.13, and they are healthy and productive enough to account for more wild harvest than any other region in the world (Beck et al., 2011). Oyster reefs provide an array of valuable ecosystem services beyond their value as seafood, including habitat structure used heavily by fish and invertebrates (La Peyre et al., 2019). Reefs differ in faunal composition from adjacent salt marshes and seagrass meadows (Hollweg et al., 2020). There have been recent concerns about oyster populations in the GoM. For instance, in 2020, Florida suspended its Apalachicola Bay wild oyster harvest for up to 5 years[28] due to droughts and lack of freshwater from rivers,[29] and the openings of the Bonnet Carré Spillway in 2019 damaged the oyster population in western Mississippi Sound due to excessive freshwater flows (Gledhill et al., 2020). Even in the relatively unaffected Suwannee River estuary in Florida, inshore intertidal oyster bars are becoming degraded and oyster counts have declined (Moore et al., 2020).

Hurricanes affect oyster stocks and reefs through direct wind and wave action, salinity changes and increased sedimentation associated with storm-related rainfall, and dissolved oxygen depletion (summarized by Mallin and Corbett, 2006). Studies have reported relatively rapid oyster reef recovery times (months), particularly in areas that are frequently disturbed by hurricanes (Livingston et al., 1999). However, oyster fisheries may take longer to recover (months to years) because of both direct impacts and public health concerns from high levels of bacteria in stormwater runoff from flooded upstream areas.[30]

Because they are such a valuable fishery, Gulf states have long-standing oyster monitoring programs that provide valuable information that can be used to assess the cumulative impacts of oyster reef restoration. Murawski et al. (2021) found that the overall abundance of natural oyster populations has been in long-term decline in areas throughout the GoM, and particularly since the mid-1990s (Tunnell, 2017; GSMFC, 2012), as reflected both in relative abundance surveys and fishery landings.

Baggett et al. (2015) exhaustively addressed oyster monitoring approaches and reported little to no monitoring of oyster reefs. In 2017, the DWH NRDA Trustees developed a strategic framework for oyster restoration that summarizes existing status and trends information for GoM oyster reefs (DWH NRDA Trustees, 2017b). Shepard et al. (2018) developed a conceptual model for assessing cumulative impacts of reef restoration efforts and provided a detailed assessment of a broad array of metrics useful as indicators of oyster and ecosystem health. At the state level, the Tampa Bay Estuary Program has an oyster habitat suitability index mapping tool to identify suitable locations for restoration.[31]

[28] See https://myfwc.com/news/all-news/oyster-commission-1220/.

[29] See https://www.npr.org/2020/07/22/894074674/floridas-oyster-beds-devastated-by-years-of-drought-other-pressures.

[30] See https://www.nature.org/content/dam/tnc/nature/en/documents/OysterRestorationintheGulf.pdf.

[31] See https://www.tampabay.wateratlas.usf.edu/oyster-habitat-suitability/.

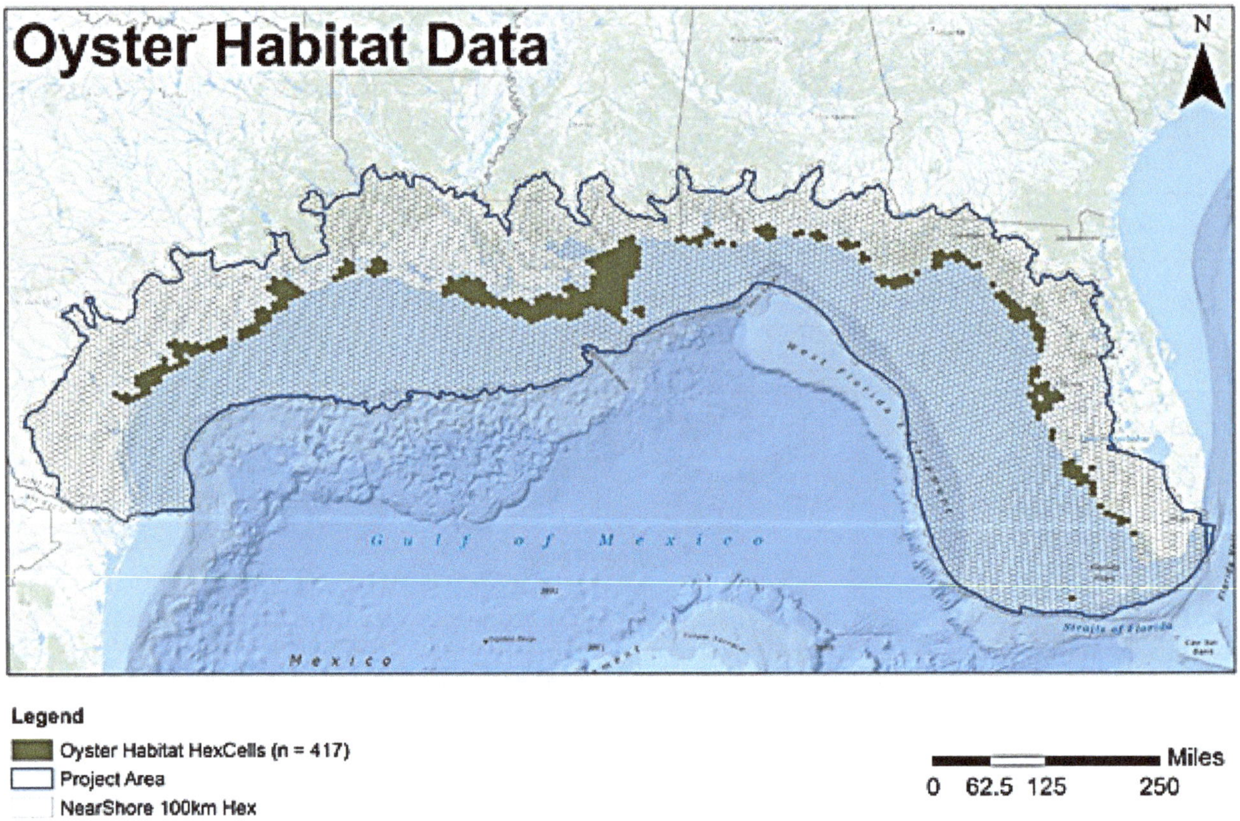

FIGURE 2.13. Distribution of oyster habitat in the northern Gulf of Mexico. SOURCE: Goodin et al., 2018.

The overall natural oyster population has been declining in monitored areas since the mid-1990s. Although Gulf states monitor oyster landings for purposes of managing their commercial fisheries, and the current locations of many oyster reefs have been mapped, Gulf-wide trends in the extent and condition of oyster reefs as habitat are currently unknown. Potential impacts from climatic drivers and anthropogenic stressors, as well as restoration design and implementation, can affect oyster reef restoration.

Barrier Islands

There continues to be considerable interest in barrier island restoration in the GoM (see Oliver and Ramirez-Avila, 2019), both to provide ecological habitat and to protect human resources on barrier islands and in the bays behind them. Studies have been conducted on barrier islands in the five Gulf states: Florida (e.g., Davis, 2016; Hine et al., 2001); Alabama (the Alabama Barrier Island Restoration Assessment,[32] which features a life-cycle response assessment model [Gonzalez et al., 2020]); Mississippi (within the Mississippi Coastal Improvements program [USACE, 2016]); Louisiana (the extensive Barrier Island Comprehensive Monitoring Program and database [e.g., Enwright et al., 2020]); and Texas (e.g., Rinaldo et al., 2021; Vinent et al., 2021).

Barrier islands themselves are dynamic features, but under the stresses of RSLR and increasing storm intensity, restoration planning entails even greater awareness of the underlying geology, geomorphology, and transport mechanisms. Justić et al. (2021) reviewed research of transport processes from GoM rivers and estuaries to the coast, shelf, and ocean—information that can be used not just in management of coastal

[32] See https://www.nfwf.org/sites/default/files/gulf/Documents/al-dauphin-assessment-14.pdf.

resources like barrier islands, but also for movement of hydrocarbons or organisms. Processes creating the Gulf barrier islands differ significantly. Rosati and Stone (2009) identified three general regions, based on sediment source, the availability of littoral and inner-shelf sediment, and the underlying substrate in the northern Gulf. They concluded that design for the barrier islands needs to include forcing processes, such as compressibility of substrates; vegetative cover; and aeolian, longshore, and onshore transport.

The barrier islands in the GoM are dynamic features that are important to both natural and human-made environments. Potential impacts on them include changes due to RSLR and storm frequency and intensity, which may affect restoration planning.

Fisheries

State and federal agencies routinely use standardized methodologies to collect data for many fish species across the GoM. When the standardized methodology is not available, one can employ well-defined formulations to convert data to a comparable format (NRC, 2000). Organizations like the Gulf States Marine Fisheries Commission's GSMFC) Fisheries Information Network[33] coordinate the standardization and collection of commercial and recreational fisheries data. NOAA's fishery monitoring programs in the GoM, especially the Southeast Area Monitoring and Assessment Program (SEAMAP)—a state/federal/university program for the collection, management, and dissemination of fishery-independent data and information in the southeastern United States, including the Gulf of Mexico—along with state fishery-independent monitoring programs--regularly provide accessible databases on fish species. NOAA also maintains databases for both commercial and recreational species for the five Gulf states.[34] The primary acquisition of recreationally sourced federal data is through the Marine Recreational Information Program, which has been in existence since 1979 (Keithly and Roberts, 2017).

Grüss et al. (2018) reviewed GoM fisheries monitoring programs and noted that they include many year-round independent fisheries monitoring programs (i.e., data are generated by direct sampling, not from those involved in the fishing activity), more than in other marine regions. Texas has a long-established program that collects independent and dependent data for both commercial and recreational fisheries and has been doing so continuously for more than 45 years (Martinez-Andrade, 2018). The Florida Fish and Wildlife Conservation Commission's Fisheries-Independent Monitoring group[35] is another such program with readily searchable data. All GoM states participate in the SEAMAP,[36] whose information is readily available to potential users, including restoration practitioners looking to assess the efficacy of restoration efforts (La Peyre et al., 2019; NASEM, 2017; O'Farrell et al., 2017).

Because fisheries data often go back decades and are likely to continue to be collected, they can be valuable for assessing long-term environmental trends. The robust nature of fisheries data allows comparison and analysis between noncontiguous databases and widely separated study sites. Fisheries data are also integrative with other data like habitat and water quality, so they can be used for filling in data gaps when assessing change. Due to the value of economic impact and regulatory requirements, collection of fisheries data (whether federal or state) will continue independently of any restoration activities that it might inform.

NOAA regularly assesses the status of fish stocks in the GoM[37] and elsewhere. In the 2019 assessment, NOAA reported on 15 GoM species and 8 highly migratory species found in the GoM. The data on which such evaluations have been based may also be useful in assessing long-term environmental trends for restoration efforts. Fishery closures in 2010 in response to the DWH spill had a significant, but in most cases short-term, impact on catches. Overall commercial fishery landings declined by 25.3 percent from 2009 to 2010 but rebounded to pre-spill levels within 2 years (Murawski et al., 2016). Murawski et al. (2021)

[33] See https://www.gsmfc.org/fin.php.
[34] See https://www.fisheries.noaa.gov/science-and-data.
[35] See https://myfwc.com/research/saltwater/reef-fish/monitoring/program.
[36] See https://www.gsmfc.org/seamap.php.
[37] See https://www.fisheries.noaa.gov/national/sustainable-fisheries/status-stocks-2019.

analyzed long-term trends for 13 key species with temporal ranges from 15 to 30 years, including 3 fish species (menhaden [*Brevoortia patronus*], spotted seatrout [*Cynoscion nebulosus*], and red drum [*Sciaenops ocellatus*]) and white and brown shrimp. Long-term trends of these key species varied in both time and space. This report emphasized the need for continued monitoring of vulnerable species in order to effectively evaluate the impacts of DWH on ecosystems and highlighted the need for continued modeling and evaluation to better understand the efficacy of habitat restoration projects in the GoM (Murawski et al., 2021).

Gulf-wide and state-level commercial and recreational fisheries data collection efforts are robust, as are fisheries-independent data collection efforts at both the Gulf-wide and state levels. The data are collected for commercial and recreational management and are useful for measuring progress toward fisheries restoration goals, and they are also useful in understanding and responding to pressures, such as oil spills.

Birds

Waterbirds use coastal habitats for forage, nesting, or both. Because of their ubiquitous presence along GoM coastal margins, waterbirds may serve as indicators of ecosystem health (Ogden et al., 2014). Similar to fish, bird life histories, including migratory patterns, integrate diverse habitats and have considerable economic (e.g., birdwatching, hunting) and ecosystem value, which has prompted extensive monitoring efforts (DeMaso et al., 2019). Burger (2017) summarized data availability, status, and trends for 15 indicator species and found that monitoring may not be robust, because of differences in data collection methods and number of surveys. In the GoM, there is a concerted effort (the Gulf of Mexico Avian Monitoring Network[38]) to coordinate the various monitoring efforts to support both management and restoration efforts. Breeding bird surveys (Sauer et al., 2011) are also a good source of data for restoration planners. The North American Breeding Bird Survey is an important resource,[39] as are shorebird annual surveys, such as those supported by Audubon.[40] In addition, citizen science efforts like eBird,[41] Partners in Flight,[42] and the Audubon Christmas Bird Count[43] may be useful, depending on location and the question being asked (Niven and Butcher, 2011).

Burger (2017) assessed 15 species to examine the health of GoM avifauna before DWH. He found habitat loss, both anthropogenic and natural, was the primary threat facing GoM birds. The results of this analysis show mixed results, with populations of a number of indicator species improving (osprey, brown pelican) or remaining stable (common loon, mottled duck), and others declining or in question (reddish egret, clapper rail). DeMaso et al. (2019) summarized waterfowl trends for many species across the northern GoM, showing similar mixed results. The DWH Trustees published the *Strategic Framework for Bird Restoration Activities* in 2017. The document outlines restoration goals and outcomes and summarizes progress to date on restoration, as well as monitoring activities underway in the Gulf.[44]

Long-term trends in waterbirds located in the GoM show mixed results, with some species improving, some remaining stable, and others declining. Habitat loss was the primary stressor. Waterbirds depend on coastal habitats for nesting and foraging and can provide indicators of ecosystem condition and restoration progress. In the GoM, the Gulf of Mexico Avian Monitoring Network coordinates the various monitoring efforts to support both management and restoration efforts.

---

[38] See https://gomamn.org.

[39] See https://www.usgs.gov/centers/pwrc/science/north-american-breeding-bird-survey?qt-science_center_objects=0#qt-science_center_objects.

[40] See https://www.audubon.org/content/audubon-coastal-bird-survey.

[41] See https://ebird.org/home.

[42] See https://partnersinflight.org/what-we-do/science/databases.

[43] See https://www.audubon.org/conservation/science/christmas-bird-count.

[44] See https://www.gulfspillrestoration.noaa.gov/sites/default/files/wp-content/uploads/Birds_Strategic_ Framework_06.23.17.pdf.

Marine Mammals

Marine mammals in the GoM area include the West Indian manatee (*Trichechus manatus*; limited to Florida) and the common bottlenose dolphin (*Tursiops truncates*; common across the region) (Ortega-Ortiz et al., 2004). Vollmer and Rosel (2013) provide a review of bottlenose dolphin research in the GoM that could be valuable in assessing this species as an integrative indicator of long-term environmental trends. Bottlenose dolphins are common in waters adjacent to highly populated and industrialized GoM coastal waters and consume large quantities of fish, which may bioaccumulate various toxins (Würsig, 2017). Bottlenose dolphins were found to be some of the most vulnerable to the DWH oil spill due to their strong affinity to specific geographic areas and their surface engagement while breathing (Murawski et al., 2021; Wells et al., 2017). In 2017, the DWH NRDA Trustees published the *Strategic Framework for Marine Mammal Restoration Activities*, which outlined restoration goals and outcomes and summarized progress to date on restoration, as well as monitoring activities underway (DWH NRDA Trustees, 2017c).

NOAA prepares regular stock assessment reports for all protected marine mammals resident in U.S. waters, including the common bottlenose dolphin.[45] *The 2018 NOAA assessment looked at 31 bay, sound, and estuary stocks across the GoM but found insufficient data to determine population trends. Status of marine mammal stocks can provide an indicator of recovery from oil spill impacts.*

Sea Turtles

There are five species of sea turtles that have been found in the GoM (Hart et al., 2020). Understanding their population abundance is important when developing conservation actions, but it is difficult because of the scale of the turtles' distribution, migratory nature, and cryptic early life stages (Ceriani et al., 2019). Most U.S. nesting beaches have programs in place to count sea turtle nests (NRC, 2010) and with appropriate coordination and cooperation can provide population estimates that are useful to restoration planners. All five species are found on Florida beaches; since 1979, the Florida Statewide Nesting Beach Survey[46] has annually surveyed 215 beaches over 825 miles. In Texas, the Padre Island National Seashore Sea Turtle Science and Recovery Program[47] has been in operation since the late 1970s and coordinates nest counts statewide. Long-term trends based on beach nesting data indicate that Kemp's ridley has made a remarkable recovery from the brink of extinction in the early 1980s and that loggerhead nesting on Florida's Gulf coast varied annually between 1979 and 2011 (Valverde and Holzwart, 2017). In Alabama, Share the Beach is a volunteer sea turtle nesting and hatching monitoring program founded in 2005, currently supported with funding from the Alabama Trustee Implementation Group Restoration Area.[48]

Satellite tracking has provided information about forage areas (Gredzens and Shaver, 2020; Shaver et al., 2017) that can be of use to restoration planners. Piacenza et al. (2019) reviewed monitoring strategies in an attempt to improve their value for conservation planning. They found that depending on objectives, 10–20 years of data were adequate for estimating population trends. Because all GoM sea turtles are protected under the Endangered Species Act,[49] jurisdiction for their recovery and conservation is shared by NOAA and USFWS, which assures ongoing monitoring to support long-term population assessments. In 2017, the DWH NRDA Trustees published the *Strategic Framework for Sea Turtle Restoration Activities*. This document outlined restoration goals, as well as progress to date on restoration and monitoring (DWH NRDA Trustees, 2017d).

---

[45] See https://www.fisheries.noaa.gov/national/marine-mammal-protection/marine-mammal-stock-assessment-reports-species-stock#cetaceans---dolphins.

[46] See https://myfwc.com/research/wildlife/sea-turtles/nesting/monitoring/.

[47] See https://www.nps.gov/pais/learn/seaturtles.htm.

[48] See https://www.gulfspillrestoration.noaa.gov/sites/default/files/2018-06_AL_RP%20II%20Sea %20Turtles%20Fact%20 Sheet%204_11_18.pdf.

[49] See https://www.fisheries.noaa.gov/sea-turtles.

Long-term trends based on sea turtle beach nesting data indicate that Kemp's ridley populations have increased since the early 1980s and that loggerhead nesting on Florida's Gulf Coast varied annually between 1979 and 2011. Because GoM sea turtles are protected, monitoring to support long-term population assessments is assured. Status of sea turtle nesting and hatching can provide an indication of recovery from long-term effects of oil spills, SLR impacts to nesting beaches, and potential effects of changes to foraging areas.

Invasive Species

A Strategy for a Healthy Gulf of Mexico (NOAA, 2015b) recognized invasive species as a significant stressor, contributing to Gulf-wide habitat loss and ecosystem decline. Florida is particularly susceptible to invasive species because of its location and climate (Doren et al., 2009; Wallace et al., 2016). All Gulf states have active invasive species identification and control programs, as well as searchable databases, and are coordinated with national programs (such as the USFWS Invasive Species Programs[50] and the USGS Invasive Species Program[51]). For example, USGS maintains a nonindigenous aquatic species database[52] and provides information on invasive fish.[53] GCOOS maintains an interactive map of the invasive lionfish and Asian tiger shrimp,[54] and the World Register of Introduced Marine Species[55] provides access to a number of databases that may be useful in tracking invasive species.

The Nonindigenous Aquatic Nuisance Prevention and Control Act of 1990 limits the introduction, spread, and impact of aquatic invasive species in U.S. waters. The Gulf of Mexico Program Invasive Species Focus Team's initial report summarized existing information. Subsequently, the Commission for Environmental Cooperation prepared a report on invasive species status, including an assessment of aquatic invasive species in the Rio Bravo/Laguna Madre Corridor (Mendoza et al., 2011). They noted 373 exotic species, mostly originating from the Atlantic coastal region, and reported that a lack of ecological and biological information hampered efforts to reduce the impacts of these invasive species on GoM biodiversity.

State agencies in the GoM maintain active invasive species programs that are closely coordinated with federal counterparts. However, states do not maintain databases of long-term trends in invasive species distribution or expansions, which would be of value for restoration practitioners. These data, if more readily available, would be useful to restoration planners or practitioners because these plants and animals can affect the success of restoration efforts.

Land Cover Changes

Naturally occurring land cover integrates a given site's climate, geology and soils, and vegetation over decades or longer. Over shorter time scales, land cover can be affected by naturally occurring disturbances, such as storms, floods, and fires, and human activities, such as population change; industrial, agricultural, or urban development; deforestation or reforestation; water diversion; and infrastructure such as roads.[56] Land cover affects or influences environmental conditions in many ways, including water quality, hydrology, and habitat and associated species composition. Changes in land cover can alter hydrologic regimes and runoff patterns, thus potentially acting as a stressor on environmental restoration projects and impacting the cumulative effects of multiple restoration projects.

---

[50] See https://www.fws.gov/invasives/programs.html.

[51] See https://www.usgs.gov/ecosystems/invasive-species-program.

[52] See https://nas.er.usgs.gov/.

[53] See https://www.usgs.gov/ecosystems/invasive-species-program/science/invasive-fish.

[54] See https://gcoos.org/invasive-species/.

[55] See https://www.marinespecies.org/introduced/.

[56] See https://www.epa.gov/report-environment/land-cover.

The Multi-Resolution Land Characteristics Consortium[57] generates land cover information at a national scale for a wide variety of environmental, land management, and modeling applications (Yang et al., 2013). Its products include the National Land Cover Database and the Coastal Change Analysis Program (Jin et al., 2013). EPA's Report on the Environment[58] uses the National Land Cover Database to track trends in and effects of land cover. Karnauskas et al. (2017) summarized land cover change between 1996 and 2010 and found that urban land cover in coastal watershed counties around the GoM increased by 15 percent during this period. While urban expansion is significant across the region, the increase in development occurred at a much higher pace in certain geographic areas, such as Houston, Texas, and Tampa, Florida (Karnauskas et al., 2017) (Figure 2.14).

Urban land cover in coastal areas around the GoM increased by 15 percent between 1996 and 2010, with some geographic areas, including Houston and Tampa, increasing at much higher rates. Data and other materials on land use change are available from the National Land Cover Database and the Coastal Change Analysis Program. Land use changes can have profound effects on coastal habitats and can interact with restoration efforts, including reduced area for restoration and increased stormwater runoff.

Oil Spills

The oil and gas industry is a strong presence in the Gulf of Mexico. There are ~1,862 platforms in the Gulf (as of April 2019)[59] and 45,310 miles of pipeline (as of 2016) (Kaiser and Narra, 2019). Production and refining is concentrated in the western Gulf of Mexico: see Figure 2.15. The history of large GoM oil spills is well known, including four very large spills—*Ixtoc* (1979; 10.2 million barrels of crude oil; Dokken, 2011), *Mega Borg* (1990; 100,000 barrels; Payne et al., 2005), *Deepwater Horizon* (> 3 million barrels,[60] 2010) and *Taylor*/MC20 (2004–present[61]). It can often be difficult to quantify exactly how much oil has been spilled; such was the case for DWH. While estimates varied from 3.26 million barrels (Fitch et al., 2013) to 5.14 million barrels (Lehr et al., 2010), the amount decided on for litigation purposes was 3.19 million barrels.[62]

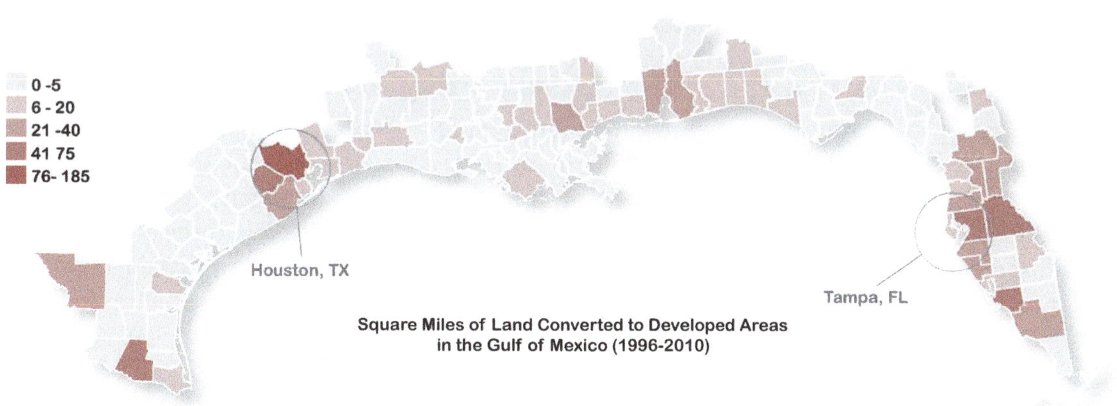

FIGURE 2.14. Total area of land converted to developed area by county, 1996–2010.
SOURCE: Karnauskas, 2017.

[57] See https://www.mrlc.gov.

[58] See https://www.epa.gov/report-environment/land-cover.

[59] See https://www.bsee.gov/faqs/how-many-platforms-are-in-the-gulf-of-mexico.

[60] See https://www.science.org/content/article/after-geoscientists-joust-judge-rules-bp-gulf-spill-totaled-319-million-barrels-oil

[61] See https://darrp.noaa.gov/oil-spills/taylor-energy.

[62] *Consent Decree for Deepwater Horizon—BP Gulf of Mexico Oil Spill*: See https://www.epa.gov/sites/production/files/2016-02/documents/deepwaterhorizon-cd.pdf.

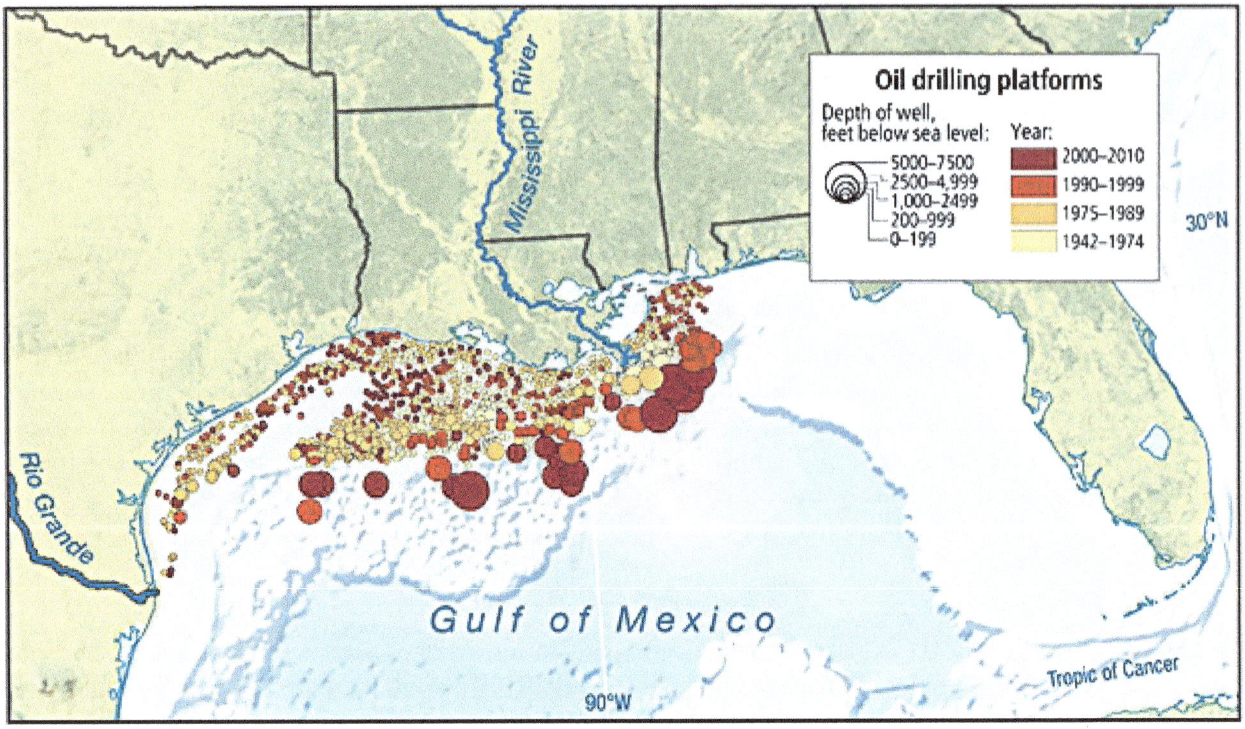

FIGURE 2.15. Oil drilling platforms in the Gulf of Mexico from 1942 to present. The size of the circle corresponds to the depth of the well, and the color corresponds to the year. SOURCES: https://www.ncei.noaa.gov/maps/gulf-data-atlas/atlas.htm?plate=Offshore%20Structures; https://www.bsee.gov/sites/bsee.gov/files/reports/shallow-water-report-01.pdf; http://www.noia.org/wp-content/uploads/2016/08/Deepwater-Gulf-of-Mexico-Report-2014.pdf.

The *Taylor* oil spill, which occurred in 2004, is still ongoing, and litigation resulted in a settlement in March 2022.[63] A NOAA technical memorandum released in 2019 estimated a daily oil flux between 9 and 47 barrels of oil per day using an acoustic survey method, or between 19 and 108 barrels per day using a bubblometer survey method (Mason et al., 2019), resulting in a very rough estimate of between 55,000 to 660,000 barrels released.[64] From an environmental management perspective, this may be viewed as a data gap in describing the location, extent, and dynamics of an acute stressor on the biota.

In addition to large-scale oil spills, there are also numerous other incidents involving hazardous materials, the cumulative impacts of which remain unknown. In 2020 and the first 6 months of 2021, there were over 200 oil and chemical spills in the GoM.[65] Oil spills in the Gulf of Mexico can occur from pipelines, wells, and transportation. In 2021, the U.S. Government Accountability Office released a report noting that updated regulations are needed to improve pipeline oversight and decommissioning: ~97 percent (18,000 miles) of decommissioned GoM pipelines have been left in place on the seafloor since the 1960s, and there is no funding for removal of these pipelines, even if there are risks of a future spill (GAO, 2021). The risks of pipelines decommissioned in place include not only the oil and gas that remain, but the hazards for commercial fishing or navigation, movement of the pipelines, and interference of other uses of the outer continental shelf.

The imprint of anthropogenic oil is evident on key components of the GoM ecosystem. A 2020 study examining ~2,500 finfish from 359 Gulf-wide locations by the Gulf of Mexico Research Initiative (GOMRI) found evidence of oil exposure in every individual examined, with especially high concentrations in some highly prized recreational or commercial species, including yellowfin tuna, golden tilefish, and red drum

[63] See https://www.doi.gov/restoration/settlement-reached-taylor-energy-oil-spill-largest-oil-spill-you-never-heard

[64] Assuming 6,129 days of total oil release (September 16, 2004, to June 28, 2021). This estimate does not account for any capping or capture of oil.

[65] Calculated from NOAA's Office of Response and Restoration raw incident data: https://incidentnews.noaa.gov/raw/index.

(Pulster et al., 2020). Organismal and community responses to oil impacts vary from species to species, with lethal or sublethal effects, including growth inhibition, changes in group behavior, decreased swimming speeds, cardiac defects, reduced immune response, and high reproductive failure rates (Murawski et al., 2021). Shallow ecosystems appear to have more resilience to oil effects than deep sea and mesopelagic systems (Halanych et al., 2021). At the landscape scale, large oil spills such as DWH can affect multiple species or ecosystems. For example, marshes provide important ecological services that can be impacted by oil spills and may be further damaged by recovery efforts, exacerbating impacts and slowing recovery of restored marshes.[66]

The GoM is also home to natural hydrocarbon seeps, which have existed long enough to be reflected in GoM phytoplankton and bacterial communities (D'souza et al., 2016). There is some evidence that, due to microbial roles in remediation, microbial communities that have evolved with natural seeps may help the GoM recover from man-made spills (ASM, 2020). It is possible that these natural hydrocarbon seeps have primed other organisms to be able to mitigate effects from oil, but that advantage may produce other potential fitness costs to the organism (Murawski et al., 2021).

Oil and other hazardous spills in the GoM are ongoing stressors: understanding of their effects is complicated by naturally occurring hydrocarbon seeps in the region. Because of the amount of oil activity in the Gulf, there are likely to be continuing impacts on organisms, communities, and habitats. NOAA's Office of Response and Restoration tracks and records oil spills and other incidents.

SUMMARY OF TRENDS AND THEIR IMPLICATIONS FOR RESTORATION DECISION MAKING

Table 2.2 summarizes each of the background trends discussed in this chapter, the implications that each trend has for restoration decision making and information needs. It also provides comments on data availability and selected GoM resources noted for each of these background trends. These trends differ across the GoM, underscoring the importance of considering trends, desired outcomes, and the likelihood of restoration success in the context of the landscape in which restoration projects are implemented. This concept is more fully discussed in Chapter 3.

[66] See https://response.restoration.noaa.gov/sites/default/files/Oil_Spills_in_Marshes.pdf.

TABLE 2.2. Summary of Background Trends

Background Trend	Summary	Implications for Restoration Decision Making	Information Needs	Comments on Data Availability	Selected Gulf-wide Resources
CLIMATE CHANGE AS A DRIVER OF LONG-TERM ENVIRONMENTAL CHANGE					
Relative Sea Level Rise	Long-term trends show an increase in RSLR in the GoM. RSLR rates in Louisiana and northern Texas are among the highest in the nation	Marsh restoration projects may not accrete vertically fast enough to keep pace with RSLR without sediment replenishment	Paucity of tide gauges and dedicated subsidence stations on the Gulf Coast inhibit the ability to accurately estimate rates of RSLR in some areas	The NOAA Tides and Currents website provides access to water levels and tides	https://tidesandcurrents.noaa.gov/
				NCCOS[1] provides a broad array of predictive tools	https://coastalscience.noaa.gov/project/predicting-impacts-sea-level-rise-gulf-mexico/
Tropical Storms and Hurricanes	Long-term data indicate that intense hurricanes are increasing in frequency	Restoration plans need to include measures to address more intense tropical storm impacts	Existing data and information sources are adequate to track long-term trends in storm patterns in the GoM	NOAA Historical Hurricane Tracks cover 150 years of GoM storms	https://oceanservice.noaa.gov/news/historical-hurricanes/
				National Hurricane Data Archive is comprehensive	https://www.nhc.noaa.gov/data/
Sea Surface Temperature	SST has gradually increased in the GoM since the early 1980s	Rising SST can increase coastal hypoxia, which can impact restoration efforts involving shellfish and other organisms	Existing data and information sources appear adequate to track long-term trends in Gulf-wide SST	The EPA SST website and program provides historic data (1880 onward) and are expected to be ongoing	https://www.epa.gov/climate-indicators/climate-change-indicators-sea-surface-temperature#ref6
				NOAA SST satellite data for GoM from 2010 to the present, expected to be ongoing	https://www.aoml.noaa.gov/phod/dhos/sst.php
Ocean and Coastal Acidification	Subsurface waters in the northwestern GoM are acidifying at a rate greater than the global surface ocean rate. Multidecadal acidification has also been reported for a majority of the Texas and Florida estuaries	Acidifying seawater can affect formation of calcium carbonate shells and skeletons in corals, mollusks, and many phytoplankton species, potentially impacting restoration projects	Especially in the southern GoM, many coastal and estuarine areas and subsurface waters currently have little or no acidification measurement	NOAA's Ocean Chemistry and Ecosystems Division makes all relevant data available on an ongoing basis; some data go back to 1957	https://www.aoml.noaa.gov/ocd/ocdweb/occ.html

[1] See National Centers for Coastal Ocean Science: https://coastalscience.noaa.gov/.

TABLE 2.2. Summary of Background Trends Continued

Tropicalization	Tropical species including mangroves and fish species have been observed year-round in the northern GoM in recent years	Expansion of tropical plant species such as mangroves can encroach upon salt marsh and other restoration efforts	Data from fisheries and wetland species monitoring programs that can be used to track tropicalization within the GoM are limited	NOAA's Atlantic Oceanographic and Meteorological Laboratory regularly produces a series of map and data products relevant to this topic	https://www.aoml.noaa.gov/phod/dhos/index.php
WATER, NUTRIENT, AND SUSPENDED SEDIMENT INFLOW TRENDS					
Freshwater Inflows	Daily streamflow discharge data (excluding the Mississippi River) generally document a downward trend in freshwater inflow to the GoM since the 1960s. Conversely, Mississippi River discharge has increased during this period	Changes in freshwater inflow impact water quality, salinity, sediment inflow, and species composition, all which can affect coastal restoration efforts	Stream gages and data collection and interpretation are limited in many areas of the GoM	Includes 40 years of data	https://pubs.er.usgs.gov/publication/wri954054
				Current water data plus historical data, varies by site	https://waterdata.usgs.gov/nwis/rt
Nutrient Inputs	Gulf-wide nutrient inputs trends are not available; trends of nutrient loading from specific rivers vary	Nutrient load reduction has been found to be the most effective restoration technique for water quality and SAV in several Florida estuaries	Long-term nutrient loading estimates are limited to several estuaries and the Mississippi River system		https://nrtwq.usgs.gov/nwqn/#/ https://sparrow.wim.usgs.gov/marb/ https://toxics.usgs.gov/hypoxia/mississippi/nutrient_flux_yield_est.html https://nrtwq.usgs.gov/nwqn/#/GULF https://www.ncei.noaa.gov/maps/gulf-data-atlas/atlas.htm https://data.amerigeoss.org/tr/dataset/gulf-of-mexico-hypoxia-dataset https://gcoos.org/hypoxia-nutrients/

TABLE 2.2. Summary of Background Trends Continued

Background Trend	Summary	Implications for Restoration Decision Making	Information Needs	Comments on Data Availability	Selected Gulf-wide Resources
Sediment Inputs	Long-term trends in sediment inflow are limited to the estuarine rivers where long-term sediment data have been collected	Declines in the Mississippi River sediment loads have contributed to marsh loss and may affect restoration plans by limiting the amount of sediment transported from the river	Less than half of the monitoring programs surveyed collect suspended solids data	Data and analysis 1950–2009	Heimann et al., 2011 Allison et al., 2012 Allison et al., 2017 Blum and Roberts, 2009
				Includes data and analysis from 1973 to 2005	Harned et al., 2009
				Provides a source of documents analyzing GoM sediment loads from various sites	https://www.usgs.gov/science-explorer-results?es=sediment+load+gulf+of+mexico
AMBIENT WATER QUALITY OF ESTUARINE AND COASTAL RECEIVING WATERS					
Salinity	Many GoM estuaries have freshened over the last 20 years, correlated with their distance from the Mississippi River system	Changes in salinity can affect plant and animal distribution and thus impact restoration efforts			Rodgers and Swarzenski, 2019 Orlando et al., 1993 https://www.ncei.noaa.gov/maps/gulf-data-atlas/atlas.htm?plate=Salinity-Zones https://gcoos.org/temperature-salinity-profiles/ https://salinity.oceansciences.org/science-results.htm
Eutrophication, Hypoxia, and HABs	Decreases in the area percentage rated "good" for an eutrophication index were observed between two periods, from 1999–2001 to 2005–2006 and from 2005–2006 to 2010. Annual assessment of the extent of hypoxia in the nearshore northern GoM show variability associated with spring rainfall in the Mississippi River watershed	Excessive nutrient loads, hypoxia, and HABs can impact the success of wetlands restoration, oyster projects, and SAV restoration	GoM-wide trends in eutrophication indicators are not readily available; NCCA plans to continue consistent sample collection every 5 years	Historical data from 2013, not current IOC-ICES-PICES Harmful Algal Event database contains bloom data from around the United States, including the GoM	https://www.ncei.noaa.gov/maps/gulf-data-atlas/atlas.htm http://haedat.iode.org/
				Data go back to 2000	https://tidesandcurrents.noaa.gov/hab/gomx.html
				Two sites provide access to data from all five Gulf states	https://www.epa.gov/waterdata/water-quality-data-wqx https://www.waterqualitydata.us/

TABLE 2.2. Summary of Background Trends Continued

OTHER SIGNIFICANT GOM ENVIRONMENTAL TRENDS					
Emergent Wetlands	Approximately 1,000 square miles of emergent wetland were lost between 1996 and 2010, primarily due to open water conversion, and, to a lesser extent, from development and urbanization	Emergent wetlands restoration is a key component of GoM recovery. Addressing issues associated with losses would benefit restoration efforts	A number of agencies collect, map, and interpret information on emergent wetlands at the Gulf-wide scale; interpretation and reporting generally lag data collection by several years	The National Wetlands Inventory includes 50 years of status and trends	https://www.fws.gov/wetlands/status-and-trends/Status-and-Trends-2004-2009.html
				The Digital Coast includes diverse data and analytical and mapping tools	https://coast.noaa.gov/digitalcoast/
				Most current and comprehensive GoM assessment	Handley et al., 2007
Seagrass/SAV	For areas where seagrass maps are available, the Gulf experienced an overall gain in seagrass extent of almost 24% between two time periods (1987–2002 and 2004–2017); gains are associated with improved water quality	Water clarity adequate to support seagrass growth and reproduction is key to seagrass restoration success	Data sources are varied, and support for future Gulf-wide assessments is not assured	Most current and comprehensive GoM assessment	Handley et al., 2020
				Analysis of data from 1987 to 2017	Handley and Lockwood, 2020
Mangroves	Gulf-wide trends since 2004 in mangrove areal extent are not known	The effects of the ongoing colonization of salt marshes in the northern GoM will be an important consideration in salt marsh restoration projects	The National Wetlands Status and Trends program estimates a 0.2% loss in areal extent between 1998 and 2004	Status and trends 1998–2004, mangroves included in the estuarine shrub classification	Dahl et al., 2011
Oyster Beds	Gulf-wide trends in the extent and condition of oyster reefs are currently unknown	Potential impacts from climatic drivers and anthropogenic stressors, as well as restoration design and implementation, can affect oyster reef restoration		Oyster landing by state and year are available	

TABLE 2.2. Summary of Background Trends Continued

Background Trend	Summary	Implications for Restoration Decision Making	Information Needs	Comments on Data Availability	Selected Gulf-wide Resources
Fisheries	Trends for fisheries species vary in time and space, with some species increasing and others decreasing over the long term	Changes in fisheries stocks and species diversity have been shown to provide indicators of restoration effectiveness	Gulf-wide and state-level commercial and recreational fisheries data collection efforts are robust, as well as fisheries-independent data collection efforts at both the Gulf-wide and state levels	States also monitor fisheries and provide useful data	https://gulfcouncil.org/
			Texas has a 45-year record of continuous monitoring and other GoM states up to 20 years		https://www.fisheries.noaa.gov/topic/recreational-fishing-data
Waterbirds	Long-term trends in waterbirds located in the GoM show mixed results, with some species improving, some remaining stable, and others declining	Waterbirds depend on coastal habitats for nesting and foraging and can provide an indicator of ecosystem condition	The Gulf of Mexico Avian Monitoring Network[2] coordinates the various bird monitoring efforts	Multiple databases, some with records over the last 50 years	https://www.usgs.gov/centers/pwrc/science/north-american-breeding-bird-survey?qt- https://www.audubon.org/content/audubon-coastal-bird-survey https://ebird.org/home https://partnersinflight.org/what-we-do/science/databases/ Fredrick and Green, 2019a
Marine Mammals	The 2018 NOAA assessment found insufficient data to determine population trends	Common bottlenose dolphin status and trends could be integrative indicators of long-term environmental health	Monitoring is ongoing and expected to continue to meet requirements of the Marine Mammal Protection Act	Regularly developed stock assessments include status and trends (if detectable) of all GoM species; began with 1994 amendment of Marine Mammal Protection Act	https://www.fisheries.noaa.gov/national/marine-mammal-protection/marine-mammal-stock-assessment-reports-species-stock#cetaceans---dolphins

[2] See https://gomamn.org/.

TABLE 2.2. Summary of Background Trends Continued

Sea Turtles	Long-term trends based on beach nesting data indicate that Kemp's ridley populations have increased since the early 1980s and that loggerhead nesting on Florida's Gulf Coast varied annually between 1979 and 2011			Because GoM sea turtles are protected, monitoring to support long-term population assessments is assured	https://www.usgs.gov/centers/wetland-and-aquatic-research-center-warc/science/distribution-and-density-sea-turtles-gulf?qt-science_center_objects=0#qt-science_center_objects https://www.ncei.noaa.gov/maps/gulf-data-atlas/atlas.htm?plate=Reptiles%20-%20Leatherback%20Sea%20Turtle https://www.boem.gov/sites/default/files/environmental-stewardship/Environmental-Studies/Gulf-of-Mexico-Region/Lamont-Hart.pdf
Invasive Species	There are no states or agencies that maintain databases of long-term trends in invasive species introductions in the GoM region	Trends, if available, would be useful to restoration planners or practitioners because these plants and animals can affect the success of restoration efforts			https://gcoos.org/invasive-species/ http://issg.org/database/reference/index.asp https://nas.er.usgs.gov https://www.usgs.gov/ecosystems/invasive-species-program/science/invasive-fish?qt-science_center_objects=0#qt-science_center_objects
Land Cover	Urban land cover in coastal areas around the GoM increased by 15% between 1996 and 2010, with some geographic areas, including Houston and Tampa, increasing at much higher rates	Changes in land cover can alter hydrologic regimes and runoff patterns, thus potentially affecting environmental restoration projects and impacting the cumulative effects of multiple restoration projects	The Multi-Resolution Land Characteristics Consortium[3] generates land cover information at a national scale for a wide variety of environmental, land management, and modeling applications	USGS National Land Cover Database includes data from 2001 onward	https://www.mrlc.gov/
Oil Spills	Oil and hazardous spills in the Gulf of Mexico are ongoing stressors, the effects of which are complicated by the naturally occurring hydrocarbon seeps in this region	Because of the amount of oil activity in the Gulf, there are likely to be continuing impacts on organisms, communities, and habitats		NOAA's Office of Response and Restoration has tracked and recorded oil spills and other incidents since 1985	https://incidentnews.noaa.gov/ https://response.restoration.noaa.gov/resources/maps-and-spatial-data https://response.restoration.noaa.gov/oil-and-chemical-spills

[3] See https://www.mrlc.gov.

NOTES: Data sources for long-term monitoring of Mississippi River status and trends are not included. Refer to text for additional references in each section.

Chapter 3
Assessing Cumulative Effects of Restoration: Current and Emerging Approaches

INTRODUCTION

It is challenging to assess the progress of ecological restoration against the backdrop of ongoing environmental change and periodic acute events. Over 50 years ago, the National Environmental Policy Act of 1969 (NEPA) offered one approach—considering the impacts of a project or multiple projects compared to a world without these efforts (the "no action alternative" or "future without project"). This concept of a future without action can also be applied to projects that have benefits—such as restoration. This method for environmental planning develops a future vision of "new" baseline conditions.

Based on understanding of environmental trends (see Chapter 2), the future Gulf of Mexico (GoM) Coast will be substantially different than it is today, with or without restoration actions. Restoration actions taken now will continue to affect the GoM Coast, and environmental changes and interacting stressors have the potential to confound future assessments of the effectiveness of these restoration efforts (Hobbs and Norton, 1996; Manning et al., 2006; Moreno-Mateos et al., 2020; Palmer et al., 2016). Using projected "future without project" environmental condition as the appropriate baseline for comparison with a future that includes large-scale restoration efforts instead of today's conditions, which are changing rapidly, can be quite valuable. Modeling related to this type of approach has started to occur in the GoM region (e.g., Meselhe et al., 2022), but it is not widespread.

Large-scale restoration in coastal watersheds, such as parts of the Greater Everglades,[1] the Louisiana Coast,[2] Tampa Bay,[3] Galveston Bay,[4] Mobile Bay,[5] and Mississippi Sound[6] are very involved undertakings. As regional restoration actions, such as those associated with *Deepwater Horizon* (DWH) settlements, grow in number and complexity, scientists and resource managers charged with restoring large ecosystems, including those outside the Gulf (e.g., Allan et al., 2013; Ortiz et al., 2018), are finding that the cumulative impacts of multiple environmental stressors need to be considered when assessing restoration success.

Restoration itself can also produce cumulative effects that interact with stressor impacts (Diefenderfer et al., 2021). Further, the effects of restoration may be beneficial, or undesirable and unplanned (Seddon et al., 2021). Failing to consider cumulative impacts of stressors may result in ecological surprises, which are the unanticipated behaviors of ecosystems. The relative occurrence of ecological surprises can increase as an ecosystem's capacity to absorb impacts diminishes (Filbee-Dexter et al., 2017; Paine et al., 1998).

---

[1] See https://www.nps.gov/ever/learn/nature/cerp.htm.
[2] See https://coastal.la.gov/our-plan/2017-coastal-master-plan/.
[3] See https://tbep.org/.
[4] See https://gbep.texas.gov/.
[5] See https://www.mobilebaynep.com/assets/pdf/FINAL-CCMP-11.25.2019.pdf.
[6] See https://www.mdeq.ms.gov/wp-content/uploads/2017/09/2016-Addendum-FINAL-10.31.2016.pdf.

As watershed- and estuary-scale restoration efforts have matured, the cumulative benefits of multiple restoration efforts of diverse types have helped to counteract the negative impacts of multiple stressors (Beck et al., 2019; CERP, 2014; Côté et al., 2016; Diefenderfer et al., 2016). Restoration and management strategies have traditionally focused on singular objectives, such as improving water quality or preventing erosion on a local or site-specific scale (Daoust et al., 2014), but it is now understood that achieving multiple restoration objectives—such as improving ecosystem structure and function, and diversifying and maximizing ecosystem services—requires a systematic and multidisciplinary approach (Diefenderfer et al., 2003; Gann et al., 2019; Hodgson et al., 2019; Moreno-Mateos and Comin, 2010; Neeson et al., 2015; Thom et al., 2011).

Since the passage of NEPA, the term "cumulative effects" has generally been defined as the collective impact of past, present, and future human activities on the environment (Spaling and Smit, 1993). The definition typically has a negative connotation because of a history of research documenting interacting human-related stressors and greatly declining ecosystem function (Darling and Côté, 2008; Halpern et al., 2015; Lotze et al., 2006; Luoma et al., 2001). However, the concept of the cumulative effects of restoration (see Box 3.1) has a positive connotation, in that the collective effects of multiple activities may contribute to a net positive change in ecosystem form or function. The approaches to evaluating the cumulative effects of large-scale restoration that have been tried include spatial analysis of big data, specialized indices, and lines of evidence (Beck et al., 2019; Diefenderfer et al., 2016; Raposa et al., 2018). Large-scale restoration efforts, consisting of formal or coordinated projects, have developed methods according to their goals and objectives and have been guided by the attributes of specific places envisioned for restoration (Achete et al., 2017; Allan et al., 2013; Konisky et al., 2006; NASEM, 2021). Overall, there is a lack of consensus about a standardized approach to evaluating cumulative effects of restoration, although there have been many calls for developing such an approach (Fischer et al., 2021; Jones et al., 2018; Love et al., 2017).

On the Gulf Coast, for example, after the DWH oil spill, extensive stakeholder engagement in the GoM identified appropriate aims for the use of settlement funds (Mabus, 2010; Walker et al., 2012). Much of the focus of Gulf Coast environmental restoration to date is on taking an ecosystem approach to the recovery of habitat conditions and associated species (DWH NRDA Trustees, 2016; RESTORE Council, 2016; NFWF, 2020). In other regions with similar aims, ecosystem restoration activities have also included intentionally facilitating the synergistic interactions of species (Eger et al., 2020; Halpern et al., 2007) or habitats (Sobocinski and Latour, 2015). Such holistic approaches present opportunities to achieve cumulative effects larger than the sum of the parts (i.e., synergistic effects). Many cumulative effects of environmental management are unplanned and/or uncontrolled (Filbee-Dexter et al., 2017), and in the worst case have negative outcomes relative to managers' aims for improved functions benefiting species and ecosystems. In some cases, negative outcomes may be addressed with well-coordinated adaptive management involving program managers, restoration practitioners, and research scientists (Littles et al., 2022; Wilber and Bass, 1998).

Approaches to assessing "cause and effect" between multiple restoration and associated management actions and their outcomes at the ecosystem scale and larger are the subject of the remainder of this chapter.

BOX 3.1
Definition of Cumulative Effects

The cumulative effects of restoration, as defined in this report, are the collective additive, synergistic, and antagonistic effects of all restoration activities that occur within a setting defined by common or connected characteristics of hydrology, geomorphology, ecology, ecological function, and biodiversity. Assessment of the cumulative effects of restoration may occur at various geographic scales, such as a wetland complex, bay, estuary, watershed, or the Gulf Coast itself. The scale of assessment may also be defined by specific interests in the outcomes, such as ecosystem processes (e.g., sedimentation), biodiversity or specific organisms (e.g., oysters), performance targets (e.g., water quality), type of restored system (e.g., wetland restoration), political or community boundaries (e.g., state boundaries), or type of restoration method (e.g., living shorelines).

Recognizing that assessment resources are limited, for instances in which relatively few or disconnected projects are being implemented in a given area (thus increasing the likelihood of the restoration signal being lost in the noise of background stressors), or in cases of restoration especially geared toward learning and experimentation (e.g., pilot projects), the evaluation of cumulative effects may be less important. For all other restoration, the approaches are described in this chapter, which includes:

- a detailed overview of antagonistic and synergistic effects of restoration actions;
- a description of modes and pathways of several types of cumulative effects;
- the use of hypotheses summarized by ecosystem conceptual models;
- an introduction to multiple lines of evidence and causal criteria frameworks and in turn, a description of the various tools needed to develop multiple lines of evidence;
- reflections on restoration planning and endpoints, including constraints; and
- a case-study discussion of cumulative effects assessment in the annually occurring GoM hypoxic zone.

ANTAGONISM AND SYNERGISM IN RESTORATION EFFORTS

Diverse pressures in estuarine and coastal waters, both natural and anthropogenic, can generate multiple stresses on an ecosystem's structure and function (O'Gorman et al., 2012). The effect of those multiple stressors can be additive (equal to the sum of their individual effects), synergistic (greater than the sum of their individual effects), or antagonistic (less than the sum of their individual effects) (Breitburg and Riedel, 2005). Antagonistic, additive, and synergistic stressors can be judged to be either beneficial or detrimental relative to program goals and objectives (Côté et al., 2016; Piggott et al., 2015). The cumulative effects of restoration efforts may also be additive, synergistic, or antagonistic and similarly judged beneficial or detrimental depending on program goals and objectives.

The idea of synergistic and antagonistic effects of multiple stressors in ecological systems and ecosystem management is well established. For example, Crain et al. (2008) synthesized 171 studies that manipulated two or more stressors in marine and coastal systems and concluded that the more stressors there were, the greater the need to account for complex interactions in both ecological studies and conservation actions. Teichert et al. (2016) evaluated the impact of nine stressor categories on fish ecology in 90 estuaries and concluded that targeting mitigation of synergistic stressors needed to be a restoration priority. In their meta-analysis of multiple stressors on seagrasses, Stockbridge et al. (2020) emphasized that understanding and accurately predicting the complex nature of stressor interaction is important in conservation, concluding that the focus needs to be on mitigating those stressors for which the greatest benefit is derived.

Antagonistic interactions may be less common than synergies, and they appear to be less understood or reported, but they are equally relevant to habitat restoration as are more additive effects (Côté et al., 2016). Understanding complex interactions at both species and community levels has been shown to enhance the effectiveness of restoration efforts in numerous habitats, from salt marshes to seagrass meadows to mangrove forests and coral reefs (Eger et al., 2020; Renzi et al., 2019; Silliman et al., 2015). Figure 3.1 shows how such complex interactions could take place among common restoration communities of the Gulf Coast, using nutrient input reduction as an example. The figure also shows how external sources affecting the four restoration communities (many of which are discussed in Chapter 2) are often less controllable than stressors local to the communities.

As discussed in the remainder of this section, exploring how to make use of ecological synergies and avoid antagonistic interactions as part of restoration efforts could improve benefits and efficacy of their implementation. This effort may also avoid costly ecological surprises following restoration investments. When assessing the impacts of synergistic or antagonistic effects of restoration activities, it is useful to separately consider two different types of restoration that occur on vastly dissimilar spatial scales—habitat and watershed.

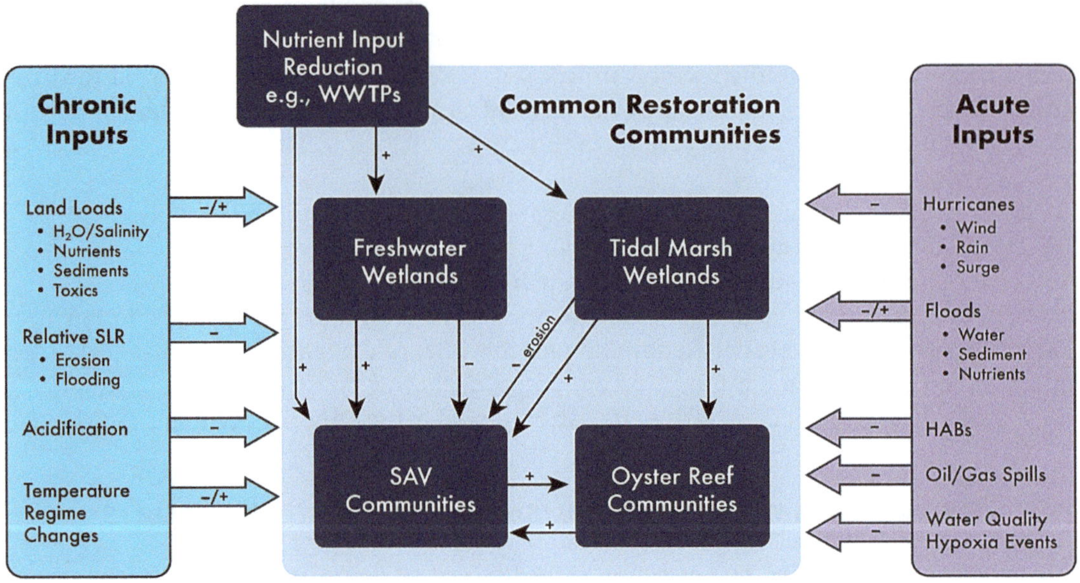

FIGURE 3.1. Chronic and acute inputs, restoration communities, and synergistic and antagonistic interactions. A box and arrow diagram showing likely and potential interactions, both synergistic (+) and antagonistic (–) between commonly used restoration communities (shown as five navy blue boxes) and two categories of inputs to these communities (Chronic and Acute). Some common antagonistic and synergistic interactions among restoration communities are also indicated. The diagram indicates important effects occurring in restoration communities from both external and often less controllable sources, as well as from interactions among restoration types. WWTP refers to wastewater treatment plants; H20 refers to water loads; SLR refers to sea level rise; SAV refers to submerged aquatic vegetation; HABs refers to harmful algal blooms.

Coastal Habitat-Scale Synergism and Antagonism

The restoration of productive coastal nursery habitats—such as oyster reefs, salt marshes, and seagrass meadows—is common. Such restoration, however, remains experimental relative to local acute and chronic inputs and still carries a risk of failure. Figure 3.1 shows some common antagonistic and synergistic interactions among restoration communities, including important effects occurring in restoration communities from both external and often less controllable sources, as well as from interactions among restoration types. Here the scale is many orders of magnitude less than that of entire watersheds, likely on the order of square meters to several hectares.

There is accumulating evidence that locating nursery habitats in close proximity to one another may positively enhance secondary productivity relative to similar habitats restored at greater distances. This has been documented most clearly for fish, shrimp, and crabs, which, because of their commercial importance, have been frequently studied. One early example is the enhanced biomass of such species in seagrass meadows adjacent to North Carolina salt marshes (Irlandi and Crawford, 1997), but a number of other examples in different locations also exist (e.g., Berkström et al., 2012; Olson et al., 2019; Sobocinski and Latour, 2015). To understand how this can occur, Gilby et al. (2018) and others have shown that many fish move daily, or with tidal cycles, between coastal nursery habitats, such as marshes, mangroves, seagrasses, and/or reefs (cf. Boström et al., 2011; Potter, et al., 2015; Olds et al., 2017). The degree to which adjacent habitats can enhance productivity is affected by the length of time in which the habitats become dry during ebb tides (Grabowski et al., 2005; Peterson et al., 2003). It is also known that better-connected ecosystems often support more fish than those that are isolated (Gilby et al., 2021; Nagelkerken et al., 2015; Olds et al., 2017) (Figure 3.2).

However, restoring habitats in close proximity could result in functional redundancy such that the combined nursery habitat benefits sum to less than those expected of the two separate habitats (Geraldi et al., 2009; McDonald et al., 2016). To resolve this issue, focused research on critical uncertainties would be needed. Functional redundancy is an example of antagonism because each hectare restored would produce less than the last. At present, there is more evidence for synergistic than antagonistic effects of adjacent nursery habitats. However, the extent to which synergism or functional redundancy occurs in adjacent nursery habitats, whether natural or restored, is a topic requiring more study and one with broad implications for developing the spatial arrangement of restored coastal habitats (Barnett and Belote, 2021).

The potential for synergistic or antagonistic interactions has implications for designing the spatial configuration of large habitat restoration projects or suites of projects. Even without considering biological effects, hydrology itself is characterized by nonlinear processes (Allan, 2004). For restoration, this is expressed by such examples as the synergistic effects of dike breaching on floodplain hydrology and consequences of the spatial position of dike breaches along the tidal–fluvial gradient, shown by Diefenderfer et al. (2012). As the spatial arrangement of restoration sites is increasingly being considered under planning processes for environmental management (Gilby et al., 2018; Lester et al., 2020; Lin and Kleiss, 2007), rigorous testing of the effects of different spatial combinations of nursery habitat restoration and those in the Gulf's different tidal regimes is now possible.

Estuary- and Watershed-Scale Synergism and Antagonism

Watershed alterations at very large scales can be expected to produce a huge number of both direct and indirect effects. Alterations of freshwater flow are one type of watershed alteration that can alter salinity regimes, nutrient delivery, water clarities, and sediment deposition rates and produce many changes in the receiving waters and their ecologies (Carle et al., 2020; Dorado et al., 2015). This type of watershed alteration has been proposed in Louisiana as a means of delivering sediments to rapidly eroding coastal wetlands, aiming to delay the loss of productive wetlands to sea level rise and land subsidence. Many additional changes occur as the diverted freshwater alters salinities, temperature, and other water quality parameters at both the origin and the destination of rerouted waters, with potentially cascading changes for the functioning of resident flora and fauna (see Figure 3.3 and Box 3.2).

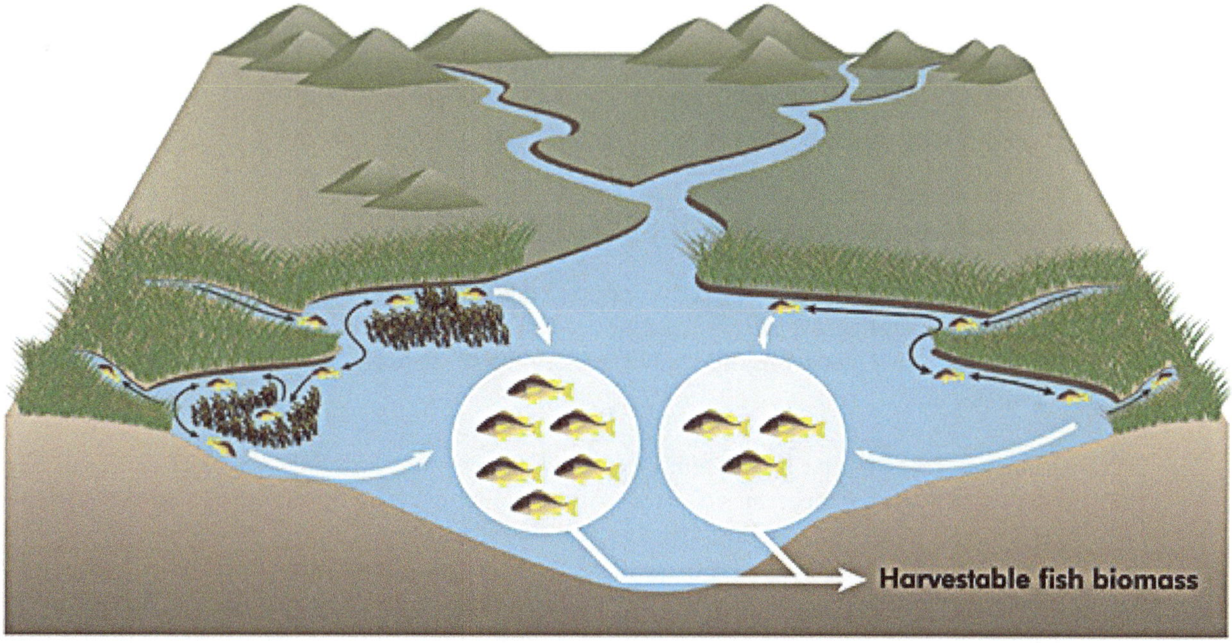

FIGURE 3.2. Example of synergism in coastal restoration. Fish move among complex habitats such as seagrass meadows and salt marshes in coastal seascapes (dark grey arrows). By restoring these habitats in close proximity to one another we might improve the habitat values, productivity and the carrying capacity of coastal seascapes for fish and fisheries (light grey arrows and circles). Symbols courtesy of the Integration and Application Network, ian.umces.edu/symbols. SOURCE: after Gilby et al. 2018, Figure 4.

Also at the watershed scale, reducing nutrient inputs via upgrades to wastewater treatment facilities and better control of non-point-source inputs from agriculture and urban landscapes can produce large changes in downstream environments by reducing algal productivity and increasing water clarity, which can facilitate the increase of submerged vegetation and its rich floral and faunal associates (shown in Figure 3.3). As described in Chapter 4, documented examples of large-scale restoration with these types of water quality and habitat goals include Tampa and Sarasota bays in Florida, and Galveston Bay in Texas, where reducing nutrient inputs led to increased light availability and seagrass extent. In these Gulf examples, multiple nutrient reduction projects implemented over several decades are associated with decreased algae concentrations (as measured by chlorophyll *a* concentration), increased water clarity, and resulting increases in seagrass acreage. Furthermore, net-beneficial interactions between nutrient reduction projects and habitat restoration projects were documented in Tampa Bay (Beck et al., 2019).

Although improved conditions associated with implementation of multiple restoration efforts have been observed for some examples at the estuary/watershed scale as described above, no documented examples of antagonistic and/or synergistic effects that may have contributed toward observed improvements in estuary-wide conditions in the GoM were identified. Restoration and management programs that were initiated years ago were not typically designed to detect synergistic or antagonistic effects, resulting in an information and data gap. At the project scale, monitoring and reporting key parameters using comparable methods would allow combining monitoring data for larger-scale analysis (for parameters, see, e.g., NASEM, 2017; DWH NRDA Trustees, 2019; RESTORE Council, 2021a).

ASSESSING THE CUMULATIVE EFFECTS OF RESTORATION

Why are cumulative effects of large-scale restoration efforts so difficult to measure and quantify? Key factors include how the magnitude of changes compares to the sensitivity of the detection method and how biological and environmental conditions can dampen outcomes (CEQ, 1997). In their recent paper on advancing understanding of the cumulative effects of large-scale restoration, Diefenderfer et al. (2021) provide a framework for assessing landscape/estuary-scale restoration progress despite inherent challenges to detecting them; in doing so, they provide an approach to also improve ecosystem outcomes beyond what might have been possible to achieve with independent, site-scale projects. To develop their cumulative effects framework, the authors modified the traditional stressor-based framework of characterizing cumulative effects discussed previously (Chapter 1) (CEQ, 1997) to identify cumulative restoration benefits. This committee has modified their work with examples from the GoM, which are detailed in the next section.

Modes of Cumulative Effects: Systemic, Spatial, and Temporal Effects

Table 3.1 presents the systemic, spatial, and temporal "modes" of cumulative effects, which categorize the general ways cumulative effects are expressed in ecosystems; the table is based on Diefenderfer et al. (2021) and modified with examples from the GoM. This table also includes the various "pathways" associated with each main mode:

- **Systemic cumulative effects:** The systemic approach to realizing cumulative effect benefits in large-scale restoration includes three means of accruing ecological benefits identified in the ecological literature: compounding or cascading, triggers and thresholds, and indirect effects.
- **Spatial cumulative effects:** Spatial approaches recognize changing spatial patterns of populations, ecosystems, and landscapes, and the cross-boundary and space crowding effects often present in complex systems and influencing habitat restoration for threatened and endangered species.
- **Temporal cumulative effects:** Temporality is built into accruing cumulative restoration benefits following restorative actions intended to catalyze natural processes to advance restoration, recognizing time lags and/or the opposite—time crowding—will occur.

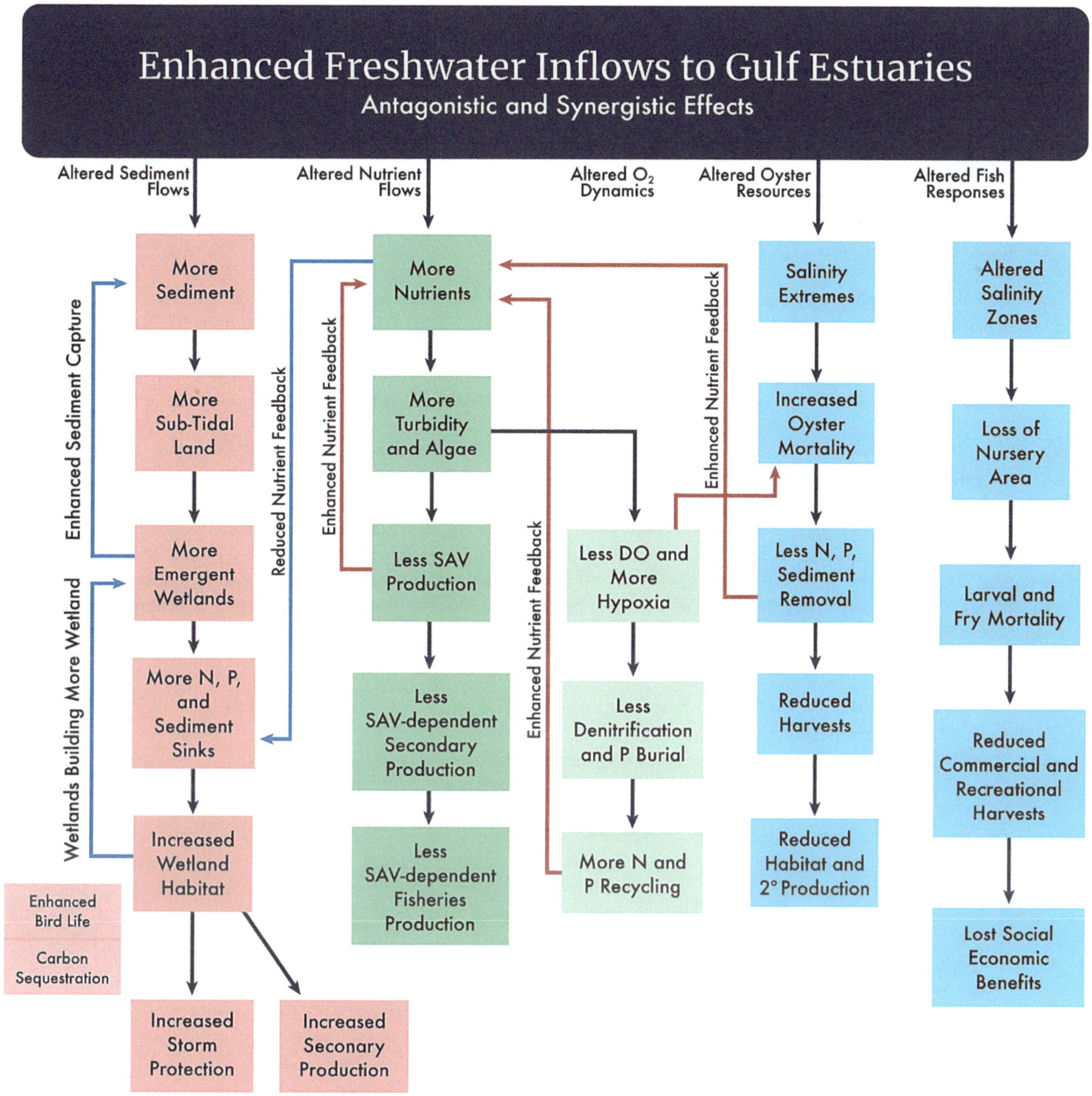

Figure 3.3. A logic-flow diagram summarizing five altered effect pathways associated with enhanced freshwater inflows to Gulf estuaries. See Box 3.2 for a detailed explanation of each pathway. SOURCE: Committee.

Box 3.2
DESCRIPTION OF PATHWAYS IN FIGURE 3.3

The altered effect pathways in Figure 3.3 include the following: (1) sediment inflows, (2) nutrient inflows, (3) O2 dynamics, (4) oyster community responses, and (5) finfish responses. The dark arrows connecting boxes in each vertical sequence represent primary expected effects. The blue and red arrows indicate synergistic and antagonistic effects, respectively. Note also that some feedback effects operate between altered vertical pathways indicating possible and complex system-level responses to changing conditions, such as freshwater inflow rates. This diagram is not all-encompassing and does not include all possible logic-flow relationships when considering synergistic and antagonistic effects.

From each effects category, a sequence of cause–effect relationships leads to ecological consequences. For example, one sequence of effects of enhanced freshwater flow indicates increased nutrient availability, leading to increased algal-induced turbidity, less submerged aquatic vegetation (SAV) production, less secondary production associated with diminished SAV community health, and, ultimately, reduced fisheries production. Each freshwater flow effect category has a cause–effect chain leading to ecological consequences. Each effect cascade is briefly described below.

Altered Sediment Flow Chain: More sediment associated with enhanced freshwater flows is shown as increasing subtidal land areas that later support increased areas of emergent wetlands. These metabolically active communities act as strong sinks for nitrogen (N) and phosphorus (P) and sediments. Continued sediment additions lead, in turn, to increased wetland habitat, other secondary ecosystem services benefits and, finally, to enhanced storm protection. In addition, several positive feedbacks indicate wetland size is enhanced as wetlands continue to grow. A negative feedback is shown interacting with the nutrient pathway, where excess N and P associated with freshwater flows can be sequestered in existing and expanding wetlands. Although not shown in Figure 3.3, autochthonous organic matter production may also contribute to marsh sediment accretion, some sediments may accumulate into channels and be removed via dredging, and enhanced turbidity could suppress primary production.

Nutrient and Dissolved Oxygen Chains: Enhanced nutrient effects are shown as promoting algal growth and water column turbidity and, in turn, stressing SAV growth and maintenance. This affects the secondary production associated with healthy SAV communities, eventually leading to less SAV-related fisheries production. There are several feedbacks within this chain and with other chains, suggesting the general importance of SAV communities in the shallow waters of the Gulf. Specifically, as SAV-related production declines, less nutrient buffering is available and more nutrients become available, leading to increased stress on SAV communities via water column turbidity increases. Two additional nutrient stress pathways that affect SAV are shown in the the oxygen and oyster logic chains, which are related to hypoxia influences on nutrient removal processes. Specifically, less filtration of algal particles by depressed oyster stock occurs along with reduced denitrification rates because of hypoxia-suppressed nitrification rates.

Altered Oyster and Fish Chains: In these chains, salinity regimes are shown as altered by changes in freshwater flows that produce stressful conditions through salinity extremes on resident finfish and oyster communities. One or both are shown as causing oyster mortalities and loss of fish nursery areas. In the oyster chain, increased mortalities result in two effects: less nutrient and sediment removal via oyster filtration (and enhanced nutrient feedback to the nutrient chain) and reduced oyster harvests and habitat availability. The impact of reduced nursery area translates into higher larval and fry mortalities and, eventually, into reduced commercial and recreational catches and their associated social and economic benefits.

Similar logic chains could be added to show both positive and negative effects of increased nutrient additions on pelagic and benthic fish production.

Table 3.1 can be used for a variety of purposes, such as considering all pathways by which long-term trends, acute events, and stressors may affect a restoration project or projects—and in turn also considering how multiple restoration efforts will affect each other via one of these pathways.

The Influence of Hot Spots, Hot Moments, and Ecosystem Control Points on Cumulative Effects

Given the importance of spatial patterns and temporal dynamics in affecting the cumulative effects of restoration (as shown in Table 3.1), the concept of hot spots and hot moments (HSHMs) is also relevant for cumulative effects assessment. This concept was originally proposed by McClain et al. (2003),who described spikes in rates and reactions in elemental cycling in biogeochemical processes, such as denitrification (Groffman et al., 2009), over space (hot spots) and time (hot moments). The applications of HSHMs in the field of ecosystem restoration to date have been limited even though HSHMs provide an opportunity to increase the benefits gained through restoration projects and programs, given that the potential returns of all locations are not equal in light of events such as natural disasters (e.g., hurricanes).

The HSHM concept was used by Kannenberg et al. (2020) who found that hot moments in ecosystems' gross primary production (GPP) constituted a significant percentage (up to 12 percent of the annual budget) of carbon (C) assimilation. HSHM dynamics occur along coastal interfaces, including the Gulf Coast, and have the potential to accelerate—or, conversely, limit—process-based cumulative effects of restoration (e.g., biogeochemical reaction rates, carbon storage, decomposition) (Ward et al., 2020). The "blue carbon" or global carbon sink functions of mangroves, salt marshes, and seagrasses are thereby modified by HSHMs (Bertram et al., 2021). Bernhardt et al. (2017), acknowledging that spatial and temporal aspects of spots and moments almost always co-occur, developed the concept of ecosystem control points. The use of HSHMs and ecosystem control points for evaluation of post-DWH restoration of the GoM coast offers an opportunity for synergistic knowledge development. Their applicability is supported by observations to date of stressor effects in Chesapeake Bay and hypoxia (discussed below). Table 3.2 provides a short description of HSHMs and related GoM examples.

Because all three previously identified modes of cumulative effects (in Table 3.1) may occur simultaneously, this committee introduces here a new mode of cumulative effects:

- **Spatiotemporal-topological cumulative effects:** The spatiotemporal-topological approach accepts that systemic, spatial, and temporal cumulative effects modes can and perhaps often occur simultaneously, as illustrated by the three modes through which they are expressed: (1) hot spots, places with anomalous ecosystem functions; (2) hot moments, times of anomalous ecosystem functions, recurring or not; and (3) ecosystem control points, which are combined "spot moments." The topological reference refers to the spatial patterns of these HSHMs and ecosystem control points, and their adjacency or feedback effects across ecosystems (Bernhardt et al., 2017). Acute events (see Chapter 2), while not synonymous with hot moments, may be causally linked to their expression.

In general, identifying HSHMs of ecosystem functions and services during the course of a restoration project could provide important clues about the environmental or climate drivers or stressors associated with these temporary events (Kannenberg et al., 2020).

The four types of ecosystem control points or integrated spot moments that are particularly important to ecosystem dynamics, based on the biogeochemical topography of a landscape, are termed permanent, activated, export, and transport (Bernhardt et al., 2017); see Table 3.2 for short descriptions and GoM examples. A variety of models, from conceptual to machine learning models, can be implemented to understand the drivers of the control points.

Measuring or monitoring these control points is likely to generate the data necessary to improve model predictions for event timing and magnitude, allowing restoration planners and managers to take advantage of beneficial outcomes and avoid deleterious effects. The contribution of these HSHMs and ecosystem

TABLE 3.1. Main Characteristics of the Systemic, Spatial, and Temporal Modes of Cumulative Eeffects with Gulf Coast Examples

		Cumulative Impacts of Stressors (NEPA Perspective, Negative Impact)		Cumulative Effects of Ecosystem Restoration and Management (Positive or Beneficial Effect)	
Modes of Cumulative Effects	**Cumulative Pathway**	**Cumulative Impacts of Environmental Stressors (Brief Description)**	**Gulf Coast Examples of Cumulative Impacts of Stressors or Degradation**	**Cumulative Effects of Large-Scale Ecosystem Restoration and Management (positive or beneficial)**	**Gulf Coast Examples of Cumulative Effects of Large-Scale Ecosystem Restoration and Management (positive or beneficial)**
Systemic	Compounding	Effects arising from multiple sources or pathways	Reef-fish food webs post-DWH (Chagaris et al., 2020) Mobile Bay habitat alterations fisheries—blue crab (Jordan et al., 2008)	In ecosystems altered by restoration, multiple internal or external drivers and stressors produce linear or nonlinear, antagonistic or synergistic effects and feedback	Tampa Bay and Galveston Bay improvements in water quality (Beck et al., 2020; Greening et al., 2014; HARC, 2020), hydrogeomorphic effects on Everglades wading bird habitat function (Beerens et al., 2015; Pearlstine et al., 2020)
	Triggers and Thresholds	Fundamental changes in system behavior or structure	Marsh edge erosion post-DWH Silliman et al., 2016) Sea level rise and coastal storm thresholds for fish, wildlife, and plant species (Powell et al., 2017)	Thresholds are points in restoration response functions at which small changes in drivers or stressors or sudden changes in state variables yield abrupt shifts between alternate ecosystems states; triggers are environmental drivers or stressors that produce nonlinear system-state responses	Tampa Bay and Florida Peninsula, threshold for light for seagrass (Choice et al., 2014; Dixon, 2000)
	Indirect	Secondary effects	Clean-up activities post–oil spill negatively impact habitat (NRC, 2013)	Restoring physical processes has biological effects, often including linkages between primary and secondary production	Oyster reef restoration aids seagrass establishment (Sharma et al., 2016)
Spatial	Landscape pattern	Change in landscape pattern	Changes in landscape patterns of Pensacola estuarine drainage area (Yang and Liu, 2005)	Reduced fragmentation, increased patch size, and restored connectivity and configuration influence ecosystem process and population dynamics	Connecting coastal nearshore and offshore habitats in GoM (Peterson et al., 2020)
	Cross boundary	Effects occur away from the source	Nitrogen export from corn belt to GoM fosters hypoxia (McLellan et al., 2015)	Restoration influences system states or processes outside of restored sites, including interactions between restoration sites	Export from upper Mississippi River Basin (Robertson and Saad, 2021)
	Space crowding	High spatial density of effects	Nonpoint source phosphorus management in Florida coastal waters (Yang and Toor, 2018)	Multiple restoration projects are implemented within the same geographic domain, with overlapping areas of influence and interaction	Mobile Bay or landbridge in Louisiana (Gregory Steyer, presentation to the committee, November 10, 2020)
Temporal	Time lags	Delayed effects	DWH impacts on coastal and nearshore fisheries—long-term implications (Murawski et al., 2021)	Important interactions and biota appear long after restoration alters drivers, stressors, or components as the system adapts of influence and interaction	Proposed sewage treatment plant and septic system upgrades (Alabama Gulf Coast Recovery Council, 2019)
	Time crowding	Frequent and repetitive effects	Pulsed inflow of diverted water timed according to river stage (Day et al., 2009; Gledhill et al., 2020)	The frequency or duration of restoration actions affects the ecosystem, or restoration alters the timing of stressors	Hydrological variability, including pulse events (Liu et al., 2021a; Montagna et al., 2018)

NOTE: The cumulative effects of stressors are shown together with the corresponding potential or actual cumulative effects of large-scale ecosystem restoration. Table adapted from Diefenderfer et al. (2021), which in turn used definitions from CEQ (1997). Gulf Coast examples of multistressor impacts and cumulative restoration effects were identified and developed by this committee.

TABLE 3.2. Conceptual Application of the HSHM and Ecosystem Control Points Paradigms to Disturbance Processes, Stressors, and Restoration on the GoM Coast

Cumulative Pathway	Short Description	Gulf Coast Natural Disturbance Process Examples	Gulf Coast Anthropogenic Stressor	Gulf Coast Restoration Examples
Hot Spots	Spatial, "patchy" Kannenberg et al. (2020)	Marsh dieback, windfalls, marsh fires	Fluid withdrawal for hydrocarbons that causes subsidence hotpots and inundation hotspots, marsh fires	Marsh creation, landbridge and barrier island restoration
Hot Moments	Temporal, "flashy" Kannenberg et al. (2020)	Riverine floods, hurricane-related wind, precipitation, and storm surges, marsh fires	Bonnet Carré release into Mississippi Sound, COVID-19 lockdown (anthropause) effect on water quality, marsh fires	Peak freshwater flow through manmade diversions and siphons delivered to marshes
Permanent Ecosystem Control Points	Sustained high biogeochemical rates relative to landscape	Describes all tidal wetland channels	Freshwater withdrawal from rivers limiting freshwater contributions to the coast	Shoreline stabilization
Activated Ecosystem Control Points	High transformation rates when conditions are optimized	The receiving area of freshwater floods	Denitrification accelerates as a result of warming temperatures	Oyster reef restoration, species-specific habitat creation (e.g., ponding)
Export Ecosystem Control Points	High accumulation capacity, with threshold for high export	Natural crevasses of the river	Failure of retention systems causes release of nutrients or harmful compounds (e.g., Lake Okeechobee)	Man-made river diversions at the source
Transport Ecosystem Control Points	High transport capacity contributes disproportionately	River-borne sediments contribute to marsh Marine-derived sediment and nutrient deposition	Excess nutrients can stress marsh health, climate change–driven saltwater intrusion	Nutrient regulation through sewage treatment

control points may not be trivial during the course of a restoration trajectory, and thus, they may need to be included while analyzing the effectiveness of restoration projects across a landscape or seascape (Kannenberg et al., 2020). Further, developing a better understanding of the combination of drivers and stressors responsible for triggering HSHMs and ecosystem control points for a particular ecosystem could help to incorporate them in a predictive modeling framework for assessing the cumulative effect of restoration projects. HSHMs and control points can be used to help address risk in projects during cost-benefit or other aspects of planning.

In summary, control points and HSHMs allow restoration planners to tailor plans with the aim of achieving beneficial cumulative effects while avoiding harmful ones. If they are ignored at the planning stage of restoration projects or programs, outcomes may differ by orders of magnitude from predictions (Petersen et al., 2008). Furthermore, understanding HSHMs is a foundation for the effective development of hypotheses and the design of monitoring systems to measure cumulative effects. HSHM and control points, while localized, can also have landscape-scale effects and measurable signals in their vicinity can overwhelm background-level annual averages over much larger areas. Awareness of HSHMs and considering them in planning efforts can also assist with avoiding ecological surprises.

THE ROLE OF CONCEPTUAL MODELS IN DEVELOPING HYPOTHESES

The ecological restoration literature includes many experiments conducted at small scales for the restoration of particular plant communities or the recovery of habitat for a species of interest (e.g., Martin et al., 2021; McDonald et al., 2016). However, relatively few coastal restoration projects are designed to test hypotheses or monitored long enough to do so (Waltham et al., 2021). Many restoration projects are initiated on the basis of a perceived shared understanding of what works, which may be based on unrelated locales or conditions. Systematic experimental restoration with replication is lacking (Howe and Martínez-Garza, 2014). In large-scale public restoration, modeling alternative restoration actions often substitutes for on-the-ground experimentation (Buenau et al., 2014; Diefenderfer et al., 2012).

The aim of evaluating progress resulting from large-scale ecosystem restoration is to determine whether, relative to chronic background trends and acute events (discussed in Chapter 2), the cumulative restoration effects are on track to meet goals or not. If they are not, the inevitable question is, "Why not?" To help inform adaptive management of the trajectory, it can be valuable to consult the conceptual model that guided restoration design.

Conceptual Models

Conceptual models are graphical representations of interrelationships between drivers, pressures, stressors, restoration actions, and ecosystem response, based on one or a series of hypotheses (Gentile et al., 2001; Suter, 1999). They are often used to represent understanding of the current and future states of the ecosystem, which are crucial for determining restoration project priorities and assessing future projections. Preparing a conceptual model can enhance understanding of the current state of the ecosystem and raise questions about underlying assumptions. The initial understanding of the ecosystem at the time of the restoration design will subsequently be tested by restoration actions (Brudvig and Catano, 2021).

The use of conceptual models is in longstanding practice in environmental management and was previously recommended for effective post-DWH GoM monitoring (NASEM, 2017). The *Monitoring and Adaptive Management Procedures and Guidelines Manual* (DWH NRDA Trustees, 2019) provides guidance on the use of conceptual models at the project level. The applicability of those procedures and guidelines to evaluation of cumulative effects of large-scale restoration bears emphasis here. For example, the model in Figure 3.1 (above) represents current expectations for potential synergistic and antagonistic effects as a result of interactions among habitats being restored—knowledge that may be tested by future studies and assessment of the effectiveness of future actions. Aronson et al. (2017) posit that expanding restoration to the landscape scale needs conceptual tools that consider not just ecological processes, but also policy and management activities that could either support or hinder the realization of restoration goals. The experience of coastal and estuarine systems with restoration programs in progress indicates the value of beginning simple modeling efforts—such as conceptual models—early on, before moving to more complex versions (Brudvig and Catano, 2021). More complexity is warranted when understanding based on related research programs, data availability from comprehensive monitoring, and management questions have been developed and evolved.

Conceptual models can help bridge the disconnect between a vision for restoration of a target species or geography and the reality of implementation that happens one project at a time across long periods of time (DiGennaro et al., 2012; Fischenich, 2008). Conceptual models are most helpful when developed in the early stages of a restoration project or program by systematically listing all the active variables, drivers, and stressors within the broader ecosystem. As new monitoring data are collected at different stages of the project or program, refining the conceptual model may become necessary (Olander et al., 2018).

Failure to achieve coastal restoration goals may occur for many reasons, including lack of understanding of initial conditions, design flaws, implementation challenges, and/or unexpected environmental changes. For example, a recent synthesis for tidal marsh restoration found that tidal marsh restoration is still fundamentally an experimental activity that is unique to each site, without universal standards (Waltham et al., 2021). Key environmental stressors of geographic variation among salt marshes include seascape configuration, the length of time when a habitat is wet or dry, riverine input, salinity, sediment supply and geomorphology, climatic region, and vegetation composition (Ziegler et al., 2021).

The Disaster-Pressure State-Ecosystem Services-Response-Health conceptual model developed by Sandifer et al. (2017) assesses ecosystem services and human health outcomes in the GoM after disasters that produce large-scale ecosystem injuries. The model connects disaster events (e.g., hurricanes, oil spills) to ecosystem pressure, states, services, and, ultimately, an array of responses. Conceptual models are also recommended for inclusion in all Natural Resource Damage Assessment (NRDA) Monitoring and Adaptive Management Plans.[7] However, O'Farrell et al. (2017) concluded that these conceptual models could be used effectively across ecosystems and projects and shared among restoration managers when prepared in a uniform framework. One such example of an explicit uniform framework is the Delta Regional Ecosystem Restoration Implementation Plan (DRERIP), which links a conceptual model to an action evaluation plan and a decision support tool (DiGennaro et al., 2012).

O'Farrell et al. (2017) also reviewed the current status of ecosystem-based fishery management modeling efforts for the GoM and identified 45 models in six classes, ranging from simple conceptual and qualitative models to models considering bottom-up and top-down interactions, whether biogeochemical based or coupled hydrodynamic and ecological model platforms. Other classes were extensions of single-species models, dynamic multispecies models, and aggregated whole ecosystem models. Another example is the Gulf of Mexico Integrated Ecosystem Assessment program of the National Oceanic and Atmospheric Administration (NOAA), which uses a simple conceptual framework to guide indicator development.[8] At the estuary/watershed scale, Tampa Bay scientists and resource managers have adopted a nutrient management strategy based on a conceptual model linking required light levels for sustaining healthy seagrass meadows with chlorophyll *a* concentrations and nitrogen loading levels (Greening et al., 2014; discussed further in Chapter 4). At regional and larger levels, the Chesapeake Bay Program and its partners have used conceptual models throughout the development and implementation of their action plan to help guide research, project implementation, and assessment of cumulative effects, as discussed in the following section (Linker et al., 2002, 2013; Shenk et al., 2012, 2013).

A potential tool for aiding the development of conceptual models is to develop a solid understanding of the restoration area by considering an ecohydrogeomorphic classification of the GoM. As described in Brinson (1993), classification simplifies the concept of a wetland or aquatic system, recognizing that while each potential area may be unique, such areas can also be placed into categories that share functional properties. The result of reducing the apparent complexity allows for improved communication among researchers and managers. Ecohydrogeomorphic classes also clarify the relationship between ecosystem structure and function. This allows for the comparison of systems with a set of common characteristics, and also acknowledges that some cross-boundary cooperation will be needed to generate useful assessments. There are several classifications in the GoM region that may be considered. For example, Pendleton et al. (2010) divided the U.S. Gulf Coast into eight regions, based on geomorphology, geology, and ecology. The definition of ecoregions led by James Omernik (Omernik and Griffith, 2014) may be useful in classifying the impacts of watershed contributions in the Gulf Region, while various older studies which presented classifications of coastal systems may still contribute much to our understanding (Lugo and Snedaker, 1974; Odum and Copeland, 1972).

How Modeling Advanced Restoration of the Chesapeake Bay and Its Watershed

The Chesapeake Bay Program focuses on nutrient load reductions, a compounding cumulative effect (see Table 3.1, above), to eliminate a spectrum of eutrophication impacts (cross-boundary cumulative

---

[7] See https://www.gulfspillrestoration.noaa.gov/sites/default/files/2018_01_TC_MAM_Procedures_ Guidelines_Manual_12-2017_508_c.pdf.

[8] See https://www.integratedecosystemassessment.noaa.gov/regions/gulf-of-mexico.

effects). During its more than 35 years of operation, there have been substantial increases in population in the 64,000 mi^2 watershed, as well as many land-use changes, such as forest losses and increases in impermeable surfaces (Clune et al., 2021). Total nitrogen and total phosphorous loads have decreased by about 27 percent and 23 percent, respectively, during the past 25-year period (Testa et al., 2018). Reductions in atmospheric deposition of total nitrogen (Eshleman and Sabo, 2016) and substantial upgrades in wastewater treatment plant operations that removed both nitrogen and phosphorous have been responsible for much of these declines (Testa et al., 2018). As a result of load reductions, water quality conditions (e.g., nutrient and algal chlorophyll *a* concentrations) have improved in many, but not all, areas of the estuary, hypoxic/anoxic conditions have decreased in size and persistence in the bay and its large tributaries, and communities of submerged aquatic vegetation (SAV) have started to recover (Testa et al., 2018). Restoration experience in the Chesapeake Bay indicates that estuarine water quality "memory" is short (seasons to years rather than decades); nutrient load reductions exhibit the strongest effects close to load-reduction locations; and season and location are also important, with high-salinity areas responding to small load reductions during the nutrient-limited summer season.

The Chesapeake Bay Program uses a variety of models, from qualitative conceptual models to complex spatially explicit simulation models. Initial understanding of nutrient dynamics and estimates of needed load reductions for the Chesapeake Bay were based on a sequence of relatively simple mass-balance computations (Boynton et al., 1995; Nixon, 1987; Smullen et al., 1982), each having limited spatial resolution and accuracy; over time, modeling efforts improved as understanding improved and better datasets became available (e.g., Lee et al., 2013). Initial coupled hydrodynamic–water quality models had limited spatial resolution and were flawed with regard to estuarine biogeochemistry (D'Elia et al., 2003; Hydroqual, 1981), but they served to stimulate testing and guide improvements (e.g., Brady et al., 2013; Cerco and Cole, 1993). The current Chesapeake Bay coupled hydrodynamic–water quality model now has considerable spatial detail and is used as a primary management tool in nutrient reduction efforts (Hood et al., 2021).

Although in a different region of the country, the management of the Chesapeake Bay Program, including its scientific and political processes, has lessons and perspectives for considering Gulf of Mexico restoration efforts, in part because of many situational similarities. Both programs also cover multiple states sharing a common resource and includes many actors working together across political boundaries, including with multiple state resource agencies and federal counterparts. Issues addressed by the Chesapeake Bay Program in common with the GoM are many and include water and sediment quality, habitat loss (especially oysters and seagrass), coastal development and fisheries disruptions, and of course climate change. Further, while the Chesapeake Bay modeling effort is more mature than efforts in the Gulf of Mexico, significant contributions toward prioritizing restoration efforts are under way. For example, the generation of the Louisiana State Master Plan, which is updated once every 5 years, is dependent on a complex modeling system that utilizes landscape modeling through an Integrated Compartment Model (ICM), surge and wave models, and risk models (White et al., 2017).

The Chesapeake Bay example suggests that value can be added by modeling early with conceptual and simple mass-balance models representing initial hypotheses about the ecosystem, then using those models as a framework for synthesizing data toward improved understanding, more comprehensive models and assessments, effective adaptive management decisions, and, ultimately, measurable estuarine restoration. This 35-year restoration effort adopted the idea of starting with initial analysis and modeling, adding complexity as scientific understanding and management needs allowed and demanded (Linker et al., 2013).

AN APPROACH FOR CONSIDERING THE CONSEQUENCES OF LARGE-SCALE RESTORATION

Evaluating the effects of a restoration effort often involves one body of water or watershed and is therefore unreplicable (Waltham et al., 2021). This means that the usual design with which most ecologists and environmental scientists are familiar, an experiment with randomly allocated treatments and replication, is not possible (Howe and Martínez-Garza, 2014). However, this does not mean that rigorous analysis of system-wide restoration projects cannot be done, only that strict assignment of causes and effects cannot be made via standard methods of statistical analysis alone.

To compensate for this inability to use traditional experimental designs, the lack of reference conditions, the lack of replication, the difficulties in establishing causality and often the shortage of appropriate data, Diefenderfer et al. (2011, 2016) proposed an evidence-based evaluation methodology that utilizes multiple lines of evidence and causal criteria. These methodologies have previously been used in biological risk assessment (Suter et al., 2002, 2010). Downes et al. (2002) recommended that causal criteria be used in ecosystem restoration, and they have since been used effectively in freshwater ecology (Norris et al., 2012; Webb et al., 2015). Although the methods have been in existence since their introduction for health research in the 1960s (Hill, 1965; U.S. Department of Health, Education, and Welfare, 1964), multidisciplinary applications like these are still expanding (Ludwig et al., 2010; Wickwire and Menzie, 2010).

Multiple Lines of Evidence

Multiple lines of evidence are intended to confer built-in redundancy, where each line of evidence functions as an "umbrella" under which relevant analyses are collected for evaluation of key hypotheses. The following discussion is built on the lines of evidence proposed by Diefenderfer et al. (2016, p. 7) for salmon habitat restoration and reconnection in the Columbia River estuary, which were intended to be "universally applicable to large-scale ecosystem restoration programs," are applied here for the more general GoM purpose of recovering aquatic species through coastal habitat restoration (see Table 3.3). The organizing principle consists of seven general lines of evidence, designated a–g, which are intended to encompass the typical kinds of indicators monitored to evaluate large-scale restoration effectiveness, and related analytical and modeling methods. As applied to large-scale ecosystem restoration, the lines of evidence were intended to be used in synthesis of the composite data and analyses to help distinguish association from causation (Diefenderfer et al., 2016). A suite of monitored indicators and analyses is developed for each line of evidence to evaluate hypotheses represented in the conceptual model.

The Chesapeake Bay example, discussed above, can be used to illustrate the concepts of multiple lines of evidence, causality, and thresholds. From the beginning of the Chesapeake Bay Program in the 1980s, extensive monitoring data analysis and various types of modeling, such as mass-balance models and large-scale numerical models, were undertaken. These tools were used to develop multiple lines of evidence, which were used to determine that excess nutrients and eutrophication were responsible for extensive loss of seagrass beds and associated habitat. Initially, it was thought that excess phosphorus was responsible for the eutrophication, but further investigation showed that nitrate was a critical component as well, showing the importance of understanding causality. Finally, it was demonstrated that the Chesapeake Bay had exhibited the characteristics of exceeding a threshold when it flipped from a benthic-oriented system to a water column–dominated system.

In the Chesapeake Bay Program example, several specific lines of evidence came up, including:

- an emphasis on long-term monitoring in tidal and nontidal waters and from atmospheric sources (line a) and of the landscape (line g);
- research aimed at improving understanding of issues related to water quality and habitat restoration (line e); and
- a coupled and evolving suite of models (line b) designed for a number of purposes.

Together, these exemplars support the needed synthesis of diverse datasets to test and forecast effects of nutrient load reductions in a complex environment (Linker et al., 2002, 2013).

TABLE 3.3. Description of Seven Lines of Evidence for the Recovery of Aquatic Species through Habitat Restoration

	Data Summary	Analysis	Synthesis	Evaluation
Line of Evidence	**Monitored Indicators**	**Analyses**	**Causal Criteria** †	**Cumulative Effects Category** ‡
a. Research on Critical Ecological Uncertainties	Various	Summarize advances in understanding cause–effect associations; iterative improvement of the conceptual model	Plausibility, temporality, specificity, coherence, exposure pathway, predictive performance	Indirect, time lags, compounding
b. Evidence-Based Review of the Literature	Species presence, residence time, survival, prey availability, diet, stomach fullness, growth	Systematic global literature search, filtering, review, and scoring based on formal criteria	Strength and consistency, plausibility, specificity, analogy, coherence, predictive performance	Not applicable to cumulative effects
c. Physics-Based and Ecosystem Models	Water-surface elevation, particulate organic matter export	Hydrodynamic modeling of inundation patterns and particulate organic matter export	Strength and consistency, plausibility, gradient, temporality, coherence, exposure pathway	Space crowding, indirect, time lags, cross-boundary, nonlinear, compounding
d. Meta-Analysis of Restoration Action Effectiveness	Water-surface elevation, water temperature, sediment accretion, vegetation similarity, species presence	Qualitative assessment of action-effectiveness studies in the restoration program; analysis of data from historically reconnected sites	Strength and consistency, gradient, specificity of association, coherence, predictive performance	Landscape, time lags
e. Analysis of Data and Modeling of Target Species	Presence, diet, stomach fullness	Comparative analysis of stomach contents; detections of migratory species	Plausibility, gradient, coherence, exposure pathway	Cross-boundary, indirect, compounding
f. Modeling of Cumulative Net Ecosystem Improvement	Prey, biomass production, prey and biomass export, area of habitat restored	Additive modeling of change in function, restored area, and probability of success	Plausibility, coherence, exposure pathway	Landscape, compounding
g. Change Analysis on the Landscape Setting	Forest cover, impervious surface	Remote-sensing data analysis of forest cover and urbanization change trajectories in watersheds	Plausibility, coherence	Landscape

† Causal criteria are described in the following section.
‡ See Tables 3.1 and 3.2.
NOTE: Causal criteria used in the synthesis are defined and described in more detail in the text below.
SOURCES: Adapted from Diefenderfer et al. (2016). Included are associated analyses of monitored indicators, the causal criteria used for synthesis (Hill, 1965; Dorward-King et al., 2001), and the cumulative effects categories used for evaluation (CEQ, 1997). Causal criteria employed in the synthesis are defined and described in more detail in the following section.

Causal Criteria

The problem of how to establish causality in coastal ecosystems is larger than statistical methods alone can resolve. Downes et al. (2002) suggest that restoration programs first define causal criteria and decide how they will be examined and measured, and then review available literature for effects of human activity and extract the information needed to evaluate response variables using each of the causal criteria. Causal criteria are aspects of the associations between two variables (Hill, 1965). Factors to be considered when deciding whether an observed statistical association is causal include:

- temporality (the effect follows the cause),
- strength of association (the magnitude of the effect),
- consistency of association (documented by multiple observers under various circumstances),

- dose–response relationships or biological gradient (gradient in the cause and response level),
- consistency of evidence through replication of findings and other knowledge,
- specificity of the association (limited to particular sites and/or effects),
- biologic plausibility (understanding of the mechanism),
- complete exposure pathway (the cause can reach the receptor),
- coherence of evidence (lack of conflict between cause-and-effect interpretation and known facts),
- experimentation (manipulation of the cause),
- analogy (comparison to similar systems),
- predictive performance (prediction of restoration outcomes), and
- consideration of alternate explanations.

Not every criterion needs to be satisfied. In fact, the only necessary criterion is temporality. A positive statistical association between an exposure and an outcome does not necessarily mean that the exposure is the cause of the outcome. Causality is more than a "link"; it is a demonstration that an exposure(s) (restoration method) is responsible for a specific outcome(s). For every exposure–outcome relationship, there will always be gradations of evidence and certainty, and observed links or associations can be due to many factors. Causal inference is not purely objective, and it always includes a subjective judgment of the degree to which the evidence satisfies each criterion, leading to the ultimate conclusion of the likelihood that a particular causal relationship exists.

Although the cumulative effects of GoM restoration are not amenable to classical statistical analysis because the coast itself or individual estuaries, bays, and watersheds are the experimental units and cannot be replicated, causal criteria may be used to help distinguish among potential causes of an observed change in the ecological system. The aim of using causal criteria would be to separate the effects of acute environmental events or chronic environmental trends from the collective actions by those implementing GoM restoration.

TOOLS FOR GATHERING MULTIPLE LINES OF EVIDENCE

This section describes the types of tools that can be applied to collect data for, analyze, and model the multiple lines of evidence described in Table 3.3 (above). As with the evaluation of the cumulative impacts of stressors on ecosystems across geographic scales (Hodgson and Halpern, 2018), multiple analytical methods are needed to encompass the complexities of the cumulative effects of ecosystem restoration (Diefenderfer et al., 2016). Many of these tools are in use by GoM restoration project and program managers. Applications of each tool have the potential to produce results that contribute to evaluation of one or more lines of evidence. This section will not cover study design methods standard to ecosystem restoration—in particular the value of reference sites versus control sites, and before versus after data, which are described in *Effective Monitoring to Evaluate Ecological Restoration in the Gulf of Mexico* (NASEM, 2017).

Research on Critical Ecological Uncertainties

Table 3.3 summarizes the use of research to understand critical ecological uncertainties as a line of evidence. As discussed in the Chesapeake Bay example, uncertainties are often acknowledged at the outset, during development of ecosystem conceptual models. They can also arise during project and program implementation, especially when things go wrong or there are unexpected outcomes (Ebberts et al., 2018). Uncertainties can be traced back to data collection and modeling methods, problems with geographically scaling knowledge, changes in trends monitored through time series, and lack of quantification of drivers and stressors (Brudvig and Catano, 2021). Any scientific method suitable to address research hypotheses falls into the category of critical ecological uncertainty research. One frequently used technique is bench- or mesocosm-scale experiments, which contribute to restoration designs by isolating various factors in the ecosystem, revealing and quantifying relationships (Fry et al., 2017; Peralta et al., 2003).

The issue of how to assess the efficacy of a system-wide manipulation carried out without replication was considered by Carpenter (1993) in a review of his and colleagues' manipulations of lakes. Carried out over many years, nutrient inputs and predator population were manipulated to alter and understand whole-lake structure and function. They intended to discover whether lakes changed nonrandomly after the manipulations, and whether the manipulations were responsible for any changes that occurred. Because of the lack of replication, standard statistical analysis could not be done and random assignment of treatments and alternative analyses were employed.

To test the first hypothesis that nonrandom changes occurred postmanipulations, Carpenter et al. (1989) used time-series analyses to test whether observed changes were outside the limits historically recorded in the manipulated lakes. If nonrandom changes were detected, the second hypothesis was tested by comparisons with similar types of lakes and model outputs, and small-scale mesocosm experimentation was employed to collect data that allowed ecological interpretations to be made. Ultimately, knowledge gained from this work was able to be used to restore damaged lakes to their original conditions.

Evidence-Based Review of the Literature

Richard H. Norris built a meta-analysis framework, EcoEvidence, to assess environmental cause and effect around causal criteria with information synthesized from multiple publications (Norris et al., 2012; Webb et al., 2015). Several quantitative methods are incorporated in the method to more objectively assess published environmental research (as shown in Table 3.3). Scores are based on the quality of the study design and the number of all available published evidence concerning a given hypothesis. The method has been applied to analyses of river flow alteration and effects on the biota from frequency, magnitude, and duration of flows (Greet et al., 2011; Webb et al., 2012). However, as in other evidence-based endeavors, such as medicine, the need to integrate the results of such weighted evidence is "relentlessly situated and contextual" (Wieringa et al., 2017). Much the same argument could be made in regard to the spatiotemporally variable conditions of the GoM.

Physics-Based and Ecosystem Models

There are many physics-based and numerical models that can be useful in the development of a modeling framework during restoration planning and implementation (Table 3.3). This line of evidence can be expanded to include ecosystem models due to increasing development and use of model linkages across physical and biological regimes, such as understanding the potential for impacts from long-term environmental trends, identifying possible synergistic and antagonistic stressors, quantifying cumulative effects, and facilitating cross-site comparative analysis. Linkages among natural resources are complex, even on small scales. Addressing these complex relationships is difficult, and even the most sophisticated modeling tools may need data that are not readily available in many areas.

Physics-based models are mathematical representations of materials and energy flow through the ecosystem and are used to capture active physical processes within a large-scale domain (e.g., Jaiswal et al., 2020). Multiple parameters or variables are often involved, with interrelationships bounded by either empirical or mathematical submodels. Physical models—such as hydrodynamic models (Hodges, 2014), sediment and nutrient transport models (Flynn, 2001; Merritt et al., 2003), and water quality models (Moriasi et al., 2015)—have been used to examine alternative coastal restoration project designs, the impact of sea level rise on coastal ecosystems, water column characteristics, soil erosion potential and soil–water dynamics, movement of riverborne sediments, carbon flux dynamics in wetlands, and other hydrological and geomorphological processes (e.g., Brown and Pevey, 2019; Burchard et al., 2006; Hiatt et al., 2018; Leach et al., 2021; Passeri et al., 2015; Rogers et al., 2012; Wassmann et al., 2006). NOAA's Ecosystems & Fisheries-Oceanography Coordinated Investigations (EcoFOCI) program is one example of a physical and biophysical modeling effort that uses mathematical modeling to synthesize physical and biological data across an ecosystem (Dougherty et al., 2010).

Physics-based models can also be a useful tool for modeling active physical processes within a restoration site at various stages of restoration activities, although they need a large amount of field or simulated data. Ecosystem models include complex interactions among ecosystem components (e.g., species, habitats), ecosystem processes (e.g., drivers, pressures, stressors), and ecosystem services (e.g., carbon sequestration, biomass, population) (see summary Table 1 in O'Farrell et al., 2017; Geary et al., 2020). The interrelationships between various components and processes are either mathematically or empirically driven (Edwards, 2001). A large amount of field and simulation data are needed to parametrize ecosystem models, which ultimately provide important clues regarding temporal trends in overall ecosystem conditions or services (Denman, 2003; Matear, 1995; Peters and Okin, 2017). Integrated modeling can be used to bring together the potential elements of conceptual, physical, and population model results (EPA, 2008; Fulton, 2010; Laniak et al., 2013).

These complex models are typically developed to help define and assess multiple factors at larger spatial scales, including synergistic and antagonistic effects and background trends (Johnston et al., 2000, 2017). As an example, community models that deal with spatiotemporal dynamics of biotic assemblies in relation to underlying ecosystem or environmental processes are a combination of ecosystem and population models. The ICM, mentioned previously, is an example of a dynamic ecosystem model that simulates changes in wetland hydrology, species cover, and elevation along the Louisiana coast (de Mutsert et al., 2021). One of the challenges with ecosystem models is that scaling across the landscape can be problematic because of the amounts and variety of data needs related to model parametrization. Without the availability of large amounts of field data, ecosystem modelers may have to spatially extrapolate the relationships established at a site, which can be prone to high uncertainty (Geary et al., 2020).

Meta-analysis of Restoration Action Effectiveness

In environmental sciences, many studies do not report the statistics needed for formal, quantitative meta-analysis (Table 3.3) of restoration efforts (Greet et al., 2011; Norris et al., 2012). In conducting a global meta-analysis of wetland restoration, researchers found only 70 studies with sufficient information in the scientific literature from 1970 to 2010, despite including estuarine, lacustrine, palustrine, and riverine wetlands (Meli et al., 2014). The design of restoration experiments and hence the use of formal meta-analysis is further limited by the impossibility of finding true replicates in nature (Howe and Martínez-Garza, 2014).

Publication in scientific literature usually lags behind the timing needed to support ongoing decision making in a given restoration program, which can prevent formal meta-analysis of published data from contributing to decisions (Diefenderfer et al., 2011). However, meta-analysis of interim reports produced by restoration projects has been done in other systems and was seen as a valuable source of information for adaptive management and course correction (Diefenderfer et al., 2016). Hence, the term "meta-analysis" is used in this report to include the assessment of interim reports and data, as well as for the traditional use of the published scientific literature. The results of many GoM restoration projects initiated after DWH for restoring coastal nursery habitats, such as oyster reefs, salt marshes, and seagrass meadows, are still under way or in the process of being analyzed, and in lieu of formal meta-analysis, a qualitative meta-analysis of interim reports could be informative.

As an example, reports developed during the initial phase of the multi-State/EPA Chesapeake Bay Program (1978–1983), some of which were later transformed into peer-reviewed publications, played an important and timely role in designing portions of the current program. Reports by Heinle et al. (1980), for example, described some early water quality trends and made clear the need for a comprehensive and long-term monitoring program; Stevenson and Piper (1979) summarized knowledge concerning the causes and consequences of SAV declines in the bay and, again, indicated the need for long-term habitat monitoring and research. Thus, gray literature—particularly rigorously reviewed federal reports—can be a very

Box 3.2
BIOLOGICAL CONDITION GRADIENT MODEL: EXAMPLES FROM TAMPA AND MOBILE BAYS

Applying the Biological Condition Gradient (BCG) model to estuarine and coastal areas relies on concepts that were originally developed for individual freshwater stream segments and necessitates a system-level view to address larger assessment scales. The BCG has been used to assess past and present conditions of coastal waterbodies and to develop numeric estuarine condition goals (Cicchetti et al., 2017). Examples of BCG use in the GoM include an estuary-wide habitat application in Tampa Bay (Cicchetti and Greening, 2011) and an ongoing effort in Mobile Bay to help guide the description of existing conditions of biological resources (Vittor, 2014, 2019; MBNEP, 2019).

To test applicability at a regional scale, the BCG approach was applied to the mosaic of habitat types in Tampa Bay. Using 1900 as the minimally disturbed condition, a stressor–response relationship for intertidal and subtidal biotopes was developed. The stressor was based on time as Tampa Bay became more developed, while response was characterized as estimated percentage change in habitat extent relative to 1900. This resulted in percent change from minimally disturbed areal extent for each habitat type and allowed translation into a BCG, thus supporting the use of the BCG approach to assist in developing restoration strategies in larger, more complex systems (Cicchetti and Greening, 2011).

The Mobile Bay National Estuary Program (MBNEP) underwent a multiyear process to develop environmental indicators to gauge progress toward meeting objectives and goals in its Comprehensive Conservation and Management Plan. As part of that effort, the MBNEP Science Advisory Committee developed a BCG framework that describes the existing biological condition of priority coastal habitats along a continuum of stress, as well as the ability of a habitat to provide ecosystem services. The conceptual framework for the BCG is based on the relative proportion of good, fair, and poor conditions for a watershed, sub-watershed, habitat type, or stream reach of interest. To test application of this approach, MBNEP is applying this BCG concept in the D'Olive Watershed Restoration program. MBNEP and partners intend to use the BCG for monitoring status and trends, communicating with the public, developing numeric criteria for condition, tracking management effectiveness, and informing coastal restoration efforts (MBNEP CCMP, 2018; Vittor, 2019).

important resource for assessing restoration effectivness in the GoM. The Strategic Conservation Assessment (SCA) tool[9] funded by the RESTORE Council has developed a conservation planning inventory tool that could potentially serve as a repository for these types of assessments and reports.

Analysis of Data and Modeling of Target Species

Often, habitats for threatened or endangered species are a focus of coastal restoration, and data on species are collected accordingly, particularly to connect species recovery with habitat restoration. Depending on the target species, a wide variety of such data may be warranted, with associated analyses and modeling (see Table 3.3, above). Population models, for example, are mechanistic models used to predict or simulate population dynamics of species within an ecosystem due to changes in habitat characteristics from a set of drivers and stressors. These models can also be used to analyze species vulnerability, movement, and individual traits. Since population dynamics are intricately linked to habitat characteristics, restoration projects can utilize population models to analyze improvement in ecosystem services in terms of species dynamics with a restored habitat. For example, a marsh restoration site can use a bird population model to examine the changes in species dynamics and individual traits before and after the restoration projects. Population models have been widely used in the GoM to project changes in plant or animal populations in response to types and severity of background trends, particularly in developing Gulf-wide or regional fisheries and endangered species management plans. Population models have also been used at smaller spatial scales to examine potential functional effects (such as changes in the level of primary and secondary productivity) associated with an oyster reef or seagrass bed restoration project.

[9] See https://www.quest.fwrc.msstate.edu/sca-project.php.

The accuracy and precision of population models depend on the quantity and quality of monitoring data collected on the species and understanding of critical factors driving population changes. A report from the National Academies of Sciences, Engineering, and Medicine (2017) linked the monitoring requirements for mobile species having large spatial ranges with monitoring requirements for cumulative effects beyond the project scale. It discussed inadvertent impacts of restoration on wide-ranging species and how some types of restoration activities might produce harmful effects, and detailed monitoring methods to help restoration planners and wildlife managers minimize impacts and benefit these species.

Modeling Cumulative Net Ecosystem Improvement

As shown in Table 3.3, one line of evidence involves estimating whether and to what degree an ecosystem may or may not have improved because of an intervention, such as restoration. This section provides a brief overview of several models commonly used by agencies for habitat change. The calculation of cumulative net ecosystem improvement (CNEI) is one recently introduced example. This model was developed for ecosystems in the Pacific Northwest (Diefenderfer et al., 2016) and is based on the earlier Net Ecosystem Improvement Index (Thom et al., 2005, 2011). Because this additive function considers multiple restoration efforts in a particular geographic area, it can be a useful tool for assessing the cumulative effects of multiple restoration projects relative to a specific target function. CNEI takes into account the project area, the number of restoration projects within that area, changes in ecological function, and the probability of long-term restoration success. In this model, examples of changes in ecological function could be an outcome such as seagrass biomass or secondary production such as the density of juvenile invertebrates in a seagrass meadow.

A strength of this model is that it can be used in any ecosystem (Diefenderfer et al., 2016). The key decisions are which metrics are most appropriate to use for assessing change in ecological function and the probability of success, based on known conditions and past responses to restoration actions. A suitable reference site and the ability to estimate success are also needed. There are techniques employing CNEI ideas that have been developed to objectively evaluate wetland functions in the northern GoM: the Wetland Value Assessment and the hydrogeomorphic approach, which are described below. For the most part, these techniques are used to evaluate the potential effects of restoration efforts, but they could potentially be used as one "line of evidence" in an overall cumulative effects investigation.

The Wetland Value Assessment, an example of metrics already assessed for other restoration tools that could be used in the GoM, was developed under the Coastal Wetlands Planning, Protection and Restoration Act (CWPPRA) program to determine the benefits of proposed wetland restoration projects. In this approach, habitat quality and quantity are measured for baseline conditions. Then, "future without project" and "future with project" conditions are predicted, based on modeled data and/or the best professional judgment of the team. This approach uses variables considered important to the suitability of a particular habitat type for supporting a diversity of fish and wildlife species. Each model consists of (1) a list of variables considered important for characterizing fish and wildlife habitat in a particular wetland type, (2) a suitability index graph for each variable, and (3) a mathematical formula that combines the suitability indices for each variable into a single value for wetland habitat quality. Modules exist for cypress-tupelo swamp, fresh/intermediate marsh, brackish marsh, saline marsh barrier islands, barrier headlands, and coastal chenier/ridge. This methodology has been used heavily in Louisiana (USFWS, 2006).

The hydrogeomorphic approach, rather than focusing only on fish and wildlife functions of wetlands, seeks to evaluate the physical, chemical, and biological characteristics of wetlands. Model development begins with the classification of the wetland system into regional wetland subclasses based on hydrogeomorphic factors (Brinson, 1993), followed by the creation of a functional profile that describes the characteristics of the regional subclass, its functions, and the ecosystem and landscape attributes that influence each function. Reference wetlands are selected from a defined geographic area, assessment models are developed, and the models are calibrated using the reference wetlands (Smith et al., 1995). Similar to the Wetland Value Assessment, the models produce a numerical value that is multiplied by area, though functions remain independent, not summed. Although the hydrogeomorphic approach is not widely used because of its detail and data requirements, such models have been developed for several GoM wetland regional subclasses, including the Northwest GoM tidal fringe wetlands (Shafer et al., 2002) and tidal fringe wetlands along the Mississippi and Alabama Gulf coasts (Shafer et al., 2007), as well as numerous interim models for the wetlands in the vicinity of Galveston, Texas.

Both of these wetland evaluation techniques are designed for individual projects, not cumulative impact assessment, where the focus shifts from functions performed at the wetland scale to the larger watershed scale. However, the functional indices in the hydrogeomorphic models may be used in conjunction with other methods designed specifically to assess cumulative impacts, and its concepts have been used in other parts of the country, such as the assessment of several of the watersheds in southern California (Smith, 2003).

The Biological Condition Gradient (BCG) approach was developed by the U.S. Environmental Protection Agency (EPA) to define and communicate existing conditions of aquatic biological resources in order to meet requirements of the U.S. Clean Water Act (Davies and Jackson, 2006; EPA, 2016), and has potential for applications in estuarine, reef, and watershed restoration (Box 3.3). Originally applied to benthic stream organisms in the U.S. Northeast, the model estimates biological conditions at a site along a continuum, from natural or undisturbed to severely altered by anthropogenic stress. Each level is defined by an empirically derived description that can be interpreted by experts and practitioners, regardless of location or habitat type (Cicchetti et al., 2017; Yee et al., 2020). Because of the consistent and structured steps defined in the model, this approach has been applied to different regions and ecosystems, including streams (Davies and Jackson, 2006), estuaries (Cicchetti et al., 2017), and coral reefs (Bradley et al., 2020).

Box 3.3
BIOLOGICAL CONDITION GRADIENT MODEL: EXAMPLES FROM TAMPA BAY AND MOBILE BAY

Applying the Biological Condition Gradient (BCG) model to estuarine and coastal areas relies on concepts originally developed for individual freshwater stream segments and necessitates a system-level view to address larger assessment scales. The BCG has been used to assess past and present conditions of coastal waterbodies and to develop numeric estuarine condition goals (Cicchetti et al., 2017). Examples of BCG use in the GoM include an estuary-wide habitat application in Tampa Bay (Cicchetti and Greening, 2011) and an ongoing effort in Mobile Bay to help guide the description of existing conditions of biological resources (MBNEP CCMP, 2018; Vittor and Associates, 2014, 2019).

To test applicability at a regional scale, the BCG approach was applied to the mosaic of habitat types in Tampa Bay. Using 1900 as the minimally disturbed condition, a stressor–response relationship for intertidal and subtidal biotopes was developed. The stressor was based on time as Tampa Bay became more developed, while the response was characterized as estimated percentage change in habitat extent relative to 1900. This analysis resulted in information about the percentage change from minimally disturbed areal extent for each habitat type and allowed translation into a BCG, thus supporting the use of the BCG approach to assist in developing restoration strategies in larger, more complex systems (Cicchetti and Greening, 2011).

The Mobile Bay National Estuary Program (MBNEP) underwent a multiyear process to develop environmental indicators to gauge progress toward meeting objectives and goals in its Comprehensive Conservation and Management Plan. As part of that effort, the MBNEP Science Advisory Committee developed a BCG framework that describes the existing biological condition of priority coastal habitats along a continuum of stress, as well as the ability of a habitat to provide ecosystem services. The conceptual framework for the BCG is based on the relative proportion of good, fair, and poor conditions for a watershed, sub-watershed, habitat type, or stream reach of interest. To test application of this approach, MBNEP is applying this BCG concept in the D'Olive Watershed Restoration program. MBNEP and partners intend to use the BCG for monitoring status and trends, communicating with the public, developing numeric criteria for conditions, tracking management effectiveness, and informing coastal restoration efforts (MBNEP CCMP, 2018; Vittor and Associates, 2019).

Change Analysis on the Landscape Setting

Data-driven models, often referred to as machine learning (ML) or deep learning (DL) models, can be quite effective in teasing apart complex relationships among ecosystem drivers or stressors and response, and in identifying patterns among different datasets. However, their performance depends on the types and amounts of training data available from a site (Clark and Gelfand, 2006; Crisci et al., 2012; Goldstein and Coco, 2015). Once these data-driven models are trained on a dataset comprising representative variables with a wide range and standard deviation, they can be used for long-term predictive modeling and monitoring. They can also be scaled up to satellite data for large-scale mapping (O'Connell et al., 2021; Sejnowski, 2020). These types of data-centric, predictive models have become a valuable tool for the ecology and environmental science community in recent years (Hampton et al., 2013; O'Connell et al., 2021; Rammer and Seidl, 2019). New refinement procedures are constantly being developed to improve their predictive ability and reduce overfitting problems, which is a common problem in these types of models when the model overperforms in the training phase and underperforms in the prediction phase (Willcock et al., 2018). One of them is ensemble modeling, for which several machine learning models are run, and their output is combined using a rule-based algorithm to produce the most accurate prediction (Sagi and Rokach, 2018).

Machine learning models are predictive models based on computational algorithms that are trained with a large number of input variables (e.g., physical, chemical, biological, ecosystem data) to disentangle complex and nonlinear relationships (see reviews by Hampton et al., 2013; Thessen, 2016). DL models often refer to a family of learning algorithms that use multiple hidden layers to build relationships for prediction, instead of using one established method to develop relationships among biophysical, biochemical, ecological, and environmental datasets (Thessen, 2016). Most ML/DL models are supervised learning algorithms that can use a variety of datasets from heterogeneous sources by compiling a large training dataset even without a set of hypotheses (Hampton et al., 2013; Rammer and Seidl, 2019). Similarly to traditional or existing physics-based and ecosystem models (discussed above), these models need a large amount of training data; however, they allow a degree of heterogeneity in the way the data were collected and processed (Crisci et al., 2012; Vinuesa et al., 2020). That characteristic brings a greater amount of flexibility to these new modeling tools compared with the rest of the modeling tools discussed above, which utilize controlled datasets and extensive parametrization. These new-generation models are gradually being used and implemented across a variety of ecosystem assessment studies (Ryo et al., 2020).

Data-driven modeling tools to assess cumulative effects of large-scale restoration projects need large amounts of data, whether field, simulated, climate, and other biophysical data; satellite-derived spatial or point data; or a combination of all data (Crisci et al., 2012; Goldstein and Coco, 2015; Vinuesa et al., 2020). Today's world of big data, pervasive sensing, massive computing capacity, ML based on artificial intelligence (AI),, and DL create an opportunity to transform ecological modeling and large-scale ecosystem synthesis studies (Humphries et al., 2018).

In recent years, there has also been a substantial expansion in federally funded ecological site networks, which provide large heterogeneous datasets for such data-driven models to be implemented. Examples of such networks include AmeriFlux,[10] FLUXNET,[11] Long Term Ecological Research Network,[12] National Ecological Observatory Network,[13] Long-Term Agroecosystem Research Network,[14] Ecosystem Phenology Camera Network,[15] and Global Lake Ecological Observatory Network[16]. These data generation networks can support data-centric modeling at a larger—that is, watershed or estuary—scale; however, their direct application to small restoration sites at a project scale could result in higher prediction uncertainty and

[10] See https://ameriflux.lbl.gov/.
[11] See https://fluxnet.org/.
[12] See https://lternet.edu/.
[13] See https://www.neonscience.org/.
[14] See https://ltar.ars.usda.gov/.
[15] See https://phenocam.sr.unh.edu/webcam/.
[16] See https://gleon.org/.

entail new site-specific calibration and validation. In addition, while many of these networks may not be currently available for Gulf sites, these examples provide program managers with options for future data collection for their sites based on their program needs. For example, the PhenoCam network[17] provides system architecture, camera specifications, mounting instructions, and source code in an open-source manner for easy implementation at any site.

In addition, satellite- and drone-based remote sensing, and a wide array of environmental citizen science projects are generating tremendous amounts of spatial and point data at different scales (Corbane et al., 2015; Ridge et al., 2020). Some of these data types are available in an open-source manner, including climate, biophysical, and satellite-derived datasets from federal agencies, such as the MERRA-2 database[18] from the Global Modeling and Assimilation Office of the National Aeronautics and Space Administration (NASA) or NASA's Earthdata database[19]. However, other data needs to be collected at an appropriate spatiotemporal scale. In the GoM, big datasets for ocean and coastal waters are available from the Gulf of Mexico Coastal Ocean Observing System,[20] which collects and archives thousands of data points from a variety of sensors each year. On a smaller scale, although it is still substantial relative to datasets for large-scale restoration across the United States, Louisiana has developed the Coastwide Reference Monitoring System[21], with 390 sites that collect data on the ecological condition of coastal wetlands.

These data-driven models can be valuable to analyze cumulative effects across multiple restoration projects at a variety of scales, provided sufficient training data are available at the initial stages. Although these models have not yet been widely adapted by researchers and restoration managers for large-scale cumulative impact assessment in the GoM, they have been increasingly popular for many types of ecosystem monitoring and predictions in the past decade, including wetland biomass, primary production, species dynamics, habitat suitability, and driver-response characterization studies (Huang et al., 2021; Michaels et al., 2019; O'Connell et al., 2021; Ridge et al., 2020; Shiu et al., 2020).

There are some examples of ML- and DL-based ecosystem models that exist for the GoM for different types of ecosystems, physical parameters, and biota. One such example is the supervised machine learning by emergent self-organization map analysis model proposed by Engle and Brunner (2019) to analyze the geochemistry of water samples collected from oil and gas wells in the northern GoM. Shiu et al. (2020) proposed a deep neural network for automated detection of marine mammal species, which enhanced the accuracy of the detection by orders of magnitude when compared with other detection algorithms. Trifonova et al. (2019) tested a data-driven model—dynamic Bayesian network models—with different levels of structural complexity and a varying number of hidden variables to predict ecosystem dynamics in GoM. Through this model, they discovered meaningful interactions among ecosystem components and their environment and examined how climate perturbations affect these relationships. One caveat is that these AI-based ML/DL models tend to indicate the correlation between ecosystem variables, not causation. One rapidly growing subfield in AI is explainable AI (xAI), which is used to decipher the complex output of ML models at various scales (Kakogeorgiou and Karantzalos, 2021; Phillips et al., 2021; Ryo et al., 2021). For example, Ryo et al. (2021) provide a summary of xAI tools available for ecologists for application in species distribution modeling at different scales. The use of models based on xAI that use field datasets or remote sensing datasets is a rapidly growing field in ecosystem modeling and environmental monitoring.

REFLECTIONS ON RESTORATION PLANNING AND ENDPOINTS

Ecosystem restoration is planned and carried out by various types of governmental, nongovernmental, and academic institutions that have differing constraints and the focus has historically been on individual project sites, ecosystems, or species (Clewell and Aronson, 2013; Roman and Burdick, 2012). The approach-

[17] See https://phenocam.sr.unh.edu/webcam.
[18] See https://gmao.gsfc.nasa.gov/reanalysis/MERRA-2.
[19] See https://www.earthdata.nasa.gov/.
[20] See https://gcoos.org.
[21] See https://lacoast.gov/crms/Home.aspx.

es to planning and evaluating large-scale restoration discussed above are not standard in the current restoration planning paradigm (Diefenderfer et al., 2021). Planning by the DWH funding entities required the invention of systems for both prioritizing restoration projects and evaluating their outcomes, with many projects proposed at ecoregional (e.g., Chenier Plain) or statewide coastal-zone scales—or focused on migratory or highly mobile species that also necessitate a broad view. Such post-DWH projects are implemented at a much larger scale than most restoration conducted anywhere previously and have the potential to be more cost-effective with good planning and management (Neeson et al., 2015). Landscape-scale restoration planning is the exception, not the rule, and in the continental United States is generally carried out by long-term partnerships led by federal and state agencies at scales, such as the Greater Florida Everglades, Missouri River system, San Francisco Bay/Sacramento Delta, or Puget Sound (e.g., Fischenich et al., 2018). In contrast, DWH funds are available for a finite period with projects implemented in a decentralized manner across multiple administrative processes.

Applied Restoration Strategies for Landscape Stressor Constraints

As discussed in Chapter 2 and seen in the acute and chronic inputs shown in Figure 3.1 (above), stressors affecting an ecosystem occur on a variety of spatial and temporal scales. These stressors (defined in Box 2.1 in Chapter 2) affect the physical and biological conditions that impact ecosystem structure and function (Groffman et al., 2004; Twilley et al., 2019). For example, changes in estuarine inflow can interfere with the sediment-trapping function of salt marsh plants, which in turn affects the physical condition, ecosystem structure, and function of the marsh. The landscape processes affecting ecosystem structure and function can be estimated and then employed to approximate the relative degrees of stress and thereby prioritize areas for particularly beneficial types of restoration actions (Diefenderfer et al., 2009; Roni et al., 2018).

The degree of stress in an ecosystem, as well as the larger landscape in which it resides, can be helpful for prioritizing of actions (NRC, 1992). Possible actions include, after Thom et al. (2005):

- restoration, which is the return of an ecosystem to a close approximation of its previously existing condition (NRC, 1992);
- enhancement, which is any improvement of a structural or functional attribute (NRC, 1992);
- preservation (or protection), which is the exclusion of activities that may negatively affect the system;
- conservation, which is the maintenance of biodiversity and natural ecosystem processes; and
- creation, which is the development of an ecosystem that is historically not present in a given geographic area.

Depending on the degree of stress both at a site and in the landscape processes supporting ecosystem functions at the site, specific management actions can be recommended to maximize the probability of success (Diefenderfer et al., 2009; Thom et al., 2005) (see Table 3.4). For example, sites with low degrees of local (site-scale) and surrounding landscape stress are the best candidates for conservation and preservation. Restoration is suitable for sites with medium to high degrees of stress, provided that landscape-scale stressors are minimal or can be treated to minimize their impact on the recovering site. Sites with high degrees of local and landscape-scale stress are unlikely to benefit from full restoration and instead the prudent action would be to focus on enhancement or creation.

Impairments at the landscape scale, which cannot be ameliorated, constrain restoration potential (NRC, 1992), as well as the types of cumulative effects (see Table 3.1, above) that may be seen after restoration efforts are completed. Outcomes of the more active land- and water-management strategies (restore, enhance, and create) may interact with the conservation and preservation of resources in the vicinity.

TABLE 3.4. Ranking the Effectiveness of Restoration Approaches Based on Site-Scale and Landscape-Scale Stress

Site-Scale Stressors	Landscape-Scale Stressors		
	Low	***Medium***	***High***
Low	Conserve Preserve	Conserve Enhance Restore Preserve	Enhance Conserve
Medium	Restore Enhance Conserve Preserve	Enhance Restore Conserve	Enhance Create
High	Restore Enhance	Enhance Restore	Create Enhance

SOURCES: Adapted from Diefenderfer et al., 2009; based on Thom et al., 2005.

Constraints of Conventional Planning on Understanding Ecological Endpoints

The expectations for potential long-term and large-scale effects of restoration need to be tempered by real assessments. For example, since the passage of NEPA, environmental planning requires documentation, including the no-action alternative also known as the "future without project" or "without condition scenario" (see discussion above). Specifically, this projected future baseline condition, accounting for the types of trends described in Chapter 2, allows for an assessment of the benefits specifically attributable to the project that are expected to accrue over the project life span and for completion of an incremental cost analysis comparing the cost of a project to the projected benefits (USACE, 2000). In the Gulf, however, due to the complexity of stressors—such as relative sea level rise, freshwater inflows, hurricane frequency, and hurricane intensity (see Chapter 2)—this standard project-planning, traditional workflow may not achieve conventionally useful outcomes.

In standard project planning, the future without project is often a relatively perfunctory part of the planning process, and this component is often not well funded; nevertheless, Yoe (2012) points out that "the preparation of the without condition scenario [future without project] [is] the single-most critical analytical task in the planning process" (p. 1). The U.S. Army Corps of Engineers has developed guidance documents detailing some of the procedures and processes that are helpful to include in such an assessment (USACE, 2014, 2018). In a changing environment like the Gulf, a complete future without project assessment, in contrast with traditional approaches, may need to entail a large part of project planning to be useful. The future conditions report in the Louisiana Coastal Master Plan, for example, considers multiple sea level rise predictions, a range of subsidence rates, and a range of hurricane impacts (CPRA, 2017).

Once a valid range of future conditions has been determined, an additional challenge to conventional restoration planning is presented by the rapidly changing ecological conditions in the GoM relative to restoration goals and objectives. Such predictions depend upon the collection and archiving of quality long-term datasets (see Chapter 2). The development and maintenance of this planning process has needed and will continue to need significant capital investment, but it is essential to determining if a proposed project will perform as designed into the future.

Traditionally, a selected restoration project yields projected benefits, or net ecosystem improvement (Thom et al., 2005) or net ecological gain (National Infrastructure Commission, 2021), often measured in habitat units or other "environmental currency" per dollar spent (USACE, 2000). In contrast, in the rapidly changing Gulf, it is possible that there may not be a net increase of benefits over the project life span. This poses a challenge to the traditional planning paradigm. Currently, there are no universally accepted guidelines for performing incremental cost analysis to compare projects that (1) maintain current conditions but yield no new benefits or (2) projects in which the action merely slows the rate of degradation. The outcome of many of the coastal land-building projects in Louisiana were designed with this second category in mind—that is, to slow, rather than stop, the rate of land loss. Acknowledgment of other social effects, including community resilience, and economic benefits of restoration projects, including avoided costs, could continue to be explored and addressed. Further work to determine restoration endpoints is a constraint for all restoration activities, particularly those in the dynamic GoM.

Importance of Detecting Mismatched Scales in Evaluating Restoration

The total area of wetlands in the GoM is about 18,261 km^2 (Turner and Rabalais, 2019). The RESTORE Council *FY2020 Annual Report to Congress*, which summarized the accomplishments for 2018–2020 from funding under the 2015 Initial Funding Funded Priorites List and State Expenditure Plans awarded to date (RESTORE Council, 2021a), reported 8.5 km^2 of wetland restoration. This is about 0.05 percent of the current wetland acreage in the Gulf. Still, as discussed relative to synergistic effects, coastal wetlands have a disproportionately high effect on fisheries production, carbon sequestration, and other beneficial functions (Bauer et al., 2013; Windham-Myers et al., 2018). Even combining land acquisition, areas that are under contract to apply best management practices, nonwetland habitat restoration, oyster habitat restored, and areas where invasive species were removed to the wetland restoration number (RESTORE Council, 2021a), the total was 234 km^2. The DWH NRDA Trustees, in summaries of data pulled from NOAA's Data Integration Visualization Exploration and Reporting (DIVER) portal, have created, restored, or enhanced 2,350 acres of marsh, beach, and dune habitats. Project summary information from the National Fish and Wildlife Foundation's Gulf Environmental Benefit Fund indicates that approximately 170,000 acres of wetlands and other coastal habitats have been protected, restored and/or enhanced. Of course, this is only the preliminary funding and granting effort, and much more will be completed in the future, but it does illustrate the difficulties of scale and points to the necessity of explicitly determining an appropriate scale for cumulative effects assessment and the need to scale restoration efforts to a size appropriate for addressing the ecosystem problem. Further, a lack of evidence of cumulative effects at the regional scales does not necessarily point to a failure of individual restoration effects, but rather may be due to an insufficient relative scale of change.

CASE STUDY OF CUMULATIVE EFFECTS IN THE ANNUALLY RECURRING HYPOXIC ZONE IN THE GULF OF MEXICO

The Mississippi Basin provides an example of the cross-boundary effects, time lags, antagonistic and synergistic effects, and time crowding discussed in this chapter. Agricultural activities in the midwestern sections of the United States exert strong cross-boundary effects on eutrophication more than 1,000 miles away in the GoM. The concept of time lags is evident in the timing of nutrients that leave the midwestern fields during snow melt, spring rains, and spring flood events and cause hypoxia in the Gulf that generally peaks a couple of months later during the summer. As an example of an antagonistic effect, freshwater inflows from the Mississippi River can have a positive effect on adjacent swamps and bottomland hardwood forests, but too much freshwater entering the estuaries can be deadly to oysters. The repeated openings of the Bonnet Carré Spillway (six times in 10 years) in the 2010s, subjecting oysters in Mississippi Sound to multiple years of injury from excess freshwater, illustrates the idea of time crowding. And, perhaps most importantly, the Mississippi River Basin is instructive of the difficulties of scale that are involved when there are many restoration efforts that, even cumulatively, are small relative to the overall area of concern.

The Mississippi-Atchafalaya River Basin is the world's fourth largest; it drains portions of 32 states and covers 41 percent of the contiguous United States. The region is home to 57 percent of U.S. farmland, including about 180 million acres of corn and soybean fields, which contributes the majority of anthropogenic nutrients (nitrogen and phosphorus) to the Gulf of Mexico. In turn, 60 percent of the nitrogen nutrient input to the GoM comes from agricultural input (farm fertilizer, agricultural inputs from legume crops, and confined animal manure).[22] An annually occurring hypoxic zone forms during the summer months as a result of these excess nutrients (Robertson and Saad, 2021), leading to algae growth and subsequent decomposition. As the decomposition process occurs, oxygen is depleted in the water, resulting in stress or death to any life that cannot move out of the zone.

Since 1985, a yearly snapshot of the size of the zone is available from a research cruise over a 2-week period each summer (see Figure 3.4). This single annual measurement of the hypoxic zone is an indicator of cumulative effects of human and natural systems on water quality and ecosystem health across large scales and long time periods. The annual hypoxic zone in the GoM illustrates a wide array of cumulative effects, which are illustrated in Table 3.5.

In this example of the traditional assessment of the cumulative effects of stressors causing environmental degradation, multiple synergistic and antagonistic effects of stressors can mask the ability to detect cumulative effects. While the quantity of nitrogen delivered to the Gulf is an important predictor of the size of the hypoxic zone, discharge rates in May of each year, as well as the presence of hurricanes and ocean currents, can generate antagonistic and/or synergistic cumulative effects (cross-boundary and compounding; see Table 3.5, which can have enormous consequences for the observed zone size in any given year. Excess nutrients are anthropogenic in source and their delivery downstream depends on precipitation upstream. A drought year can cause the following year's delivery of nutrients to be unusually small, even if the total nutrient usage was large. Likewise, years of flooding and extensive rainfall can result in large nutrient deliveries the following spring, despite little or no increase in the application of nitrogen and phosphorus upstream (in Table 3.1, "Time Lag"). In addition to precipitation upstream, hurricanes and ocean currents can significantly impact the magnitude of the measured hypoxic zone. Given that precipitation is more variable and hurricanes are intensifying with climate change, it seems likely that these effects will continue and become more pronounced. Because of the presence of these significant confounding effects, rather than

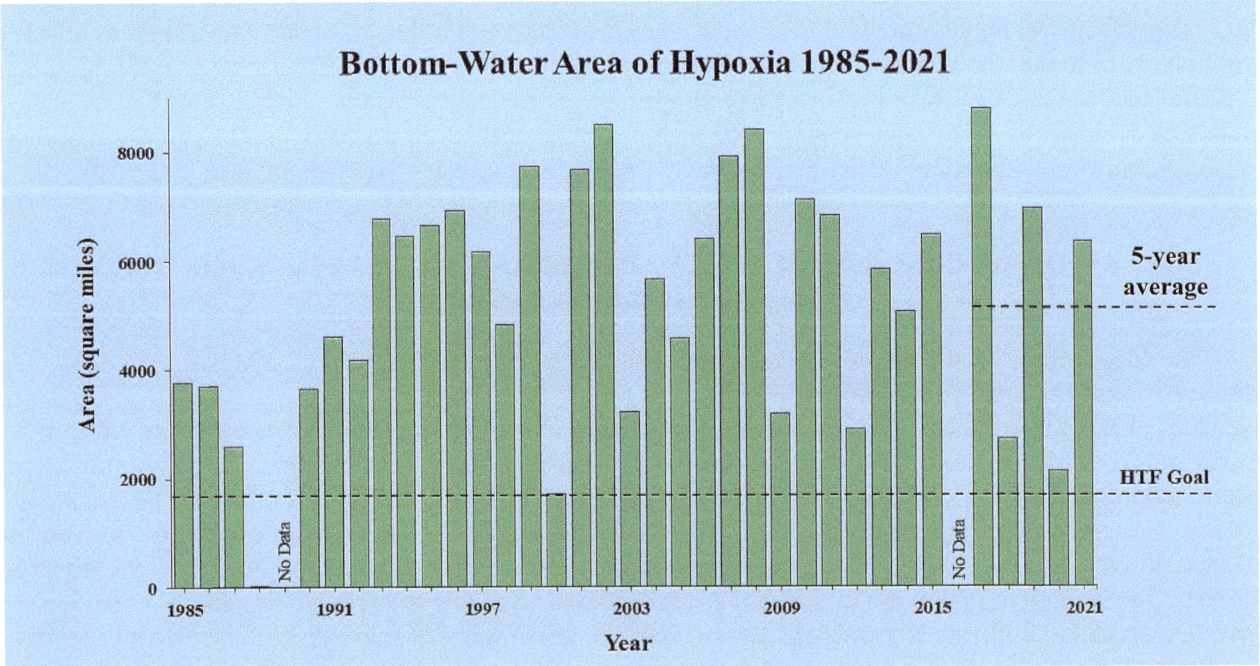

FIGURE 3.4. Size of GoM hypoxic zone. The year-to-year variability is due to a variety of synergistic and antagonist effects. HTF refers to the hypoxia task force. SOURCE: https://www.noaa.gov/news-release/larger-than-average-gulf-of-mexico-dead-zone-measured.

[22] See https://www.usgs.gov/media/images/sources-nitrogen-delivered-gulf-mexico-0.

focusing on a single-year measurement of the hypoxic zone, most analysts consider the 5-year running average in order to smooth out some of these effects.

The second key point of this example is that multiple modes and pathways of cumulative effects of restoration activities will often occur simultaneously and can reduce or augment the effects of individual restoration projects. In the context of the hypoxic zone, restoration activities are those that change land use, farming practices, and/or water delivery across the Mississippi River Basin with the goal of reducing nutrient inflows to the GoM; examples are provided in Table 3.5.

The third key point of this example is that, in the absence of a sufficient scale of restoration activities, cumulative effects of restoration activities may not be measurable and/or will be easily obscured by background variability. In 1997, the national Mississippi River/Gulf of Mexico Watershed Nutrient Task Force was created to coordinate activities, set targets for reductions in the size of the zone, and support a variety of related actions. In 2001, an Action Plan was released and over the next two decades, a number of reassessments and updates to it have been published.[23]

The official goal for the 5-year average size of the zone is depicted in Figure 3.4 and lies at the line labeled "hypoxia task force (HTF) goal." In this figure, the goal is not currently being met and has not been met in any single year. Further, every 5-year average far exceeds it. Is the failure to meet this goal evidence that individual restoration efforts were not effective? To answer this question, it is first necessary to determine whether the set of restoration projects and activities that have been implemented since 1985 are of sufficient scale and/or duration to have had a measurable effect at the large-watershed scale. Second, it may be that adequate change in farming actions has occurred, but there may be significant lag times between changes in farming practices throughout the watershed and a clear, measurable signal that these changes are reducing the average size of the hypoxic zone, particularly considering the confounding effect of weather variability that contributes to high variability of the zone's size.

TABLE 3.5. Role of Cumulative Effects in the GoM Hypoxic Zone

Cumulative Effect (from Table 3.1)	Examples Relevant to Restoration Efforts to Reduce the Hypoxic Zone
Compounding	Multiple wetland restoration projects upstream can generate additive nutrient reductions.
Triggers and Thresholds	Once the temperature is warm enough, harmful algae blooms (HABs) grow and bloom.
Indirect, Secondary Effects	Land use and agricultural practice changes that generate nutrient reductions have co-benefits in the form of carbon sequestration in soils and provision of habitat for pollinators, birds, and other wildlife—but some may have negative secondary effects too.
Landscape Pattern	Restoration of a wetland in ideal topographical locations can act to catch water flowing from many agricultural fields, allowing nutrient recycling from a broad area.
Cross Boundary	One watershed flows into the other. Nitrogen, phosphorous, and sediment flow downstream, but at different rates and paths.
Space Crowding	Locations with multiple conservation practices in place, such as vegetative drainage ditches that feed into a retention pond, provide multiple opportunities for nutrients to settle or be cycled.
Time Lags	Phosphorus moves more slowly than nitrogen in some systems, so while there are time lags associated with both nutrients, they should differ from each other.
Time Crowding	One large rainfall event followed soon by another will have nonlinear erosion effects.

[23] See https://www.epa.gov/ms-htf/history-hypoxia-task-force; https://www.epa.gov/ms-htf/looking-forward-strategy-federal-members-hypoxia-task-force.

Insight on this question comes from work done by the state of Iowa to document the extent of conservation actions that have been taken to address nutrient loss from the state to the Gulf.[24] The Iowa Nutrient Reduction Strategy (INRS, 2017) estimated that over 90 percent of Iowa's 22 million acres of row crop would need to be treated with conservation actions to reach the state's goal of a 41 percent reduction in nitrogen and 29 percent reduction in phosphorus loads leaving the state. Currently, 3 percent of Iowa's farmlands are treated. There is no evidence to suggest producers in other states have done more than Iowa farmers to address nitrogen losses from their fields.[25] With this context in mind, it is not surprising that there is no measurable signal in the annual size of the hypoxic zone of these minor changes implemented to date.

A LONG-TERM VIEW

Traditional large-scale environmental management frameworks, derived from legal and regulatory requirements, were focused on remediating human activities with the capacity to harm whole regions. However, activities at this scale, such as restoration, may have positive effects on an ecosystem (Daoust et al., 2014; Diefenderfer et al., 2011, 2016, 2021). In light of the United Nations Decade on Ecosystem Restoration, which began in 2020, this type of widespread investment in restoration per se may become more commonplace (Aronson et al., 2020; Waltham et al., 2020). The assessment of cumulative restoration effects at regional or other intermediate scales depends on the ability to match the scale of improvements with the scale of detection methods.

The effort to maximize successful large-scale restoration efforts is a daunting task not only because of the normal complexities of these coastal ecosystems, but also because of trends in important chronic and acute influences on these ecosystems. As shown in the case studies, these complexities are present in the long-term and ongoing Chesapeake Bay Program and in the development of hypoxia in the GoM. Likewise, these complexities are present and have posed formidable challenges for the collective restoration investments across the Gulf Coast since the DWH. Approaches to meet the challenges detailed in this chapter include a framework of the modes of cumulative effects in ecosystems, causal criteria used in judgment, and the evaluation of analyses representing multiple lines of evidence. Currently, the comprehensive restoration framework on the Gulf Coast subsequent to DWH represents a unique and crucial opportunity to experiment with, develop, and demonstrate measurable restoration outcomes at a regional scale.

---

[24] See https://store.extension.iastate.edu/Product/15915.

[25] See https://www.epa.gov/ms-htf/report-nonpoint-source-progress-hypoxia-task-force-states.

Chapter 4
Applications of Synthesis and Cumulative Effects Assessment in the Gulf of Mexico

INTRODUCTION

The projects funded by the legal settlements of the 2010 *Deepwater Horizon* (DWH) oil spill (RESTORE Council, 2016) represent an opportunity and a challenge for assessment and learning, because of the unprecedented number and diversity of projects and the 5- to 30-year timeframe of payments from the settlements. Individual project monitoring can inform efforts to ensure that what is learned from early projects benefits successive projects. The large-scale, long-term restoration ongoing in the Gulf of Mexico (GoM) is also an opportunity to assess cumulative effects of multiple projects on a landscape/seascape scale and to lay a foundation for future restoration efforts that will extend beyond the DWH funding timeframe.

Common methodologies and guidance for monitoring individual GoM restoration projects have been established, first by the trustees of the DWH Natural Resource Damage Assessment (NRDA) (DWH NRDA Trustees, 2016; NASEM, 2017), with the most recent monitoring guidance from the RESTORE Council (RESTORE Council, 2021a, b). Shared protocols, when implemented, allow for project comparisons and facilitate adaptive management. A common methodological approach also allows for assessment of cumulative effects at broader spatial scales. However, uncertainty remains about the aggregate benefits or effects of the unprecedentedly large number of restoration projects being concurrently implemented through settlement funds and how to assess them.

The RESTORE Council and NRDA Trustees recognize the need for assessing cumulative effects of multiple restoration projects[1] and have signaled that this task will be addressed in future versions of program guidance documents (DWH NRDA Trustees, 2019). The National Fish and Wildlife Foundation (NFWF) has not made a decision on whether a global synthesis of the Gulf Environmental Restoration Fund will be done.[2]

Some of the major challenges and barriers to assessing cumulative progress in restoration activities have been noted previously (NASEM, 2017). They include:

- inadequate funding and incentives at the project scale for short- and long-term monitoring of appropriate data and lack of human resources support to archive and document collected data,
- the lack of consistent protocols across projects to assure that monitoring and data collected are comparable and useful for cumulative effects assessments,
- the absence of a single freely accessible data repository that receives and archives all types of data from restoration projects in the Gulf, and

[1] See https://www.gulfspillrestoration.noaa.gov/sites/default/files/wp-content/uploads/Final-Phase-III-ERP-PEIS-Record-of-Decision_FINAL.pdf.
[2] Personal communication from Jonathan Porthouse, NFWF (September 16, 2021).

- the lack of a single entity or authority tasked with designing and undertaking a cumulative effects assessment to provide information needed for long-term planning at multiple spatial scales and the implementation of adaptive management.

In addition, the lack of standardized methods for assessing cumulative effects (as discussed in Chapter 3) hinders progress in evaluating restoration efficacy at larger scales, as summarized in the DWH NRDA report *Deepwater Horizon Oil Spill: Final Programmatic Damage Assessment and Restoration Plan and Final Programmatic Environmental Impact Statement* (DWH NRDA Trustees, 2016, p. 5-370):

> *Even for restoration approaches and/or techniques that are relatively well established (e.g., coastal habitat restoration), uncertainties about the aggregate benefits and/or impacts of restoration projects will be higher as the total number of projects implemented, size of individual projects, and extent to which projects are concentrated in particular geographic areas increases. As restoration scale (i.e., number and size of restoration projects, both independently and within a particular geographic area) increases, it will be more important to ensure that the information about aggregate restoration benefits and potential unintended consequences are incorporated into the monitoring and adaptive management framework.*

The challenges identified highlight the need for projects to initiate consistent monitoring for enough time following construction to enable future cumulative effects of multiple projects or large-scale assessments (NASEM, 2017).

This chapter discusses the scales at which cumulative effects of multiple restoration projects have been observed in the GoM, considers cumulative effects assessment methods suitable for different spatial and temporal scales, and identifies elements of a process that could be used to assess the effects of multiple restoration efforts given the potential confounding effects of environmental background trends. The chapter also examines how long-term, larger-scale restoration efforts are suited for analysis by multiple lines of evidence and demonstrates on-the-ground results and techniques that support management decisions.

KEY CONSIDERATIONS AND INFORMATION FOR GULF COAST CUMULATIVE EFFECTS ASSESSMENT

To accurately assess cumulative effects beyond the project scale in the GoM, a deep knowledge of Gulf ecosystems and restoration goals is needed. The challenge of undertaking cumulative effects assessment has been tackled in other locations, where beneficial cumulative effects of multiple restoration projects have been observed at different spatial and temporal scales. Well-documented restoration examples illustrate methods for evaluating large-scale water quality, seagrass, and marsh habitat restoration—the types of projects now being funded in the GoM. They include reductions in nutrient inputs from peak loads in the Baltic Sea, leading to improvements in ecosystem condition (Reusch et al., 2018); nutrient reduction projects that led to submerged aquatic vegetation (SAV) recovery in northern Chesapeake Bay (Gurbisz and Kemp, 2014; Lefchek et al., 2018); dam removals throughout the United States, which catalyze sedimentary changes in coastal landform, habitats, and biota (Foley et al., 2017); and the reconnection of floodplains to deliver food resources that support juvenile salmon in the Columbia River (Diefenderfer et al., 2016). Other evaluations are currently being initiated, such as the cumulative effects of two decades of habitat restoration on Chinook salmon recovery in Puget Sound's Whidbey Basin (Trujillo et al., 2021). A 2021 special session[3] at the National Conference on Ecosystem Restoration, which featured talks from across the United States, indicates the strengthening of a national community around this topic.

As noted throughout this report, detecting the effects of restoration efforts at any spatial scale will likely be hampered by background environmental variation and trends, such as those associated with climate change, as well as anthropogenic impacts, such as overfishing, pollution, and habitat damage. The degree to which background trends prevent detection of restoration benefits will depend on the assessment

[3] "Applying a Beneficial Cumulative Effects Paradigm to Advance Large-Scale Ecosystem Restoration"; see https://conference.ifas.ufl.edu/ncer2021/detailed-agenda.html.

methods used; the magnitude of the trend relative to combined characteristics of the restoration efforts themselves (the number, areal extent, and location of projects); the average effectiveness of those projects in achieving their goals; and the possibility of experiencing antagonistic and synergistic effects from other restoration activities.

The spatial scales at which beneficial cumulative effects from multiple GoM restoration efforts could potentially be observed range from project level (hectares) to estuary/watershed, or even larger scales. Detecting cumulative effects necessitates predictions about the improvements expected at the desired scale, often based on conceptual modeling (see Figure 3.1, in Chapter 3), and sufficient data to accurately determine whether those improvements have been achieved. Key considerations and information needed to perform an assessment of cumulative effects of restoration efforts include an understanding of the following broadly defined factors:

- current conditions and status of the watershed, waterbody, habitats, or animal populations;
- type and severity of background trends (such as climate change effects) and stressors (such as water quality degradation) affecting the ecosystem, and relevant future predictions based on these factors;
- a conceptual or other type of model that predicts how cumulative effects of restoration efforts can be expected to influence ecosystem form, function, or condition;
- restoration efforts, including types, proximity to other restoration projects, potential synergy and antagonistic effects, and spatial and temporal scales of implementation; and
- results of the restoration efforts that can be accessed for analysis.

PRIOR ASSESSMENTS OF THE CUMULATIVE EFFECTS OF ECOSYSTEM RESTORATION AND MANAGEMENT IN THE GULF OF MEXICO

Restoration efforts associated with some of the examples discussed below were initiated before DWH restoration efforts began and provide insight into how assessments of the cumulative effects from multiple projects have occurred to date.

Meta-Analysis and Synthesis of Restoration Projects by Habitat Type

Quantitative analysis and synthesis[4] of the effectiveness of different types of restoration, such as the effects of seagrass or oyster reef restoration on a local bay or statewide basis, remain uncommon in the GoM. Several syntheses determined whether restored areas, such as revegetated seagrass plots, remained present at varying numbers of years after the initial revegetation. In a synthesis of seagrass restoration worldwide, van Katwijk et al. (2016) found survival rates of only 22 percent for relatively small-scale seagrass restoration sites and 42 percent of larger sites 2 years after planting. In contrast, in a meta-analysis of Florida seagrass restoration projects, Rezek et al. (2019) found that if a restoration site was still in existence 3 years after installation, almost 90 percent of restored sites contained seagrass; however, the average densities were 37 percent lower than reference areas. In an assessment of seagrass coverage from 2007 to 2017, Handley and Lockwood (2020) compared seagrass gains and losses for each of the Gulf states and found that, for areas where data are being collected, the five Gulf states experienced an overall cumulative gain in seagrass extent of almost 24 percent (127,910 ha) during the 10-year period (see Table 2.1 in Chapter 2). However, the degree to which seagrass meadows are providing ecosystems services at the Gulf scale remains unknown.

Challenges to the validity of these trends in seagrass gains and losses include the lack of a Gulf-wide monitoring strategy, with each state currently using its own methods, as well as the lack of an identified

[4] Synthesis in ecology aims to discover new knowledge by bringing information together and has been defined for estuarine and coastal science as "the inferential process whereby new models are developed from analysis of multiple data sets to explain observed patterns across a range of time and space scales" (Kemp and Boynton, 2012, p. 3).

entity charged to regularly analyze and report trends. Although remote sensing data and models have been used successfully at many U.S. coastal sites to map spatial coverage, density, and biomass of submerged vegetation (Hestir et al., 2016; Meyer and Pu, 2012; Moore et al., 2008; Pu et al., 2014), remotely sensed datasets have not been used to assess the condition of seagrass on a Gulf-wide scale because of the complexities in assessment and modeling that arise from site-specific heterogeneity in water quality and geomorphic settings. Emerging sensing technologies and advanced data analytics and classification methods capable of ingesting heterogenous spatial datasets (discussed in Chapter 5) could make reliable, comprehensive Gulf-wide seagrass monitoring feasible in the future (Wang and Furrer, 2021).

Increasingly, assessments of restoration projects include a comparison of whether restored habitats are providing similar functions, and at the same level, as reference areas (i.e., have demonstrated functional equivalency). Simenstad and Thom (1996) explored functional equivalency trajectories, defined as the time it takes for restored wetlands to equate to natural references using appropriate parameters. They evaluated 16 attributes, finding most inconclusive, and they recommended approaches to improving such assessments. They further noted that understanding the source and extent of natural variability of reference sites is key to understanding constructed sites. Kentula (2000) expanded consideration of functional success to landscap-scale restoration, focusing on the contribution to ecological integrity (e.g., Karr et al., 2022; Rohwer and Marris, 2021; Wurtzebach and Schultz, 2016) of a region or landscape in maintaining biodiversity.

Kuhn et al. (1999) showed that some marsh management methods impaired function relative to reference sites in coastal Louisiana. Dillon et al. (2015) compared restored and natural oyster reefs in Grand Bay Mississippi over 2 years and found that constructed reefs performed better in several metrics. A recent meta-analysis of nekton (e.g., shrimp, crabs, fish) abundance in restored versus reference salt marsh areas in the northern Gulf used a functional type of comparison and found that nearly 13 years were needed for the densities of nektonic organisms in restored marshes to approximate those in reference marshes (Hollweg et al., 2020). Suir et al. (2020) found that restored wetlands reached structural and functional equivalency to reference wetlands 3–10 years after construction. They also found that with maturity, restored wetlands outperformed reference wetlands in such metrics as land aggregation, vegetative productivity, and floristic quality.

DWH-funded restoration projects have programmatic guidance available for identifying restoration objectives, monitoring certain performance-based metrics throughout the project, and assessing whether restoration objectives have been achieved by the end of the project. Table 4.1 illustrates the main DWH funding entities and their data reporting requirements. Specific guidance regarding project monitoring and performance assessment can be found in *Monitoring and Adaptive Management Procedures and Guidelines Manual Version 1.0* (MAM Manual) for NRDA (DWH NRDA Trustees, 2019) and the Observational Data Plan Guidelines for the RESTORE Council (RESTORE Council, 2021b).

Synthesis of changes in project-level metrics for multiple restoration efforts within a defined geographic area theoretically could provide evidence of cumulative effects (as discussed in Chapter 3), though this effort could be complicated because similar project types could conduct postimplementation monitoring for varying lengths of time, depending on the project sponsor. Synthesis of analytical results within lines of evidence has been identified as an important step in assessing the cumulative effects of large-scale restoration for at least a decade (Diefenderfer et al., 2011, 2016). To examine the feasibility of synthesizing DWH-funded project-level monitoring data and information regarding attainment of restoration objectives, the committee attempted to obtain relevant data online or through direct contacts for a cumulative effects assessment. It was not successful in obtaining adequate information to do an analysis, due to the inability to access data without extensive outreach to individual project contacts.

Despite repeated calls for consistent and transparent monitoring and assessment (summarized in NASEM, 2017), barriers continue to exist for accurate collation of monitoring and assessment of successes or failures in meeting restoration objectives. Funding streams have varied monitoring specifications and for inconsistent durations, and not all funding sources post final reports or have readily accessible monitoring data. Many of the records in both the Council Monitoring and Assessment Program (CMAP) and

TABLE 4.1. Main DWH Funding Entities Responsible for Providing Data Reporting Guidance at the Project and Programmatic Scales

DWH Funding Entities	Project-Level Monitoring Guidance	Project Monitoring Data Location	Program-Level Monitoring Guidance
National Fish and Wildlife Foundation (NFWF) Gulf Environmental Benefit Fund	Not publicly available. [a]	Project monitoring data is not publicly available based on searches. NFWF encourages, but does not require, recipients to make their data public in some way. [a]	Not publicly available.[b]
DWH National Resource Damage and Assessment (NRDA) Trustees	The Monitoring and Adaptive Management Manual provides "the TIGs [trustee implementation groups] with detailed information on recommended MAM [monitoring and adaptive management] procedures and guidelines, as well as additional guidance for the development of MAM Plans and the implementation of MAM at the project, [r]estoration [t]ype, and programmatic levels" (p. 2).[c]	The DIVER Repository contains MAM parameters. Additional project data may also be available in some cases.[d]	DWH NRDA Trustees indicate in the Monitoring and Adaptive Management Manual that "further [r]estoration [t]ype and program-level guidance will be developed in future versions of the MAM Manual" (p. 2).[c]
RESTORE Council	Observational Data Plan (ODP) Guidance: "Data should be submitted to a publicly available repository within two years of data collection, unless otherwise specified in the ODP Plan. Annual reporting on data collection is required during the Council's annual performance reporting period" (p. 26).[e] The Council also requires annual reporting and project closeout reporting.	Council MERLIN site contains metadata for Bucket 2 projects, as well as links and downloads to project-specific monitoring data that may be housed in other locations.[f]	None at this time. The Council's Comprehensive Plan Update notes, "The Council will utilize its staff, CMAWG [Council Monitoring and Assessment Work Group], CoP [Gulf of Mexico Alliance Monitoring Community of Practice], and coordination with other entities as a means to develop common standards and monitoring protocols for Council projects and programs; indicators and metrics of restoration and conservation success (including ecological function, benefits, and services) by project, region and/or watershed; identify data gaps in the assessment of the success of Gulf-wide restoration; and evaluate tools to measure Gulf-wide benefits" (p. 28).[b]

[a] Personal communication from Jonathan Porthouse, NFWF (March 25, 2022).
[b] See https://www.restorethegulf.gov/sites/default/files/CO-PL_20161208_CompPlanUpdate_English.pdf.
[c] See https://www.gulfspillrestoration.noaa.gov/sites/default/files/2019-08%20MAM_Manual_FULL_Updated%202019.pdf (p. 2).
[d] DIVER, Data Integration Visualization Exploration and Reporting; see https://www.diver.orr.noaa.gov/web/guest/diverexplorer?sqid=642&subtitle=DWH+Restoration+Project+Data.
[e] See https://restorethegulf.gov/sites/default/files/20210520_Council_Observational_Data_Plan_Guidelines_ Version%20 2.0_508.pdf.
[f] See https://restorethegulf.gov/merlin/srv/eng/catalog.search#/home.

the Gulf of Mexico Alliance (GOMA) project tracker[5] databases lack links to posted results or points of contact. The NRDA Trustees recognize these barriers[6] and have made recent progress toward reporting on restoration projects implemented to date, including publication of their 2021 programmatic review (DWH NRDA Trustees, 2021). Future MAM manuals are anticipated to include larger-than-project-scale guidance (DWH NRDA Trustees, 2019), and the LA TIG issued MAM guidance for Louisiana in 2021 (DWH NRDA LA TIG, 2021).

When more data are made available, it may be possible to "roll-up" project-level information into an assessment of potential cumulative effects (as defined in Box 3.1, in Chapter 3) within a geographic area or for specific outcomes of interest, ecosystem types, or restoration methods. The lines of evidence used by a program manager to assess cumulative effects will depend on the type and number of projects being implemented, the scale and relative impact of the projects, the proximity of projects to each other, the geographic extent of the program area being assessed, and the type and extent of data available to support the assessment. Examples of syntheses in the GoM that have attempted to detect beneficial cumulative effects of multiple restoration efforts at the estuary/watershed- and Gulf-wide scales, though limited, are provided in this section as references for future assessments.

Estuary- and Watershed-Scale Program Assessment Examples

Existing examples of cumulative effects of restoration at the estuary and watershed scales share the characteristic of an intermediate geographic scale, between an individual restoration project and the Gulf Coast, and are hydrogeomorphic units with high internal connectivity. This connectivity of hydrological and sedimentary processes that are foundations of and controlling factors on ecological processes and species' habitats (cf. Groffman et al., 2004) makes them a natural unit of management and assessment (Li et al., 2020a).

At the estuary/watershed scales, several long-term, science-based resource management programs located in the GoM offer examples of the application of multiple lines of evidence (see Chapter 3) to assess cumulative effects of the many restoration actions implemented in each estuary over the last 3 decades. Although not designed to assess cumulative effects, the Tampa Bay Estuary Program (TBEP) in Florida and its partners, the Galveston Bay National Estuary Program in Texas in partnership with the Galveston Bay Foundation, and the Comprehensive Everglades Restoration Plan each used tools and techniques similar to the theoretical approaches outlined in Chapter 3. Each program developed and now applies site-specific approaches to assessing cumulative effects of restoration efforts within their study areas, as outlined below.

In addition, each of these examples (initiated decades ago) did not originally incorporate explicit considerations of long-term environmental trends in their conceptual models and restoration designs. However, as these trends (especially those associated with climate change) have become obvious, each program has used an adaptive management approach to revise restoration plans, goals, and project designs. For example, TBEP has recently updated their habitat masterplan with a new conceptual approach, which accounts for future stressors (including sea level rise, climate change, and watershed development) in the target setting process (Robison et al., 2020).

Tampa Bay Estuary Program

Located on the west coast of Florida, Tampa Bay is a shallow, 1,035 km^2 estuary with a population of 3 million in the 6,735 km^2 watershed.[7] In the 1970s, Tampa Bay was significantly impacted by nutrient overenrichment and eutrophic conditions, including degraded water quality, persistent algal and macroalgal blooms, depleted oxygen levels and fish kills, and an estimated 50 percent loss of seagrass between 1950 and the mid-1970s (Greening et al., 2014).

---

[5] See https://dwhprojecttracker.org/.

[6] See https://www.gulfspillrestoration.noaa.gov/sites/default/files/wp-content/uploads/Final-Phase-III-ERP-PEIS-Record-of-Decision_FINAL.pdf.

[7] See https://tbep.org/estuary/.

In response to local citizens' calls for action, state regulatory restrictions on wastewater treatments plants located in the Tampa Bay watershed were put in place in 1978. This marked the beginning of Tampa Bay's restoration and recovery. Nearly 900 projects for improving water quality have since been completed by the public/private Tampa Bay Nitrogen Management Consortium (TBNMC), representing regulatory compliance-driven activities (e.g., water treatment upgrades, stormwater retrofits, power plant scrubber updates, industrial discharge reductions, residential fertilizer use ordinances) (Raulerson et al., 2019) and voluntary actions (e.g., habitat acquisition and restoration, education and social marketing campaigns) (Beck et al., 2019). The cumulative effects of nutrient reductions from these actions—implemented by local governments, industries, utilities, and homeowners—have reduced total nitrogen loads to half of those estimated in the mid-1970s, and water clarity conditions were improved roughly to those observed in 1950, when human population in the watershed was less than 20 percent of what it is today. In 2018, 164.4 km^2 of seagrasses were reported throughout the bay, exceeding the 1950 areal extent target established by TBEP partners in 1995 by about 10.6 km^2 (Beck et al., 2020). Between 2018 and 2020, however, mapped seagrass extent was reduced by almost 16 percent, possibly associated with recent seasonal algal blooms[8] (Beck et al., 2022).

Key elements of TBEP's assessment of cumulative effects associated with these multiple restoration efforts include:

- annual synthesis, analysis, and reporting of ambient water quality monitoring data collected by the three counties surrounding the bay;
- seagrass extent estimates every 2 years by the Southwest Florida Water Management District; and
- comparison to numerical targets for both water quality metrics and seagrass extent (Beck et al., 2022).

A publicly accessible database of restoration projects[9] is maintained by TBEP, updated by the TBNMC partners, and used to synthesize the types, locations, and estimated nutrient reductions from projects implemented in the bay and the watershed every 5 years as part of demonstration of attainment of water quality regulatory thresholds. If thresholds are not met, the TBNMC has developed a decision matrix to identify and address potential factors impacting water quality and/or seagrass extent (TBNMC, 2017).

A timeline of the Tampa Bay nutrient management strategy demonstrates the use of lines of evidence, including those described in Chapter 3 and those developed by TBEP and its partners (see Table 4.2). In addition, a recent paper (Beck et al., 2019) applied a cumulative effects approach to an empirical modeling-based reassessment of data on the interactions between water quality improvement projects and seagrass recovery projects in Tampa Bay.

Galveston Bay National Estuary Program

Galveston Bay is the seventh-largest estuary in the United States, encompassing ~1,600 km^2, with a 62,200 km^2 watershed and a population of 5.4 million.[10] It is a subtropical estuary with an average depth of 2.1m, fed primarily by the San Jacinto and Trinity Rivers. Houston, the fifth-largest U.S. city, sits at the northwest end of the bay. The port of Houston ranks second nationally by tonnage and first in foreign traffic, primarily petroleum based. The Houston area produces about 44 percent of U.S. petrochemicals (42.6 million tons per year) and 14 percent of U.S refining capacity.

Galveston Bay has faced many challenges due to coastal development. Approximately 182 km^2 of wetlands were lost over 60 years (Texas Environment, 2011), and seagrass beds have mostly disappeared over that same period. Seafood consumption advisories were issued for the bay starting in 1990, and approximately half of the bay's wetlands have been lost due to relative sea level rise, diminished sediment supply, and human activities (GBEP, 2018). Freshwater diversions for municipal, industrial, and agricultural uses, particularly in the Trinity River watershed, have negatively impacted the Bay (GBNEP, 1995). Loss of freshwater has altered salinities, diminished sediments, and negatively affected water quality.

---

[8] See https://tampabay.wateratlas.usf.edu/seagrass-monitoring/.

[9] See https://apdb.tbeptech.org/.

[10] See https://gbep.texas.gov/galveston-bay-101/.

TABLE 4.2. Assessment of Cumulative Effects of Environmental Restoration Efforts in Tampa Bay and Galveston Bay (by decade). Lines of evidence include those defined in Chapter 3 (Table 3.1)

	Tampa Bay			Galveston Bay		
Decade	**Conditions and Changes: Adaptive Management**	**Site-Specific Assessment Techniques**	**Lines of Evidence**	**Conditions and Changes: Adaptive Management**	**Site-Specific Assessment Techniques**	**Lines of Evidence**
1970s	Ecosystem degradation: clear water seagrass-dominated system to macroalgae- and phytoplankton-dominated system	Water quality monitoring data collection initiated (monthly at 60 stations in the bay)	Research on critical ecological uncertainties		Water quality monitoring data collection initiated	Research on critical ecological uncertainties
1980s	Wastewater treatment plant upgrades stipulated to reduce by 90%; implemented by 1982 (state regulation)	Seagrass areal extent monitoring initiated (every 2 years)	Analysis of data and modeling of target species	Galveston Bay designated an estuary of national significance, 1987		
1990s	Tampa Bay National Estuary Program formed; conceptual model and simple hypothesis developed Tampa Bay Nitrogen Management Consortium agrees to collaborate to meet nutrient load target, multiple nutrient reduction and restoration projects implemented Consortium develops total nitrogen (TN) loading caps for all sources	Numeric targets approved and adopted for seagrass acreage, light attenuation, chlorophyll *a* concentration, nutrient loading caps Tampa Bay Nutrient Management Strategy	Review of literature: *Conceptual model*: nutrient loading drives chlorophyll a concentration and light attenuation, which affect seagrass growth *Hypothesis*: managing nitrogen loads at appropriate levels will result in adequate water clarity to support seagrass recovery *Modeling of drivers and associated response*: empirical model and mechanistic model agree on TN load targets Analysis of data on target species: light conditions of seagrass at deep edges Mechanistic and empirical models	1995 Comprehensive Management Plan adopted Texas Estuaries Act (1999) funds restoration actions and coordinates state agency actions Total maximum daily load (TMDL) and implementation plans established for lower Galveston Bay watershed; Watershed Protection Plans established for lower Galveston Bay		Review of literature: *Hypothesis*: A collaborative public–private effort to restore habitat, conserve fish and wildlife populations, improve water quality and sediment contamination, and harmonize multi-use conflicts will improve and sustain bay health *Conceptual model*: modeling of drivers and associated response based on EPA TMDL water quality models.
2000s-ongoing	Assessment of cumulative effects on nutrient loading, ambient water quality and clarity, seagrass; reevaluation of models every 5 years; tracking restoration efforts Active adaptive management	Water quality and seagrass monitoring with annual reporting, assessment, and adjustment; land use and land cover change analysis every 5 years Tampa Bay Nutrient Management Strategy Assessment	Change analysis on landscape setting Meta-analysis of restoration action effectiveness Synthesis of cumulative effects from multiple restoration efforts Modeling of cumulative net ecosystem improvement	Multiple restoration efforts on water quality, wetlands, and oysters were implemented		

TABLE 4.2. Assessment of Cumulative Effects of Environmental Restoration Efforts in Tampa Bay and Galveston Bay (by decade) Lines of evidence include those defined in Chapter 3 (Table 3.1) Continued

2010s-ongoing		Tampa Bay Nutrient Management Strategy Assessment		Galveston Bay Program sets goals for improving water quality, habitat conservation, species conservation, sustaining freshwater inflows, collaborative monitoring Active adaptive management 1995 Comprehensive Management Plan updated based on annual Galveston Bay Report Card findings Charting the Course to 2015: Galveston Bay Strategic Action Plan (2009) and The Galveston Bay Plan, 2nd Edition (2018)	Galveston Bay Report Card analyzes data and trends from 19 metrics for water quality, pollution events and sources, wildlife, habitat, human health risks, and coastal change; report cards continue annually	*Analysis of data on target species* *Synthesis of cumulative effects from multiple restoration efforts* *Physical modeling of ecosystem controlling factors:* hydrodynamics/ freshwater inflows
2020s-ongoing	Evaluation of alternative hypotheses for seagrass recovery backsliding in some areas	Tampa Bay Nutrient Management Strategy Assessment	Physical modeling of ecosystem controlling factors: hydrodynamics Research on critical ecological uncertainties: algal blooms	Watershed and bay assessment of nitrogen phosphorus concentrations, dissolved oxygen Synthesis of sediment pollution loads and sources related to seafood consumption. Fifteen years data for organics and metals Assessment of oil spill impacts since 1998 Freshwater inflow assessment of change from 2003 to 2016 data Assessment of relative sea level change, pH and water temperature related to climate change between 2005 and 2018	Galveston Bay Report Card	Synthesis of cumulative effects from multiple restoration efforts *Research on critical ecological uncertainties*: Synthesis of finfish, shellfish, and water birds data and trends—2005–2010 Synthesis of habitat (emergent and submergent) status and trends 1953 to 2019; oysters, 1995 to 2019 Analysis of data on target species Meta-analysis coastal change from 2010 to 2019 Change analysis on landscape setting

The formation of the Galveston Bay Council[11] in 1995 brought together a 42-member group of federal and state agencies, local government, nongovernmental organizations, businesses and industry, and private citizens to guide plan implementation and future strategies. Several tools were developed to monitor environmental trends and the progress of the plan's implementation, including the Galveston Bay Report Card.[12] Since 2015, the Galveston Bay Report Card has generated an annual analysis of the bay's environmental health. Led by the Galveston Bay Foundation and Houston Advanced Research Center, the report card assesses six different metrics of bay health: water quality, pollution events and sources, wildlife, habitat, human health risks, and coastal change. Data sources include hydrologic monitoring and various state and federal agency data sources compiled into a regional monitoring portal.[13] Table 4.2 outlines the process used in Galveston Bay including lines of evidence described in Chapter 3 and those developed by the Galveston Bay partners.

The Galveston Bay Report Card has proven an effective tool for guiding restoration efforts, as a communication tool, and as an implicit assessment of the cumulative effects of restoration within the Bay and its watershed. Since 2000, 117 km^2 of coastal and watershed habitat have been protected, restored, or enhanced (GBEP, 2018). Water quality monitoring data collected between 1970 and 2017 (HARC, 2020) indicate that, despite a declining trend in bay waters, dissolved oxygen levels appear healthy and remain above 6 mg/L. Total phosphorus and ammonia concentrations have declined as well. However, after the sharp decline in chlorophyll *a* concentrations in the 1980s and 1990s, peaks were seen in 2003, 2007, and 2009. Since then, concentrations have remained elevated. Changes in freshwater inflows have the greatest effect on parameters like chlorophyll *a* (Pinckney et al., 2017), but the complex interaction of multiple stressors within the bay system preclude a current understanding of why the chlorophyll *a* levels are elevated.

Key characteristics of the methods and processes used by estuary-scale efforts have been associated with the attainment of goals and ecosystem improvements (Greening et al., 2014; Gross and Hagy, 2017; Sherwood et al., 2016; Tomasko et al., 2018), thus providing an indication of the cumulative effects of multiple restoration efforts. These characteristics include:

- targeted resources and metrics are recognized as important by both the public and scientists;
- the community is willing to work together toward common goals;
- science-based measurable goals and targets have been adopted;
- multiple tools are considered, such as regulation, public–private collaborative actions, and citizen actions;
- long-term monitoring, synthesis, and reporting mechanisms are in place;
- a convener to track, facilitate, and report progress is recognized;
- program assessment and adjustment (an adaptive management approach) is used;
- a link to the economic value of a healthy ecosystem has been established; and
- large (size) or a large number of projects addressing common stressors have been implemented.

Comprehensive Everglades Restoration Plan

The Comprehensive Everglades Restoration Plan (CERP) is a watershed approach for restoring, protecting, and preserving the greater Florida Everglades ecosystem. The plan aims to restore the hydrology, improve water quality, restore natural habitats, and protect native species.[14] Monitoring the numbers of nesting birds over three decades indicated that the cumulative effects of extensive water management efforts and restoration projects within the Everglades system have contributed to restoring contiguous aquatic habitat and prey populations, as well as supporting increased nesting success of several wading bird populations (Beerens et al., 2015; Pearlstine et al., 2020). Another metric of cumulative effects of multiple restoration efforts in the greater Everglades ecosystem is water quality. Over the last 20 years, $1.8 billion

[11] See https://gbep.texas.gov/the-galveston-bay-council/.
[12] See https://www.galvbaygrade.org/.
[13] See https://www.waterdatafortexas.org/coastal and https://gbep.texas.gov/regional-monitoring-portal/.
[14] See https://www.sfwmd.gov/our-work/cerp-project-planning.

has been invested in phosphorus control programs that have significantly improved Everglades water quality, with average phosphorus concentrations reduced from 24 to 9 ppb; at least 90 percent of the Everglades now meets water quality standards of 10 ppb or less for phosphorus concentration.[15] Similar improvements were noted by NASEM (2018b), which found that completed CERP components were beginning to show benefits, including creating hydrologic conditions increasingly similar to predrainage flows in some areas. However, as noted in *Progress Toward Restoring the Everglades: The Eighth Biennial Review* (NASEM, 2021), the ability to detect restoration progress is limited by monitoring, analysis, and results communication.

Despite multiple water management projects and observed improvements within the Everglades system, widespread downstream effects in the receiving waters of Florida Bay have not yet been realized. A major CERP objective is to manage the volume, distribution, and timing of freshwater inflow to Florida Bay, thereby returning the bay to historical ecological conditions (Cole et al., 2018). Changes in water usage in the Everglades have caused a 60 percent decline in freshwater inflow into Florida Bay, with resultant increases in hypersalinity (Herbert et al., 2011). Cole et al. (2018) examined year-to-year and basin-scale interactions between water quality and seagrass across 15 transects in Florida Bay between 2006 and 2013. They noted that, although restoration actions related to increasing freshwater inflow have been initiated by CERP, this intervention has, at this point, not led to significant decreases in salinity in parts of Florida Bay. They concluded that without further efforts to release freshwater during droughts, seagrass die-off events may recur.

Over the past few decades, the three estuary/watershed-scale examples highlighted above implicitly incorporated key elements of ecosystem restoration (Gann et al., 2019), including a basis in conceptual ecosystem models from which hypotheses and questions about ecosystem functions and the steps necessary to achieve them were derived. These questions were addressed and refined through long-term (multidecadal) monitoring and research, and adaptive management approaches to considering effects of long-term trends continue to be used to refine restoration plans.

Gulf-Wide and National-Scale Programs

National programs have contributed to significant improvements in U.S. waterbodies, including those in the GoM. For example, significant improvements in many U.S. waterbodies are the result of projects implemented to meet Clean Water Act provisions, which are aimed at reducing pollution from industry, sewage treatment plants, and other point sources. Between 1962 and 2001, many metrics of water pollution, such as biochemical oxygen demand, fecal coliform counts, total suspended solids, and water bodies considered "not swimmable," collected at 240,000 monitoring sites nationwide showed improvement (Keiser et al., 2018). Federal Clean Water Act (CWA) grants to install nutrient reduction technology from wastewater treatment plant discharges made rivers significantly cleaner for up to 25 miles downstream (Keiser et al., 2018).

Similarly, from 1970 to 2017, implementation of the national Clean Air Act resulted in a reduction of 73 percent of the six common air quality pollutants nationally.[16] In the GoM watershed, trends in atmospheric wet nitrogen deposition (from National Atmospheric Deposition Program data) from four sites in the Mississippi River watershed show decreasing or relatively steady nitrogen deposition since the early 1980s. As noted by Whitall (2008), these modest improvements may be related to the Clean Air Act Amendments of 1990. The Clean Air Act may have resulted in a reduction in the amount of nutrients deposited via atmospheric deposition to the GoM watershed and directly to the Gulf's surface waters (Whitall, 2008).

The scale of effort to improve clean air and clean water through the programs of the Clean Air Act[17] and Clean Water Act was significant and sustained (Keiser et al., 2018). In each case, adequate monitoring and data collection to measure progress and assess the performance of the programs occurred over a long period of time. It is notable, however, that while some datasets are tightly quality controlled, data on water

[15] See https://www.sfwmd.gov/our-work/cerp-project-planning and https://www.sfwmd.gov/our-work/wq-stas.
[16] See https://www.epa.gov/clean-air-act-overview.
[17] See https://www.epa.gov/sites/default/files/2015-07/documents/summaryreport.pdf.

quality have been collected under the auspices of federal agencies, states, nongovernmental organizations, and academic researchers, and these collection methods and measurements are not all consistent. In addition, data accessibility remains inadequate, so it has been difficult for analysts to use these data in a rigorous and consistent manner (Keiser et al., 2018).

Despite significant reductions in pollutant sources from national water quality programs and demonstrated improvements in upstream reaches, Gulf-wide water quality assessments from the National Coastal Condition Assessment of the U.S. Environmental Protection Agency (EPA) showed increases in the percent of areas rated "poor" for the eutrophication index from three periods between 2005 and 2015 (EPA, 2021). Several factors have likely contributed to the lack of observed water quality improvements at the Gulf-wide scale. Most experts agree that non–point-source pollution, which the Clean Water Act does not directly address, is the most pressing and challenging water quality problem in the northern Gulf of Mexico (NRC, 2012). Robertson and Saad (2013) estimated that 60 percent of nitrogen and 49 percent of phosphorus entering the GoM are from agricultural sources (e.g., cropland, fertilizers, manure, nitrogen fixation). Although management of agricultural practices has resulted in nutrient reduction from individual fields throughout the Mississippi River basin, (reviewed in Daniels et al., 2018), collectively these reductions contribute only a very small percentage to the overall nutrient budget of the Mississippi River basin.

In addition, lag times between field-scale management and response of water quality metrics in streams often mask any immediate gains in reduction and may take years of in-stream monitoring to detect (Sharpley et al., 2009). Variability in rainfall and rainfall patterns result in large differences in nutrient delivery and transport at both annual and larger time scales, increasing the difficulty in detecting changes in the receiving waters (streams, river, or Gulf) associated with nutrient management practices on land and other long-term environmental trends.

The five Gulf states had large-scale water-quality improvement proposals awarded in 2020–2021 through the RESTORE Council.[18] These included Louisiana's River Reintroduction into the Maurepas Swamp, Alabama's Perdido River Land Conservation and Habitat Enhancements, the Coastal Alabama Regional Water Quality Program and Perdido Watershed Water Quality Improvements and Restoration Assessment Program, the Florida Water Quality Improvement Program and Florida Gulf Coast Tributaries Hydrologic Restoration Program, the Water Quality Improvement Program for Coastal Mississippi Waters, and the Texas Coastal Water Quality Program. With water quality projects being implemented at the multistate scale, cumulative beneficial effects may occur, some of which may cross state lines through aquatic connectivity (such as Mississippi/Alabama and Alabama/Florida). This may also be an opportunity for states to collaborate on sampling methods and designs to capture larger-scale responses to collective restoration efforts and assess effects of background trends.

Living Resources: Oysters, Fisheries, Turtles, and Marine Mammals

Gulf-wide assessments of coastal living resources and target species, including oysters, nearshore fisheries stocks, turtles, and marine mammals, can produce quality data that may be able to detect cumulative effects Gulf-wide in the future.

Currently, and as identified in Chapter 2, there are no accurate Gulf-wide assessments of oyster abundance. However, although landings data cannot provide accurate estimates of population size, they provide some indication of a resource's distribution (GSMFC, 2012). Annual combined commercial oyster landing statistics for the five Gulf states show yearly fluctuations over the period 1950–2019, with a possible decline from a peak in the mid-1980s.[19]

Significant funds and effort have been dedicated to restoring oyster abundance and habitat in the GoM. An inventory of oyster restoration projects by the DWH funding entities identified a total of 67 approved

[18] See https://www.restorethegulf.gov/funded-priorities-list-3b.

[19] See https://www.fisheries.noaa.gov/national/sustainable-fisheries/commercial-fisheries-landings.

oyster-related projects (for a total of ~$175M) distributed throughout the states over the last decade.[20] Project types include research, aquaculture, decision support, shell recycling, and planning, and 30 projects are focused on restoration or enhancement of oyster reefs. As of January 2021, completed restoration and enhancement projects have resulted in a total of 2,522 acres of oyster reefs, with a goal of an additional ~2,000 acres from active and awarded projects (Brooke, 2021).

Stock assessments for a limited number of inshore and nearshore fish species, such as Gulf menhaden, are prepared by the Southeast Data, Assessment, and Review process (SEDAR).[21] SEDAR is a cooperative program between state and federal resource managers to improve the quality and reliability of fishery stock assessments. Their procedures and experiences can inform efforts in monitoring and assessment efforts related to restoration assessments across disciplines and political boundaries. SEDAR cooperators include the South Atlantic Fishery Management Council and the Fisheries Southeast Fisheries Science Center and Southeast Regional Office of the National Oceanic and Atmospheric Administration (NOAA), the Gulf of Mexico and Caribbean Fishery Management Councils, the Atlantic States and Gulf States Marine Fisheries Commissions, and NOAA's Fisheries Highly Migratory Species Division. Advantages of SEDAR stock assessments include the use of consistent methodology, the likelihood that SEDAR will continue to be supported by the cooperators over time, and the accessibility of data and analyses.

There have been recent efforts to develop more holistic ecosystem-based approaches to fisheries management—for example, transitioning from single to multispecies assessments in the GoM. Dell'Apa et al. (2020) suggest that tools conducive to the effective integration of ecosystem considerations into fisheries management are needed. These tools could inform and guide the work of regional managers, decision makers, and stakeholders in the GoM by allowing them to include the effects of other fisheries sectors, as well as environmental and ecological background trends. Development and applications of ecosystem-based methods could assist in separating background variation from the cumulative effects of restoration on fisheries in the GoM. Comprehensive data on additional fish stocks in the GoM other than those tracked by SEDAR and bird rookeries were not collected in regular intervals before DWH, and thus it would be difficult to assess longitudinal impacts as a result of DWH.

Sea turtles and marine mammals move among many locations in the GoM and are affected by a variety of environmental stressors (Carmichael et al., 2012; Frasier et al., 2020; Kellar et al., 2017; Kucklick et al., 2011), making it difficult to assess stressors that can lead to declines in individual and population health. Focused restoration efforts for marine mammals were developed by the DWH NRDA Trustees (2017c) and considered the long time frames needed for these populations to be restored. For example, it is estimated that, without targeted restoration, Barataria Bay dolphins need 39 years and and offshore cetaceans need from 10 to as many as 105 years for their communities to be restored from DWH effects.[22]

As demonstrated by the examples included in this chapter, the successful detection of cumulative effects of GoM restoration projects have to include consideration of a number of factors throughout the program's duration (see Figure 4.1). Key factors include the availability of adequate baseline monitoring to assess the effects of background trends, the scale of restoration projects, the collection of consistent project-level data for sufficient time to allow cumulative effects assessment, and the use of multiple lines of evidence and models to synthesize results.

MEETING THE CHALLENGES OF GULF COAST–SCALE ASSESSMENT

Programmatic Goals of the DWH Funding Entities

There is a wide array of restoration programs within the GoM, but all are working toward beneficial effects on the Gulf Coast. DWH settlement–funded programs have developed complementary goals related

[20] See https://www.gsmfc.org/ann_mtgs/2022-03/MSSC/4.%20Gulf%20oyster%20restoration%20projects%20Brooke%20FSU.pdf.
[21] See http://sedarweb.org/.
[22] See https://www.fisheries.noaa.gov/national/marine-life-distress/sea-turtles-dolphins-and-whales-10-years-after-deepwater-horizon-oil.

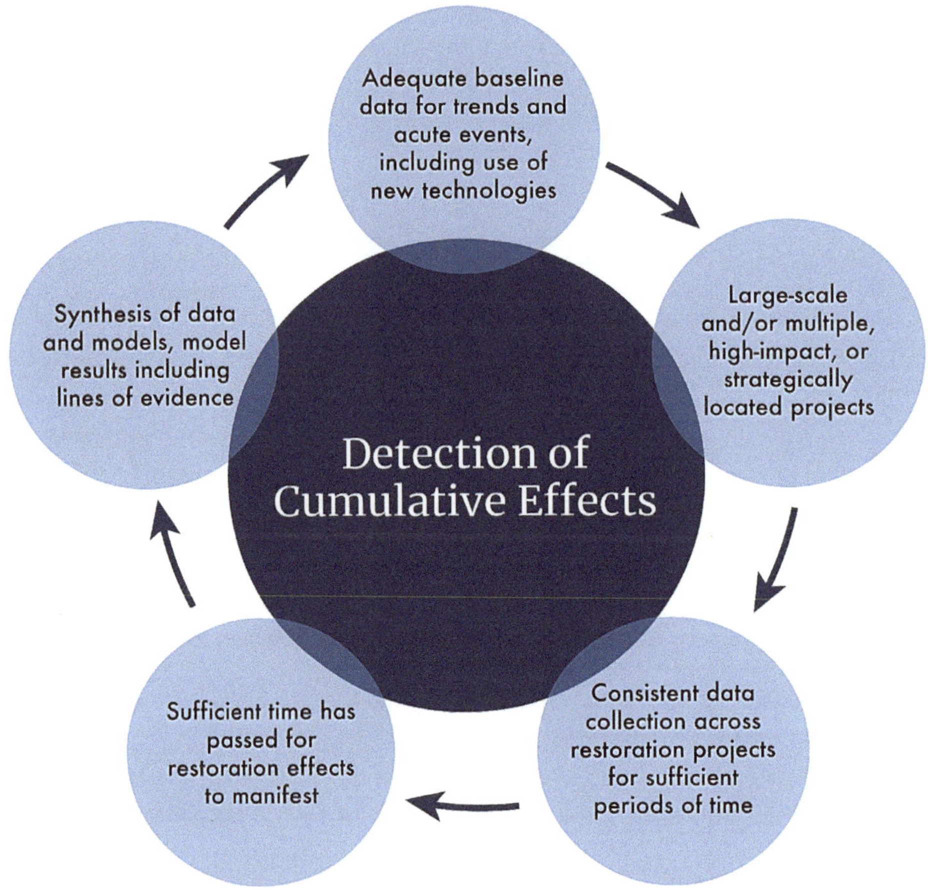

FIGURE 4.1. Key factors supporting cumulative effects assessments for large-scale and/or multiple restoration projects in the Gulf of Mexico. SOURCE: Committee.

to ecosystem restoration, generally focusing on four areas: habitat, water quality, living coastal and marine resources, and resilience (see Table 4.3). The DWH NRDA Trustees goals are to understand the GoM ecosystem and the injuries sustained from the DWH oil spill, developed through a scoping process that included public involvement (DWH NRDA Trustees, 2016). The RESTORE Council updated its overall goals in 2016 (RESTORE Council, 2016). The NFWF goals are consistent with the terms of the plea agreements and support projects that remedy harm to habitats and species where there has been injury resulting from DWH.[23] The U.S. Fish and Wildlife Service North American Wetlands Conservation Fund, with their focus on wetland conservation and bird habitat, also has similar aims. These four focus areas provide the foundation for a general hypothesis framework for the synthesis of programmatic outcomes (see Table 4.3).

Assessing Potential Nonlinear (Synergistic and Antagonistic) Cumulative Effects of DWH Program Approaches

Chapter 3 introduced the potential for nonlinear effects, either synergistic or antagonistic, to occur in multiple restoration activities being conducted to improve the condition of ecosystems, water bodies, and species. Table 3.1 in that chapter presents examples of practical methods for predicting, measuring, and evaluating certain nonlinear effects in several Gulf Coast subregions. This section further specifies the ecosystem pressures and stressors, potential synergies and antagonisms, and measurement and modeling methods associated with these restoration types.

[23] See https://www.nfwf.org/gulf-environmental-benefit-fund/priorities.

TABLE 4.3. Restoration Focus Areas and Hypotheses to Support Synthesis of Programmatic Outcomes and Corresponding Goals and Objectives of the DWH Settlement–Funded Programs

	Restoration Focus Areas			
	Habitat	**Water Quality**	**Living Coastal and Marine Resources**	**Resilience**
	Hypothesis: Through restoration, coastal habitats will maintain or increase areal coverage, function, and ecosystem diversity	Hypothesis: Through restoration, water quality will be restored or protected in coastal and marine areas	Hypothesis: Through restoration, species populations will be maintained or replenished	Hypothesis: Through restoration, coastal ecosystems and communities will be resilient to changing conditions
Entities Controlling DWH Ecosystem Restoration Funds	**Habitat**	**Water Quality**	**Living Coastal and Marine Resources**	**Resilience**
National Fish and Wildlife Foundation ($2.544B) Source: https://www.nfwf.org/gulf-environmental-benefit-fund/priorities	Restore and maintain the ecological functions of landscape-scale coastal habitats, including barrier islands, beaches, and coastal marshes, and ensure their viability and resilience against existing and future threats, such as sea level rise.		Replenish and protect living resources, including oysters, red snapper and other reef fish, Gulf Coast bird populations, sea turtles, and marine mammals.	
DWH Natural Resource Damage Assessment ($8.1B) Source: DWH NRDA Trustees 2016. pp. 5-15 to 5-17, Figure 5.4-1	Restore and conserve habitat (includes wetlands, coastal and nearshore habitats)	Restore water quality (includes nutrient reduction and water quality)	Replenish and protect living coastal and marine resources (includes fish and water column invertebrates, sturgeon, sea turtles, submerged aquatic vegetation, marine mammals, birds, mesophotic and deep benthic communities, oysters)	
RESTORE Council ($1.56B) Source: RESTORE Council (2016)	Restore and conserve the health, diversity, and resilience of key coastal, estuarine, and marine habitats	Restore and protect the water quality and quantity of the Gulf Coast region's fresh-, estuarine, and marine waters	Restore and protect healthy, diverse, and sustainable living coastal and marine resources	Build on and sustain communities with capacity to adapt to short- and long-term changes

Understanding the linkages between ecosystems is essential for predicting and designing appropriate measurement methods for nonlinear effects. The potential for synergistic effects on secondary production to be achieved through the restoration of primary production (as shown in Figure 3.2, in Chapter 3) has been understood for decades (Peterson and Lipcius, 2003), yet verification of this relationship still necessitates a time-consuming suite of field and laboratory procedures (Sather et al., 2016; Sobocinski and Latour, 2015). Understanding the drivers, pressures, and stressors on ecosystems (see definitions in Box 2.1, in Chapter 2) is a primary step that facilitates the ability to predict nonlinear effects and design methods to detect them.

Examples of pressures and stressors on six primary GoM restoration types are shown in Table 4.4. The interactions between pressures and stressors are evident throughout the table. Connectivity among aquatic habitats, for example, stands out as a primary mechanism for interactions with nonlinear effects having to do with water quantity and quality, plant detritus, and biota. For another example, oyster reefs help mitigate the negative effects of nitrogen (Arfken et al., 2017) that over-fertilizes coastal waters (see case study in Chapter 3; Boynton and Kemp, 2008), thus having the potential for synergistic effects with DWH-funded projects aimed at water quality (shown in Table 4.4). Furthermore, intensifying trends in both chronic and acute stressors affecting SAV, emergent marshes, and mangroves (see Chapter 2) enhance the urgency for determining which restoration actions—or suites of actions—will have positive or negative effects.

TABLE 4.4. Potential Synergistic and Antagonistic Effects of Restoration Approaches

Restoration Approaches and Anticipated Outcomes		Ecosystem		Synergistic and Antagonistic Considerations	
Restoration Type	**Example Restoration Approaches/Techniques**	**Example Restoration Outcomes/Goals**	**Example Pressures/Stressors**	**Potential Synergies with Other Restoration Types**	**Potential Antagonisms with Other Restoration Types**
Oysters	Restore or create oyster reefs through placement of cultch in nearshore and subtidal areas Construct living shorelines Enhance oyster reef productivity through spawning stock enhancement projects Develop a network of oyster reef spawning reserves	Enhanced ecosystem services Enhanced spat settlement and recruitment Enhanced broodstock development Positive rates of shell and reef accretion Self-sustaining populations	Physical disturbance or removal Sedimentation and burial Pollution, eutrophication Hypoxia Disease, predation, and competition Hydrologic alterations Climate change (air temperature, sea surface temperature changes in precipitation/extreme events, increased storm intensity, sea level rise)	Water quality (dissolved oxygen, turbidity, salinity, nutrients) Submerged aquatic vegetation (SAV) (seagrass) Wave attenuation/shoreline erosion (oyster reefs protecting marsh restoration) More species diversity of fish (better than either marshes or oysters separately) Improvements in water clarity Improvements in denitrification	Alteration of hydrology, including salinity/freshwater shocks generated from river diversions Bringing sediments to build wetlands Displacement of benthos and bottom-dwellers
Submerged Aquatic Vegetation (SAV)	Propagation and transplanting Enhancing beds through nutrient addition Protective measures to limit disruption of existing beds Protection of beds through wave attenuation structures Natural increase in extent in suitable areas as a result of water quality improvements (assuming viable seed bank or rhizome expansion from existing plants)	Improvement of water clarity/ reduced turbidity Increase in vegetated area Carbon sequestration Increase in nursery habitat for commercially important species Denitrification Increase in pH	Light attenuation/turbidity (and effect on epiphytes) Physical impacts (e.g., boat wakes) Salinity fluctuations Eutrophication Formation of dense monoculture (competitive exclusion)	Possible positive interactions with both adjacent marshes and oyster reefs Water quality/nutrient reduction projects	Processes such as marsh erosion due to relative sea level rise or hurricanes could have negative influence on SAV communities Construction of projects within footprint of known SAV habitats

TABLE 4.4. Potential Synergistic and Antagonistic Effects of Restoration Approaches Continued

Wetlands (Salt Marshes and Mangroves)	Create restore and enhance coastal wetlands Restore oyster reef habitat Restore and preserve Atchafalaya River (Mississippi) processes Create, restore, and enhance barrier and coastal islands and headlands Restore and enhance SAV Protect and conserve marine coastal estuarine and riparian habitats Provide detritus for food web enhancement	Additional acreage of functional habitat under conservation and management Reduced rate of erosion of shorelines protected by oyster reef living shorelines Increased bird nesting and resting areas. Improved water quality conditions through enhanced denitrification and reduced sediment loads	The trends in acute and chronic environmental inputs are key issues here and most would seem to have negative consequences for restoration projects Tropicalization Sediment deposition (may be beneficial or detrimental, depending on the quantity, type, location, and duration)	Oyster restoration Habitat enhancements for the benefit of birds Improvement of conditions for SAV restoration (e.g., reduced turbidity) Diversions could be considered synergistic by rebuilding land and enhancing fisheries productivity Oyster reef positioning and hydrology could also be synergistic if planned appropriately	
Water Quality and Nutrient Reduction	Reduce nutrient loads to coastal watersheds Reduce pollution and hydrologic degradation to coastal watersheds Create, restore, and enhance coastal wetlands Protect and conserve marine coastal estuarine and riparian habitat	Reduce pollutant loadings, including nutrients and pathogens, to priority watersheds Improved water clarity	Excess nutrient inputs from watershed sources Bacteriological sources Water temperature	Very possible positive interactions of oyster reef and wetlands with SAV Where appropriate, colocate pollutant reduction projects with other restoration projects to enhance ecological services provided by other restoration approaches	Careful attention needs to be paid to acute and chronic foundational inputs; many have trends associated with them at present (e.g., sea level rise)
Birds	Restore and conserve bird nesting and foraging habitat Establish or reestablish breeding colonies Prevent incidental bird mortality Create, restore, and enhance wetlands and SAV, dunes and beaches, barrier and coastal islands, and headlands Protect and conserve marine, coastal, estuarine, and riparian habitats	Increased extent and quality of foraging and nesting habitat Increased population of target species	Habitat loss Wetland fragmentation and conversion of open water as a result of sea level rise and subsidence Nesting and foraging area disturbances, including anthropogenic sources (e.g., light pollution, dogs on beaches) Depredation (e.g., foxes, coyotes) Food source availability	Wetlands SAV Water quality (pH, organic material, nutrients)	Creating bird habitat can minimize shallow water fish habitat Anything that increases anthropogenic disturbances or conversion of natural habitats

NOTE: Many of these pressures and stressors, their trends, and interacting effects are detailed in Chapter 2.
SOURCES: The example restoration approaches noted in the second column are based on information found in https://media.fisheries.noaa.gov/dam-migration/pdarp_2016.pdf and https://www.gulfspillrestoration.noaa.gov/sites/default/files/DWH-ARZ008721.pdf.

Scaling Data and Information to the Gulf Coast

The challenge of evaluating cumulative effects of restoration at the Gulf-wide scale necessitates scaling up data from both representative restoration projects and intermediate-scale hydrogeomorphic units like estuaries or watersheds. As discussed in Chapter 3, avoiding scale mismatch is necessary for defensible assessment of environmental change. When scaled-up information is evaluated against monitoring and analysis of long-term trends and acute events (see Chapter 2), restoration trajectories are likely to be better understood and future conditions predicted with a reasonable level of certainty.

The results of intermediate-scale assessments—whether qualitative, quantitative, and/or using lines of evidence—can be evaluated Gulf-wide through comparative cross-system analyses. This is one of five categories of methods used in coastal synthesis studies, as discussed in Kemp and Boynton (2012). The others are analysis of time-series data, balance of cross-boundary fluxes, system-specific simulation modeling, and general systems simulation modeling. In the past, multiple methods of analysis have been applied to strengthen the conclusions drawn about the factors controlling the behavior of coastal systems. These analytical and simulation methods are categorized under lines of evidence in the cumulative effects approach (see Table 3.3, in Chapter 3). Each of these methods could conceptually be applied in the GoM. The use of comparative cross-system analysis (e.g., Boynton and Kemp, 2008) among the 34 U.S. GoM estuaries could provide an initial step to bridge the spatial gap between existing estuary-scale assessments of cumulative effects and Gulf-wide-scale assessment.

As Kemp and Boynton (2012) noted, many factors influence selection of types of analysis and simulation to effectively address hypotheses about controls of ecological processes. Different forms of synthesis need different types of data and some types (routine monitoring variables, such as temperature, salinity, nutrient concentrations) are more available than others (process and rate measurements, such as primary production, respiration, and various nutrient fluxes). Because every synthesis effort will be constrained by the data available, ensuring that appropriate monitoring efforts are in place will determine the basis for synthesis ultimately carried out. It is for this reason that this report emphasizes the importance of monitoring efforts. The unavailability of adequate monitoring data prevented the committee from carrying out a rigorous evaluation of the effectiveness of GoM restoration efforts (see Chapter 1).

More specifically, cross-system comparison analysis involves the development of a quantitative statistical model (often structured as a linear regression) using similar data from different systems to examine how key properties or processes vary with inputs from external drivers or other internal properties of an ecosystem (Kemp and Boynton, 2012). Within the GoM, limited examples of cross-system comparison analysis include seagrass areal extent response to hurricane disturbance across three estuaries in southeast Florida (Tomasko et al., 2020), the influence of tidal and nontidal wetlands on nitrogen export from estuaries to coastal waters (White et al., 2019a), decapod crustacean assemblages associated with northern GoM oyster reefs (La Peyre et al., 2019), ecosystem response to climate variability (Trifonova et al., 2019), and DWH impacts on nearshore and coastal living marine resources (Murawski et al., 2021). McKinney et al. (2019) used cross-systems analysis and analysis of time-series data and general systems modeling to compare living resources, water quality, and habitat across regional estuaries.

A number of estuaries within the GoM have established programs that maintain extensive water quality, freshwater inflow and nutrient loading, high-frequency water quality monitoring, and coastal habitat extent databases. Primary among these are the five NOAA-supported National Estuarine Research Reserves, the seven EPA-supported National Estuary Programs, and state-supported monitoring programs. These databases could be used to develop and test hypotheses using cross-system comparison analyses among estuaries across the GoM. Because each of these long-term programs and their partners have implemented a variety of restoration projects and efforts within their respective study areas, an opportunity exists to use cross-system analysis to ask questions addressing cumulative effects of implemented projects. For example, these questions could include:

- What lessons have been learned from previous projects?
- When beneficial environmental effects been observed following restoration efforts, what are the characteristics of those estuaries?
- What types, extent, and/or frequency of restoration efforts are associated with various observed beneficial effects?
- Has any restoration project adversely affected another project?
- What is the lag time between restoration projects' implementation and observable, detectable effects, especially due to effects of background environmental trends?

USING AN ADAPTIVE MANAGEMENT APPROACH TO ASSESS CUMULATIVE EFFECTS

Adaptive management is a structured process that considers uncertainties and allows for flexibility in management activities to address these uncertainties (Holling, 1978; Pastorok et al., 1997; Williams, 2011). When environmental restoration efforts, monitoring, and evaluation throughout the planning, implementation, and tracking stages are integrated, knowledge gained at each step can be considered to improve the chance of successful restoration in the following stages. Adaptive management has been implemented in various ways, but generally includes assessment and setting restoration goals, planning restoration efforts (including identifying critical uncertainties that may affect restoration success), implementation, monitoring, evaluation of progress toward restoration goals, and adjustments throughout the process (summarized in NRC, 2004).

The DWH NRDA Trustees recognized that, because of the large temporal, spatial, and funding scales associated with the DWH oil spill, the restoration effort warranted a robust framework to support restoration decisions. In 2016, an adaptive management approach (MAM Plan) to guide DWH restoration efforts was identified, as an element of the *Final Programmatic Damage Assessment and Restoration Plan* and *Final Programmatic Environmental Impact Statement* (DWH NRDA Trustees, 2016) (see Figure 4.2). In 2018, the trustees developed a MAM manual, which included a commitment to monitor and evaluate restoration outcomes and provide feedback to inform decision making for current projects and refine the selection, design, and implementation of future restoration actions (DWH NRDA Trustees, 2017a, 2019). The trustees note that

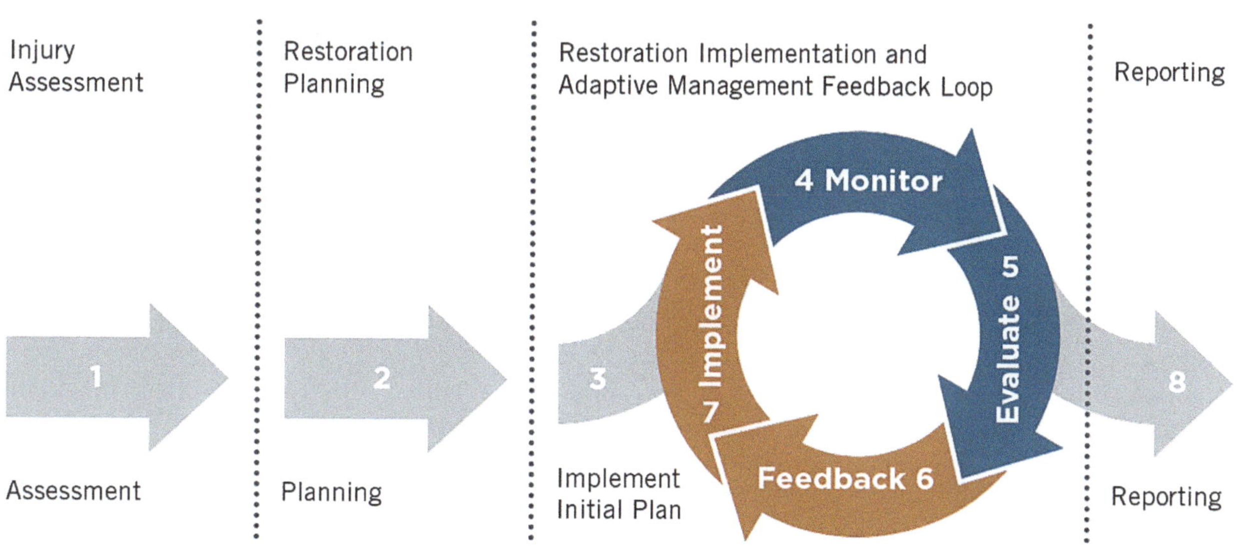

FIGURE 4.2. Monitoring and adaptive management framework guiding Deepwater Horizon restoration. SOURCE: DWH NRDA Trustees, 2016.

the MAM framework "may be more robust for some elements of the restoration effort with higher degrees of uncertainty, or where large amounts of restoration are planned within a given geographic area and/or for the benefit of a particular resource" (DWH Open Ocean Trustee Implementation Group, 2020, p. 4; see also DWH NRDA Trustees, 2016). Importantly, the adaptive management feedback loop provides the trustees with the opportunity to adjust restoration actions as needed, based on monitoring and evaluation of restoration outcomes (DWH NRDA Trustees 2016, 2017a, 2019, 2021). Data from completed projects can inform future restoration efforts. The adaptive management approach developed by the trustees provides the ability to make refinements to wetland restoration projects and programs as more information becomes available over time. The approach supports using the latest scientific data to inform how to improve future wetland restoration decision-making processes, including predicting and/or measuring the influence of external factors (e.g., sea level rise, tropical storms or hurricanes) on restoration outcomes, characterizing interactions among restoration actions, and collecting additional data needed to support regional-scale restoration (Hijuelos and Hemmerling, 2015; Steyer et al., 2003).

Key considerations in an effective adaptive management approach when assessing the cumulative effects of multiple restoration efforts include monitoring that is sufficient to detect environmental change associated with the restoration effort (Westgate et al., 2013) and to assess the potential effects of long-term environmental trends; evaluation and synthesis of monitoring data from multiple projects and background environmental trends to assess impacts on cumulative effects; feedback and learning by which the information obtained through the synthesis of the monitoring data is used to adapt future project plans; and implementation to determine and administer an appropriate remedial action or course correction. The cycle is then repeated as needed.

When applied to individual restoration projects, such as those associated with DWH, use of an adaptive management monitoring framework allows restoration to proceed in the face of uncertainty. Earlier sections of this report highlight the many uncertainties restoration efforts face within the GoM, such as effects from climate change, relative sea level rise, long-term environmental trends, and variability of freshwater discharges. Adaptive management provides a critical framework in which these uncertainties are integrated into assessment of restoration success within the Gulf and provides for adjustment in restoration approaches if the initial plan does not make progress in attaining restoration objectives. NASEM (2017) found that adaptive management is most suited for situations where environmental conditions may respond to management actions (restoration projects meet this qualification), where there is considerable uncertainty regarding the response, where reducing that uncertainty could improve decisions regarding the project, or where stakeholders are committed and have the capacity to sustain an adaptive management approach.

To date, guidance for individual restoration projects supported by DWH recovery funds include monitoring plans that incorporate adaptive management as part of their approved project plans and monitoring to be conducted in accordance with those plans. However, as noted above, data and results from the projects completed to date are sparse or not currently accessible, and the evaluation step of the cycle does not appear to have been implemented or disseminated beyond the individual-project scale.

At the project scale, the use of adaptive management for DWH-funded projects has been infrequently documented. In one example, when evaluating the progress of a seagrass planting restoration project in Santa Rosa Sound, Florida, project scientists determined that it was necessary to erect temporary fencing to exclude rays and blue crabs that were uprooting newly planted seagrass as a result of their feeding on benthic infauna. From studies in other regions (Thom et al., 2005), crabs are known to disrupt seagrass plantings through foraging and burying. The fencing allowed the plants to become established, although disturbance by tropical storm activity also hindered the overall success of the project (Heck and Byron, unpublished report). The potential for applying an adaptive management approach to larger spatial scales (including Gulf-wide) is discussed in Chapter 5.

GULF-WIDE SYNTHESIS: AN EXAMPLE

During information-gathering sessions for this report, several GoM stakeholders and project managers noted that while there is a need for Gulf-wide cumulative assessments of restoration efforts, incentives and resources are not currently in place to achieve this goal.[24] Gulf-wide synthesis activities have been recommended since at least 2017 (NASEM, 2017).

A recent assessment of meta-analyses of lag times between restoration implementation and the development of associated benefits (summarized in Carle et al., 2020) provides an example of how synthesis can inform future restoration and monitoring efforts. The duration of the monitoring period needed to evaluate the effectiveness of a restoration project varies depending on the habitat type(s) involved, the restoration outcomes that are being evaluated, and factors specific to each location. Table 4.5 presents representative time periods that have been reported by literature reviews and meta-analyses of restoration projects in the GoM (e.g., Armitage, 2021; Baumann et al., 2020; Carle et al., 2020; Ebbets et al., 2020; Hollweg et al., 2020; Rezek et al., 2019).

In restored coastal marshes, attributes such as plant cover and use by fauna as spawning and nursery habitat can become well established within 1–5 years following restoration, while other features, such as below-ground biomass, soil organic content, and trophic functions (such as benthic infauna densities), may need a decade or more to approach levels typically observed in natural marshes (Baumann et al., 2020; Carle et al., 2020; Ebbets et al., 2020). In contrast, restored oyster reefs "tend to demonstrate rapid development of both ecosystem structure and function, as long as they are located in environmental settings with suitable salinity, water quality, and oyster larval transport" (Carle et al., 2020, p. 1682).

In the case of SAV restoration, time lags and outcomes can vary depending on the scale of the restoration effort. At the project scale, where the goal is to establish individual SAV beds using transplanting or seeding techniques, outcomes tend to be somewhat binary, either failing rapidly at sites where water clarity or other environmental factors do not support SAV survival or showing long-term persistence and growth at sites where SAV needs are adequately met (e.g., Rezek et al., 2019). In successful project sites, nekton and epifauna can colonize quickly and reach abundances and species composition that resemble natural SAV beds in 1–2 years (Carle et al., 2020). At larger sub-estuary or estuary scales, SAV restoration has typically been achieved by improving water quality and water clarity to levels that meet SAV needs, resulting in natural SAV recruitment or expansion from existing beds (see Table 4.4 for examples). In such cases there can be multiyear time lags between the initial implementation of water quality improvement projects, typically focused on nutrient load reductions, and the establishment of water quality conditions adequate for SAV restoration. Once the necessary water quality conditions are met, there can be additional multiyear lags before seagrass cover and habitat function increase to target levels (e.g., Greening et al., 2014; Tomasko et al., 2018).

Following project implementation, long-term monitoring and assessment will be needed to track and (if possible) correct for impacts due to climate change; relative sea level rise (RSLR); and changes in hydrologic, sediment and nutrient loads. The information and references summarized in Table 4.5 can be used to inform long-term monitoring protocols needed to track restoration progress, cumulative effects of multiple projects, and assess impacts of background environmental trends.

Living shorelines are designed to provide shoreline stabilization and typically incorporate native materials, such as marsh vegetation, oyster, or mussel beds, or SAV, alone or in combination with some type of harder shoreline structure for added stability (NOAA, 2015a; Smith et al., 2020). After construction, living shoreline projects can reach ecological equivalency with reference marshes for habitat use by snails and herons within 2 years (Chambers, 2021; Guthrie et al., 2022) and can establish nekton within 3 years (Gittman et al., 2016). However, soil composition equivalency with natural marsh organic matter lag several to many years behind those associated with habitat use (Chambers, 2021; Isdell et al., 2021).

[24] G. Grandy, J. Henkel, A. Hunter, G. Leonard, P. Mickle, J. Porthouse, and B. Sutter, presentations to committee, August 12, 2020.

TABLE 4.5. Observed Lag Times for Various GoM Habitat Restoration Approaches

Habitat Type	Restoration Approach	Typical Spatial Scale	Outcomes Monitored/ Assessed	Months	1 Year	2 Years	3 Years	5 Years	10 Years	20 Years	30 Years	References
COASTAL MARSH	Bathymetric modifications +/- planting	Project	Plant cover; above-ground biomass and faunal spawning; nursery habitat									Carle et al., 2020, and references therein
		Project	Below-ground biomass; soil organic matter, benthic infaunal and epifaunal density, production, and diversity; trophic support for nekton growth and development									Armitage, 2021, and references therein; Carle et al., 2020; Baumann et al., 2020, and references therein; Hollweg et al., 2020, and references therein; Ebbets et al., 2020, and references therein
OYSTER REEF	Placement of cultch/ substrate	Project	Oyster density and reef structure; macrofaunal production and diversity; food web complexity									Carle et al., 2020, and references therein
SAV	Planting (transplants/seeds)	Project	Plant cover and condition; above-ground biomass, nekton production, and diversity; benthos production/diversity									Carle et al., 2020, and references therein; Rezek et al., 2019
	Water quality restoration and management	Sub-estuary/ Estuary	Plant cover, condition, and biomass; nekton production and diversity; benthos production and diversity									Greening et al., 2014; Sherwood et al., 2016; Beck et al., 2020; HARC, 2020; GBEP, 2018; Tomasko et al., 2018

NOTE: Bathymetric modifications include dredged sediment placement, thin-layer sediment enhancement, and excavation of upland areas. Lag times between restoration implementation and functional equivalency of 3 years or less are shown in yellow shading; green shading indicates lag times of more than 3 years. GoM, Gulf of Mexico; SAV, submerged aquatic vegetation.

INTEGRATING CUMULATIVE EFFECTS AND ADAPTIVE MANAGEMENT INTO RESTORATION: NEXT STEPS

As noted above, cumulative effects at the estuary and sub-estuary scales have been observed in the GoM, but extensive and prolonged monitoring efforts were necessary for detection. Additionally, the strength of large external environmental drivers, pressures, or stressors (such as seal level rise or hurricanes) and the relatively small amount of area restored may have prevented the detection of beneficial effects of cumulative restoration at the Gulf-wide, regional, or state levels.

To better ensure detection of cumulative effects in restoration, several steps could be taken. At the project level, they could include identifying project-level metrics to support assessment of effects on a broader scale, utilizing consistent methodologies for collection and reporting of project-level metrics and ensuring that data are freely accessible and that metrics are available to facilitate assessment. At other levels, steps could include identifying other metrics or datasets that could support cumulative effects assessments, but may not be readily collected at the project scale. These include non-DWH restoration programs that are collecting data across large spatial and temporal scales.

Figure 4.3 outlines an approach for planning, implementing, assessing, and synthesizing cumulative effects of multiple restoration efforts in a defined geographic area. The use of an adaptive management strategy allows progress on implementation of restoration efforts to continue, even if all necessary information is not immediately known. For example, Step 2 asks whether adequate baseline environmental conditions are sufficiently known to assess change; if not, this step provides for the development of preliminary baseline estimates using existing data sources (Step 2a) that can be used to continue toward implementation. The adaptive management feedback loop allows reevaluation and adjustment of the restoration strategy as additional data and information are collected on environmental conditions, performance metrics, and comparison with baseline conditions.

Key elements of the multiple lines of evidence discussed in Chapter 3 are used throughout this approach. In Step 1, the conceptual model incorporates potential synergistic and antagonistic effects, hot spots and hot

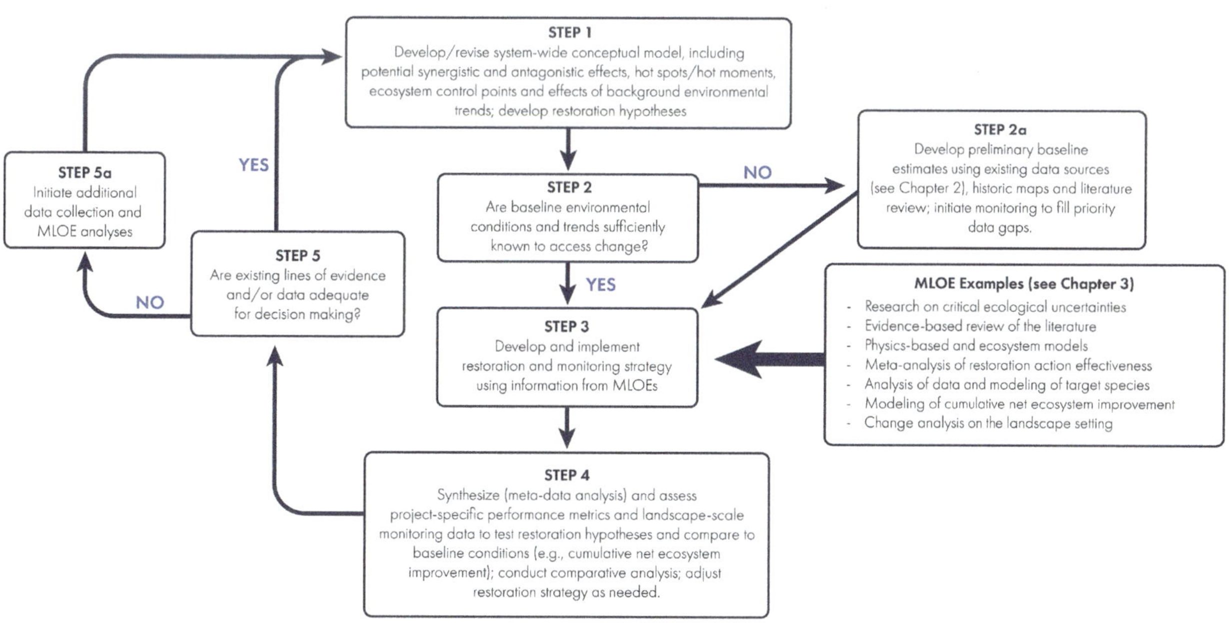

FIGURE 4.3. A flow chart approach for environmental restoration to assess cumulative effects of multiple restoration projects, using multiple lines of evidence (MLOEs) (see Table 3.3) and an iterative adaptive management approach. The large arrow between MLOE examples and Step 3 indicates that one or more lines of evidence can be used to develop the restoration and monitoring strategy. The steps in this approach may not be applicable for all existing and planned large-scale or multiproject restoration in the GoM, and other information and lines of evidence could be included at each step.

moments, and background trends, and develops restoration hypotheses. Step 2a uses historic data sources and literature review to help fill information gaps and allows progress to continue. Development of the restoration strategy in Step 3 can take into account one or more lines of evidence. Synthesis of changes in landscape-scale monitoring data to test restoration hypotheses and the use of data-driven models (see Chapter 3) are key elements in Step 4. Step 5 assesses the adequacy of existing lines of evidence and data to make restoration strategy decisions. The cycle is then repeated if necessary, starting with revising the original conceptual model with new information.

Chapter 5
Moving Forward

INTRODUCTION

Calls for providing the scientific efforts (monitoring, modeling, and research) needed to support an adaptive management process for *Deepwater Horizon*–funded restoration efforts are not new. In 2011, the Gulf Coast Region Ecosystem Task Force developed a shared vision for the Gulf Coast, to "achieve a resilient, healthy Gulf of Mexico ecosystem that supports the diverse economies, communities and cultures of the region" (Gulf Coast Ecosystem Restoration Task Force, 2011, p. 10). The authors note that achieving this vision has to include alignment of activities among the federal government and states, as well as collaboration in order to create realistic goals for restoration. In 2012, the Gulf Coast Ecosystem Restoration Task Force (GCERTF) Science Coordination Team developed a science plan (Walker et al., 2012) for cross-cutting monitoring, modeling, and research efforts needed to provide the basic science infrastructure to support the overall vision for the Gulf of Mexico (GoM) restoration program. A specific objective of the plan called for the development of an iterative and flexible approach to allow adaptive management and decision-making. In addition, Brown et al. (2011) recognized that, because the GoM ecosystem contains many political boundaries, coordination across geographic and organizational lines would be needed, and recommended that strategies be implemented at the largest scale possible.

As discussed in Chapters 2–4, significant progress has been made since 2012 in advancing monitoring and modeling capabilities[1] (*Deepwater Horizon* NRDA Trustees, 2017a; NOAA and USGS, 2019). Still, new technologies (such as sensor networks) and involvement of nontraditional partners (such as citizen/community scientists) offer opportunities for further advancement. Long-term environmental trends for some environmental parameters and species are now available at local and regional scales, though data collection, analysis, and reporting are often inconsistent, and existing efforts are not adequate to detect all important Gulf-wide trends.

Significant challenges remain in the development of the critical analysis and synthesis of the cumulative effects of *Deepwater Horizon* projects. Assessment of cumulative impacts—additive, synergistic, and possibly antagonistic effects—of multiple restoration projects of similar or diverse nature over spatial and temporal scales beyond that of an individual project are uncommon in the GoM. Development of synthesis capacity that can support an adaptive management process to integrate diverse restoration projects over significant spatial scales is recognized as a need by the *Deepwater Horizon* funding entities,[2] but has not yet been widely initiated.

[1] See https://www.gulfspillrestoration.noaa.gov/sites/default/files/2019-08%20MAM_Manual_Attachment_ E_07_Marine%20 Coastal%20Estuarine%20Riparian_Habitats_%20Monitoring_Guidance%202019.pdf.

[2] As described in Chapter 1, when used in this report, the "*Deepwater Horizon* funding entities" are the RESTORE Council state and federal members (States of Alabama, Florida, Louisiana, Mississippi, and Texas; U.S. Department of Commerce; U.S. Department

This chapter discusses critical aspects that underpin progress in moving forward: (1) data needed to assess long-term environmental trends and their impact on restoration activities; (2) emerging monitoring strategies for large-scale multiproject restoration efforts; (3) how cumulative effects analyses and adaptive management approaches can be utilized in planning, implementing, and assessing future restoration efforts in the GoM; and (4) barriers and opportunities for synthesizing the large amount of data and information already collected from *Deepwater Horizon* projects to maximize the probability of successful ecosystem restoration in the GoM.

DATA RESOURCES FOR ASSESSING LONG-TERM ENVIRONMENTAL TRENDS

Possessing baseline and trend data for important environmental variables when evaluating restoration efforts provides fundamental support for the synthesis activities needed to inform adaptive management actions (Ellison et al., 2020; LoSchiavo et al., 2013; Thom, 1997). The committee's assessment of the lack of data availability for monitoring background environmental indicators and trends is similar to that summarized previously by Love et al. (2015) and more recently by Carle et al. (2020). For example, Love et al. (2015) identified gaps in long-term monitoring that can be summarized in three overarching findings:

- There are many existing monitoring efforts that restoration decision-makers can use to track the recovery of injured natural resources. Building on these existing efforts will improve consistency, efficiency, and coordination.
- There are gaps in monitoring and in the understanding of natural resources in the Gulf that need to be filled to effectively evaluate recovery and thus the success of restoration programs in the Gulf ecosystem.
- Addressing the currently disjointed monitoring system and moving toward a Gulf-wide ecosystem monitoring network will provide a more efficient, integrated, and accessible tool for ecosystem information.

A collection of papers focused on the GoM, introduced by Carle et al. (2020), found that monitoring of restoration projects was often limited to a few years and focused on meeting specific construction targets (e.g., acres of marsh) or features (e.g., percentage of vegetation cover). In a specific example, they stated that "without more long-term monitoring data, natural variability makes it difficult to accurately predict restoration trajectories for some characteristics of restored marshes, particularly the extent to which slower developing functions might emerge over time" (Carle et al., 2020, p. 1685; see Table 4.5, in Chapter 4).

Although several national programs do provide consistent and accessible data, most environmental indicators are not monitored consistently, and datasets are fragmented spatially and/or temporally (GOMA Water Quality Team, 2013; Love et al., 2015). State and federal resource agencies collect long-term environmental monitoring data specific to their agency and state missions, and this information could be useful for assessing cumulative impact of multiple restoration projects at estuarine or larger scales. The RESTORE Council Monitoring and Assessment Program (CMAP) developed an inventory of existing habitat and water quality monitoring, and mapping metadata for the GoM programs, which includes links to electronically available datasets (NOAA and USGS, 2019). However, efforts to access and synthesize data are hampered by the lack of unifying GoM analysis and synthesis activities for many key stressors (e.g., ocean and coastal acidification, hypoxia, tropicalization). No entity that the committee could identify has the resources and the explicit responsibility to synthesize data from multiple sources. This type of analysis and synthesis activ-

---

of the Interior; U.S. Environmental Protection Agency; U.S. Department of Agriculture; U.S. Coast Guard; and U.S. Department of the Army); the *Deepwater Horizon* NRDA Trustees (National Oceanic and Atmospheric Administration, U.S. Environmental Protection Agency, U.S. Department of the Interior, U.S. Department of Agriculture, and the five Gulf states); the National Fish and Wildlife Foundation Gulf Environmental Benefit Fund; the U.S. Fish and Wildlife Service North American Wetlands Conservation Fund; the Centers of Excellence; the NOAA RESTORE Science Program; and the Gulf Research Program.

ity, so essential for accurate assessments of cumulative effects of large-scale restoration activities, remains to be undertaken.

Monitoring Challenges Faced by Program Managers

Chapters 3 and 4 discuss a detailed framework with case studies on cumulative effect assessment of multiple GoM restoration projects. Such a framework, a foundation for Gulf-wide synthesis studies, is built on the assumption that comprehensive monitoring datasets are acquired regularly during and after restoration, and data are freely available and accessible. However, establishing a comprehensive monitoring program is not easy and can face logistical, technical, and economic challenges (e.g., Biber, 2013). Increasingly intense hurricane seasons (e.g., Holland and Bruyère, 2014) can disrupt established monitoring programs, destroy expensive field equipment (as shown in the example in Figure 5.1), deny access to field sites, and distract program managers with other urgent priorities (Kozlov, 2021). After a hurricane, program managers have to partially or entirely resurrect the monitoring effort with new locations, equipment, and sampling protocols (e.g., Burke, 2018).[3] For example, in 2018, the Gulf Research Program of the National Academies of Sciences, Engineering, and Medicine funded 22 small grants to help with repair, replacement, or recovery of equipment, data, and research material incurred from Hurricanes Harvey and Irma (NASEM, 2019).

As documented in Chapter 2, access to long-term monitoring data is limited, and the data are often dispersed among many agencies, states, and local entities. CMAP compiled an inventory of 544 water quality monitoring, habitat monitoring, and mapping programs operating in the GoM (NOAA and USGS, 2019). The authors found that accessible metadata was commonly missing from monitoring programs and recommended monitoring programs invest in making metadata web accessible (NOAA and USGS, 2019).

The following sections examine emerging technologies as potential solutions to avoid data discontinuities. These technological solutions are in addition to traditional field-based data collection discussed in Chapter 2. These technologies are not meant to replace well-established field-based monitoring programs developed through decades of research, but to complement them. Some of these technologies are already being used in the GoM. Many of the proposed sensing frameworks and sensor devices, towers, and networks

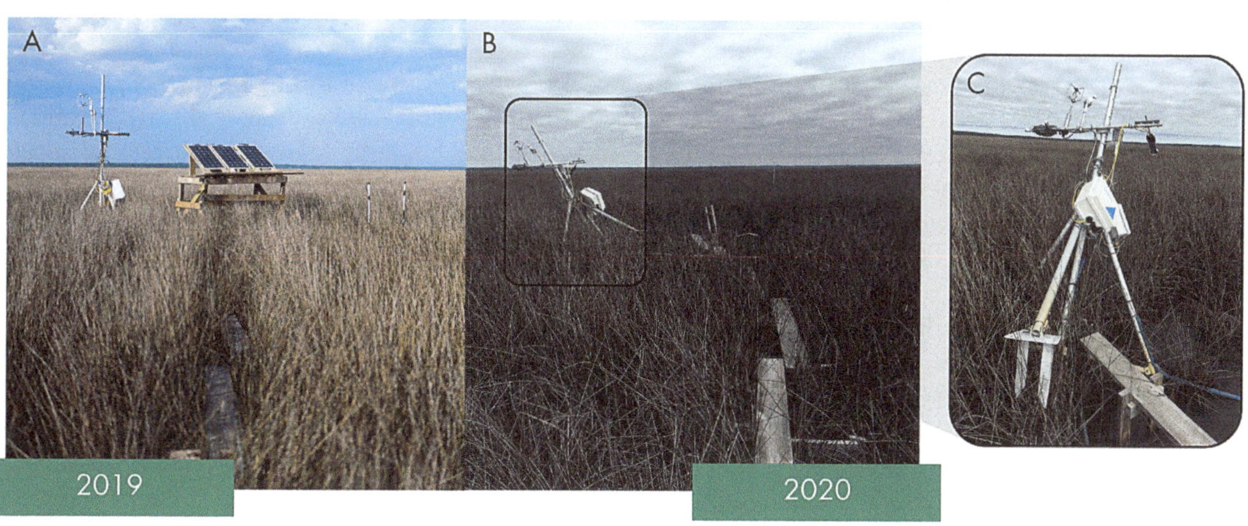

FIGURE 5.1. The Eddy Flux Tower at the Grand Bay, Mississippi, National Estuarine Research Reserve. (A) shows the tower in 2019. (B) and (C) show the tower after the 2020 GoM hurricane season. The solar panels and the underlying structures were not found. The tower itself was broken off from the base. SOURCES: (A) Peter Hawman, University of Georgia; (B) and (C) Hailong Huang, Gulf Coast Research Laboratory, University of Southern Mississippi.

[3] This phenomenon was also described by Gregory Steyer of the U.S. Geological Survey (personal communication, October, 11, 2021).

are also prone to technical and communication failure, miscalibration issues, and damages by natural disasters, but repairing, replacing, and reestablishing these monitoring devices may be easier, faster, and more economically viable than the replacements needed in field-based sampling programs (Biber, 2013).

Emerging and Novel Monitoring Strategies and Techniques

Recent advancements in data-driven techniques—such as artificial intelligence (AI), machine learning (ML), and deep learning (DL)—as well as cloud and edge computing are expected to fundamentally transform many domains of human endeavor, including post-restoration monitoring.[4] Traditional remote sensing combined with new sensing technologies and AI-driven techniques can generate high-quality long-term monitoring data across terrestrial and coastal ecosystems[5] (Corbane et al., 2015; Ridge et al., 2020). Although some large-scale remote sensing studies to monitor GoM-wide water quality and wetland habitats have been conducted in the past (Chen and Hu, 2017; Ghosh et al., 2016; Hu et al., 2011; O'Connor et al., 2016), few examples of integrated remote sensing and emerging technology-driven monitoring studies (e.g., using AI, ML, or DL) are being adopted for Gulf restoration efforts: for example, the CMAP inventory assessment found that only 7 percent of oyster restoration efforts used AI or ML (NOAA and USGS, 2019).

It is clear that transformational changes in capturing, recording, transmitting, synthesizing, archiving, and extracting meaningful physical, chemical, biological, and ecological data and information about restoration projects across GoM are needed from multiple sources, users, and scales. As environmental monitoring data generated using these techniques will most likely be heterogeneous—textual, numerical, or spatial—models or synthesis tools will need to be capable of handling a variety of data for predicting ecosystem responses to restoration activities (Peters et al., 2014; Peters-Lidard et al., 2017; Zhong et al., 2021). These newer data-driven frameworks are not expected to replace traditional restoration science–driven ecosystem monitoring, but complement and strengthen them. The ecosystem restoration and monitoring community could benefit from using a hybrid method, combining traditional approaches with new advanced tools, techniques, and partners. Such changes will likely apply to all aspects of the long-term trend monitoring framework, including data collection, modeling (making predictions, analyzing trends, and detecting anomalies), and synthesizing and disseminating information.

Field-based monitoring and sensing technologies have grown substantially in the past decade, driven by advancements in cloud computing, crowdsourcing or crowdsensing, AI, plug-and-play sensors, low-cost computer boards, drones, and small satellites.[6] Use of these advanced multiplatform sensing technologies and data analytics and visualization methods can make post-restoration monitoring targeted, effective, inexpensive, and sustainable over long time frames (Corbellini et al., 2018; Mayton et al., 2017; Rundel et al., 2009). To employ these methods of data acquisition effectively, a comprehensive, multifaceted approach, such as those that already exist and are being increasingly implemented in ecosystem monitoring studies, will need to be adopted. For example, the GoM restoration science and practitioner community could adopt the AI-integrated Cyber-Physical Systems (CPS) framework of the National Science Foundation (NSF) to develop innovative ways to acquire, process, visualize, and share data for long-term monitoring of restoration sites and to assess the cumulative impact of projects. CPS is defined as "engineered systems that are built from, and depend upon, the seamless integration of computation and physical components."[7] The physical component in CPS could be a restored site, such as a salt marsh, bay, seagrass meadow, or oyster reef, or include a larger area, such as multiple restored sites across a landscape. The three components of long-term monitoring and assessment would involve acquiring multiple heterogeneous environmental data from the site; transmitting and processing those data; and modeling, visualizing, and disseminating map

[4] See https://www.microsoft.com/en-us/ai/ai-for-earth.

[5] See https://coastalresilience.org/project/ai-conservation/; https://salo.ai.

[6] See https://www.nsf.gov/cise/harnessingdata.

[7] See https://www.nsf.gov/pubs/2017/nsf17529/nsf17529.htm.

products to end-users or data contributors (in the case of crowdsensing) (e.g., Mishra et al., 2020; Nundloll et al., 2019). Once set up, a CPS system can be automated to save time and reduce costs.

Implementing such a framework across restoration sites and states is predicated on a willingness to adhere to the open data concept, such as the White House's Climate Data Initiative[8] and the Open Data Portal of the National Oceanic and Atmospheric Administration (NOAA)[9] (Peters, 2010; Schmidt et al., 2016), which means data are accessible and available to be repeatedly used in modeling and analysis. Another integral part of such a framework is the use of AI-based data-centric environmental models based on ML and DL (LeCun et al., 2015; Zhong et al., 2021). As noted in the Executive Order on maintaining American leadership in AI,[10] ML and DL have the potential to drive growth and innovation in many fields, including coastal ecosystem science. However, to date, the coastal environmental restoration field has only minimally integrated or adopted ML and DL in modeling and data synthesis activities because of such challenges as large training datasets, data labeling, and computation needs (Lamba et al., 2019).

AI-based models capable of highlighting the complex nonlinear relationships between restoration activities and ecosystem response (structure and function) need to be at the forefront of tool creation because they can isolate the cumulative impacts of long-term restoration projects and can be used for large-scale synthesis across sites in the GoM that are under the constant influence of broad-scale natural and anthropogenic drivers, pressures, and stressors. AI-based modeling with existing monitoring data has been increasingly used in the past few years but primarily for smaller-scale projects (Beijbom et al., 2015; Norouzzadeh et al., 2018; Parashar et al., 2021; Weinstein, 2018; Williams et al., 2019). Its use in large-scale, long-term monitoring projects has been rare.

The sections below describe the types of data, ecosystem variables, models, and visualization tools that could be helpful in advancing a cost-effective, easily implementable, easily adoptable monitoring framework in the GoM.

Multisource Monitoring Data

Remote Sensing Satellites

Monitoring coastal environments using point-based field data can be time-consuming, expensive, and spatially unrepresentative or inadequate to characterize long-term trends. To overcome these difficulties and conduct site- and regional-scale monitoring and mapping, remote sensing satellite data are often used (NASEM, 2015, 2018c). Depending on the environmental variables being monitored, restoration practitioners have access to open-source satellite data ranging from 10 m to tens of km in resolution. Very high-resolution[11] multispectral image data (1–3 m) from commercial satellites—such as PlanetScope (3–5 m), IKONOS (4 m), QuickBird (2.4 m), and WorldView-2 and -4 (1.24–4 m)—have been available since the 2000s. However, they are expensive and often cost-prohibitive for long-term time-series analysis, and the highest-resolution data is not always best suited to the research question (Ghosh et al., 2016). National Aeronautics and Space Administration (NASA)[12] and European Space Agency (ESA)[13] satellite imageries are available in an open-source form and represent a wide range of data from passive and active sensors, including multispectral sensors (e.g., Landsat, MODIS, Sentinel-2 and -3), Light Detection and Ranging (LiDAR) sensors (e.g., ICESat-2), and Interferometric Synthetic Aperture Radar (InSAR) sensors (e.g., Envisat). These are typically considered moderate-resolution (10 m to 1 km) satellites useful for various local and regional analyses.

[8] See https://www.data.gov/climate.
[9] See https://odimpact.org/files/case-studies-noaa.pdf.
[10] Executive Order 13859 (February 11, 2019): https://www.federalregister.gov/documents/2019/02/14/2019-02544/maintaining-american-leadership-in-artificial-intelligence.
[11] See https://www.satimagingcorp.com/satellite-sensors.
[12] See https://eospso.nasa.gov.
[13] See https://earth.esa.int/eogateway.

In the past few years, there has been substantial growth in the number of small satellites, or CubeSats, launched to the lower earth orbits that provide remote sensing images at various resolutions[14, 15] (NASEM, 2016; Stephens et al., 2020). One of the prominent examples is the Planet Lab's CubeSat constellations consisting of roughly 150 CubeSats with the goal of imaging the whole Earth daily at a spatial resolution of 3 m.[16] These kinds of high-temporal-frequency data have the potential to be valuable in monitoring the impact of acute disturbance events, such as hot spots and hot moments (see Chapter 3), on restoration sites in the GoM. There are now numerous satellite options for acquiring imaging data from a specific restoration site or for multiple restoration projects, and the choice of satellite sensor and data depends on the nature of the investigation, type of data needs (e.g., spatial resolution or pixel size; optical imagery, thermal imagery, or elevation or height data), frequency of acquisition (daily, weekly, monthly), and environmental constraints (e.g., cloud cover, tide height). When multiple datasets from different satellite sensors are available, cross-calibrated models can be developed (Liu et al., 2021b; Page et al., 2018), which means the general model will work for multiple satellite datasets without requiring satellite-specific calibration. These cross-calibration techniques can help maintain long-term trend monitoring, even if a specific satellite used in the analysis gets decommissioned.

Uncrewed Aerial Systems

In the past few years, uncrewed aerial systems (UASs) have become ubiquitous in environmental data collection (NASEM, 2020). The most common types of UASs for remote sensing observations offer multispectral (tens of bands) imaging and thermal imaging.[17] Although hyperspectral (hundreds of bands) UASs exist, they tend to be expensive and are not now commonly used, though applications for their use in post-restoration monitoring of habitat and fisheries are in development (Roegner et al., 2019). UASs offer several advantages that satellites currently do not, such as the flexibility of when and how to fly, which is essential to control the data quality in terms of scale, resolution, and environmental constraints. UASs can fill in the temporal gaps created by satellite-based monitoring, offering very detailed mapping of ecosystem structures that satellites often miss (Emilien et al., 2021; Evans Ogden, 2020). They are often used to scale-up ground-based models to an intermediate scale before implementing them on satellite data (Emilien et al., 2021). The use of UASs in environmental monitoring has increased in the past few years, but standardization in data processing and analysis has not caught up.

Autonomous Underwater Vehicles

The use of autonomous underwater vehicles (AUVs) has increased dramatically in the past decade as a way to collect essential ocean variables, such as temperature, salinity, oxygen, nutrient, pressure or depth, and benthic cover or topography, at a fine spatiotemporal scale (Barrett et al., 2010; Joint Nature Conservation Committee, 2018; Whitt et al., 2020). Advancement in technologies in communication and operation autonomy has made it possible for these AUVs to acquire comprehensive environmental datasets, which can be ingested by existing marine observation systems, such as the Global Ocean Observing System (GOOS)[18] and the Gulf of Mexico Coastal Observation System.[19] ML, DL, and intelligent remote sensing systems are now commonly embedded into AUVs, enhancing their environmental perception (e.g., DeeperSense Project[20]) and allowing them to optimize their path to collecting data from areas with environmental anomalies (e.g., algae blooms, thermal hotspots, hypoxic zones, coral die-off areas) (Paull et al., 2013; Wynn et al., 2014; Whitt et al., 2020). However, operations in nearshore coastal waters are affected by such factors as shallow depths, turbidity, tides, and submerged and emergent vegetation: AUVs have not necessarily been engineered or widely adopted for these applications (e.g., Watson et al., 2020).

---

[14] See https://www.jpl.nasa.gov/cubesat.
[15] See https://www.nasa.gov/mission_pages/cubesats/overview.
[16] See https://www.planet.com.
[17] See https://conservationdrones.org.
[18] See https://www.goosocean.org.
[19] See https://www.deepersense.eu/www.
[20] See https://www.deepersense.eu/www.

Ground Sensor Networks

Ground- or field-based imaging and nonimaging sensors can be an accurate source of field monitoring data, which can be integrated and matched with drone and satellite data (Babaeian et al., 2019; Mishra et al., 2020). Field sensors are becoming increasingly cheaper and versatile (Mao et al., 2019). They can be seamlessly integrated into inexpensive computer boards, such as Raspberry Pi or Arduino (Mao et al., 2019; Ojha et al. 2015; Prasad, 2015) and solar panels to automate data collection, storage, and transmission (Boddula et al., 2017). The imaging or optical or wireless sensors can provide multi- and hyperspectral remote sensing reflectance data, which can be used to monitor wetlands' biophysical characteristics, soil organic matter, water quality, seagrass status, and several other coastal habitats or ecosystem indicators (Babaeian et al., 2019; Boddula et al., 2017; Geller et al., 2017; Mishra et al., 2020; Ojha et al. 2015; Prasad, 2015; Quinn et al., 2010; Rundel et al., 2009; Sadinski et al., 2018). Field-based cameras, such as the PhenoCam or wildlife cameras, can be valuable tools to provide geocoded image data from a site[21] (O'Connell and Alber, 2016; Richardson et al., 2018). Other sensors—such as CO_2 sensors mounted on eddy flux towers, nutrient sensors, soil sensors, floating water quality sensors (for conductivity, temperature, depth) and other buoy-based sensors, and temperature and water-level sensors—can also be useful in monitoring coastal estuarine ecosystem properties (Jones et al., 2019; Moore et al., 2009; Nehir et al., 2021; Yan et al., 2008).

Crowdsourcing or Crowdsensing

Crowdsourcing/crowdsensing[22] has been an increasingly effective tool used in many types of environmental monitoring (Ghermandi and Sinclair, 2019). Inviting members of the public to contribute meaningful environmental data and disseminating the spatial or map products generated from those datasets back to the people as information create meaningful incentives that can keep community scientists motivated to sustain the data collection and sharing feedback loop (McKinley et al., 2017). In recent years, many projects have successfully used crowdsourced data and online social media for targeted and effective environmental monitoring: examples include the Enviro-Net project from the University of Alberta and IBM, the Aqueduct Project from the World Resources Institute,[23] the Secchi Disk Project (Seafarers et al., 2017), the CyanoTRACKER project (Mishra et al., 2020), and Australian Fishes project.[24] There is considerable interest in leveraging community scientists or the public at large, including residents, tourists, and environmental advocates, by enabling and encouraging them to contribute observational data (de Sherbinin et al., 2021). The proliferation of smart mobile devices, online social media, and cloud computing has significantly lowered the barrier for nonscientists to participate in environmental monitoring (LeCun et al., 2015). Since the advent of social media platforms and microblogs, crowdsourcing and community science activities have been revolutionized, and researchers have used the crowd-sensed data for detecting all sorts of ecological, environmental, and geological phenomena, such as vegetation health, coastal flooding, plant and animal identification, earthquake and landslide damages, and urban temperatures (e.g., Earle et al., 2012; Mishra et al., 2020; Scott et al., 2016). Community science and crowdsourced data could be valuable in providing first-hand information about coastal habitats, species, or disturbance impacts if there is proper infrastructure in place, such as reporting mobile apps, information about participation on- and off-site, and incentivizing participation (Cigliano et al., 2015; Kelly et al., 2020).

Automated Data Collection and Transmission

New avenues for environmental data collection discussed above can create an opportunity to supplement traditional approaches to environmental monitoring. Once set up, these systems can be relatively

[21] See https://phenocam.sr.unh.edu/webcam.

[22] Crowdsourcing is data that is collected by nonprofessionals, such as community members and stakeholders, instead of scientists and government agencies (e.g., Conrad and Hilchey, 2011), while crowdsensing uses devices to collect field data.

[23] See https://www.wri.org/aqueduct.

[24] See https://australian.museum/get-involved/citizen-science/fishes.

autonomous and inexpensive to operate and eliminate the need for frequent field travel or physical sample collection. Being able to collect automated data from relatively isolated restoration sites by remotely operating the sensors from a laboratory using AI technologies and guaranteeing reliable transmission of monitoring data from these environments (e.g., coastal marshes) is one of the most important features of the NSF CPS infrastructure. Restoration project managers could further explore automated data collection using feasibility studies with quality assurance and quality control procedures in place for cross-validation with occasional field data, as recommended in Louisiana's 2015 System-Wide Assessment and Monitoring Program (SWAMP) report (Hijuelos and Hemmerling, 2015) and currently under way in the Columbia River estuary restoration (Roegner et al., 2019). It can ultimately be a cost-effective way to address postrestoration monitoring challenges. Various data transmission approaches are available, such as data hopping via WiFi, 4G/5G mobile network–based data transfer, and transmission via low-earth-orbit satellite communication datalinks, such as the Iridium satellite constellation[25] (Alpert et al., 2016; Hart and Martinez, 2015; Zhang et al., 2019).

Data-Driven Modeling to Ingest Heterogenous Monitoring Output

Environmental models, in the form of conceptual or physical models or ecosystem, population, or mechanistic models, and landscape or rule-based models (discussed in Chapter 3) have been used traditionally to model cumulative impact assessment in terms of environmental biophysical and biochemical characteristics and ecosystem responses. They can be effective when accurately parametrized with high-quality field data derived from field sample analysis or controlled monitoring. However, with the new technology-driven sensing and monitoring techniques discussed above, the available data will be heterogeneous and inherently noisy (e.g., multisource sensing data, spatial and remote sensing data, and numerical and contextual data) (Clark and Gelfand, 2006; Goldstein and Coco, 2015; Peters et al., 2014; Peters-Lidard et al., 2017; Zhong et al., 2021). Using these new datasets, modeling the complex and nonlinear relationships between ecological variables and post-restoration ecosystem responses over space and time within the traditional modeling approaches can produce large uncertainties (Clark and Gelfand, 2006; Zhong et al., 2021). ML- and DL-based models (e.g., convolutional neural network or recurrent neural network) are increasingly widely used alternatives that can handle heterogeneous monitoring imaging and non-imaging data for model training and tuning (Lamba et al., 2019; Zhong et al., 2021). They can be highly effective in teasing apart the complex relationships between ecosystem drivers, pressures, and stressors and response and patterns between different datasets and environments, although their performance depends on the types and amounts of training data available from a site (Goldstein and Coco, 2015; Lamba et al., 2019). Once trained for a site across a wide range of variables, they can be used for a long time without needing frequent re-training (Sejnowski, 2020). They can also be scaled up to airborne and satellite sensors for large-scale and long-term monitoring and modeling (Lapenna and Soldovieri, 2021). The data-driven models will not replace traditional environmental models, such as physical or mathematical or simulation models, but will strengthen them when used in a hybrid framework (Clark and Gelfand, 2006).

Ensemble ML is a modeling framework in which several ML models or coupled physical, ecosystem, and ML models are run using environmental input variables either obtained from monitoring processes (as described above) or estimated using traditional environmental modeling, and their output is combined using a rule-based algorithm (e.g., stacking or voting; sometimes referred to as meta-learning) to generate the most accurate prediction (Finn et al., 2017; Sagi and Rokach, 2018). These types of data-centric models are progressively becoming a go-to modeling tool for the ecology and environmental science community for building assessment and predictive models (Beijbom et al., 2015; Norouzzadeh et al., 2018; Parashar et al., 2021; Peters et al., 2014; Weinstein, 2018; Williams et al., 2019). New refinement procedures for improving the predictive ability of these models are available in an increasingly open-source manner (Ghannam and Techtmann, 2021). This modeling framework can contribute to assessment of the long-term cumulative effects of large-

[25] See https://www.iridium.com.

scale GoM restoration projects, along with other methods that support synthesis, such as literature review and simulation modeling (Chapter 3) and comparative analysis and time-series analysis (Chapter 4).

Open-Source Data for Visualization and Sharing

Accessibility to data, maps, and information is at the heart of the concept of open science.[26] Access to data can be the biggest bottleneck for large-scale synthesis studies or comparative studies across restoration sites (e.g., de Groot et al., 2010; Farley et al., 2018; Palomo et al., 2018). Data transparency (such as types, collection and analysis procedure, uncertainty) and access and sharing mechanisms (e.g., open-source archive or web mapping tools) have to be determined in the early stages of each restoration project to facilitate future modeling and synthesis studies. An open-science framework adopted by the restoration practitioners and government agencies across the GoM in the form of openly accessible and comprehensive monitoring databases could help transform the post-restoration monitoring process to study long-term cumulative impacts. The primary objective would be to create a centralized online location where any user can visualize and download relevant data and products (e.g., imaging, nonimaging, community science). A diverse set of end users (such as environmental managers, researchers, nongovernmental organizations, policy makers, and commercial and recreational users) could visit the site to explore the available products and write custom code, if needed, to use these products in further analysis. There are several efforts under way across the Gulf states to either enhance existing open-source databases and online mapping interfaces or establish new ones capable of synthesizing monitoring data. Some examples include the Gulf of Mexico Research Initiative (GoMRI)'s Information and Data Cooperative (GRIIDC),[27] NOAA's Gulf of Mexico Data Atlas,[28] the Gulf of Mexico Coastal Ocean Observing System (GCOOS),[29] the U.S. Army Corps of Engineers' Institute for Water Resources Gulf of Mexico Regional Assessment for the National Shoreline Management Study,[30] the hydrologic and dnvironmental databases of the U.S. Environmental Protection Agency (EPA) and the U.S. Geological Survey (USGS), and the Tampa Bay Estuary Program's reporting and assessment methods.[31] These databases are important sources of input data, but they lack the ability to integrate models to synthesize, process, and display outputs about restoration projects across the GoM. Recent growth in cloud-based platforms—such as Google Earth Engine,[32] which is open to anyone with an internet connection—can be an interface for users to model and visualize data, and carry out further processing (e.g., Boothroyd et al., 2020; Campbell and Wang, 2020; Mishra et al., 2020; Vos et al., 2019). The Google Earth Engine platform does not need local computing resources or specialized software, and users can perform several analyses, which may include empirical or data-driven modeling, mapping, and time-series trend analysis of multiple restoration sites, at once. These analyses are important components contributing to the lines of evidence and synthesis used for postrestoration cumulative effects (see Table 3.3, in Chapter 3).

Data Repositories

It is important that all data and information regarding restoration be available to all users, regardless of where data are deposited. FAIR principles—Findable, Accessible, Interoperable, Reusable (e.g., Wilkinson et al., 2016)—are widely accepted and provide guidance for data management and stewardship. In addition, many interim products and gray literature reports are scientifically peer reviewed and often contain more extensive datasets than those that are published in scientific journals. Including these products in data repositories allows them to be available to the greater community. To ensure that data and information

---

[26] See https://en.unesco.org/science-sustainable-future/open-science.
[27] See https://data.gulfresearchinitiative.org.
[28] See https://gulfatlas.noaa.gov.
[29] See https://gisdata.gcoos.org.
[30] See https://www.iwr.usace.army.mil/Missions/Coasts/National-Shoreline-Management.
[31] See https://tbep.org/our-work/data-visualization.
[32] See https://earthengine.google.com.

generated by *Deepwater Horizon*-funded restoration projects do not become lost or unavailable over time, data repositories need to ensure the following:

- FAIR principles are followed,
- multiple data formats and types (including video and audio formats) are accepted,
- data and metadata submissions are reviewed for completeness,
- an easy-to-use submission process is available,
- training and assistance for data providers (including planning, organizing, documenting, and submitting datasets) is provided,
- detailed geographic data are available to enable spatial searches,
- Digital Object Identifiers (DOIs) are assigned to datasets so they can be cited,
- datasets linked to publications are easily identified,
- data are freely available and downloadable,
- data are retained permanently in an archiving database,
- metadata are discoverable through common search engines, and
- data reporting formats are machine readable.

The broad set of restoration efforts envisioned by *Deepwater Horizon* funding entities and others, both coastal and oceanic, will generate significant quantities of data over the coming years. The amount of data that already exists is not insignificant, but it can be difficult to find and access and may need curation to meet the standards listed above. Much of the data is in diverse formats, which need to be stored appropriately. As ML, DL, and AI advance problem-solving approaches across both research and industry (e.g., Kersting, 2018), it is important to ensure that all restoration data are accessible to innovative approaches and contribute to adaptive management needs.

Table 5.1 summarizes example project-level parameters that could be useful for cumulative effects assessment for key restoration types being implemented since *Deepwater Horizon* (including oysters, submerged aquatic vegetation, salt marshes and mangroves, water quality and nutrient reduction, and birds). Both traditional parameters (discussed in Chapter 2) and novel techniques and emerging technology (discussed in this chapter) are included, as well as examples of cumulative effects assessment tools (discussed in Chapter 3). Key considerations regarding scale that might be important when designing and assessing project-level restoration efforts are also noted, for use by restoration managers.

KEY METRICS NECESSARY FOR ASSESSMENT BEYOND THE PROJECT LEVEL

Better understanding of the cumulative effects of multiple restoration projects is needed to ensure effective use of funding, sustainable restoration results, and demonstrable ecosystem benefits, along with other desirable societal goals. Assessing cumulative restoration is a focus of growing concern and analysis by the restoration community (Cooke et al., 2018, 2019; Ladouceur and Shackelford, 2021), as well as by funders and practitioners looking to advance restoration beyond project-level goals. Drivers and pressures (discussed in Chapter 2) affect restoration at both project and larger scales. Climate change generates pressures that, along with acute events such as hurricanes, affect all scales of restoration.

In the GoM, as with other developed coasts, anthropogenic drivers generate pressures (e.g., land use change, water management strategies) that can negatively affect restoration at all levels, further confounding synthesis activities (Magliocca et al., 2015). Monitoring the landscape-scale changes generated by these diverse drivers and pressures needs to be a priority moving forward, as those observations will support synthesis activities that inform adaptive management. This will include creating the metrics needed to assess restoration and the data needed to develop the metrics, and ensuring that data are compiled and made available to restoration planners in a useable form.

Key Indicators of Long-Term Trends Needed to Assess Restoration Efficacy

Numerous approaches and techniques are available for assessing the efficacy of GoM restoration activities, and clear and useful examples of synthesis have been completed or are under way in the Gulf region (e.g., Fennel and Laurent, 2018).[33] Specific approaches and examples of synthesis activities completed in coastal and estuarine areas have been described in Chapters 3 and 4. The available approaches noted in Chapters 3 and 4 include mass balances (often called budgets), comparative analyses (use of data from many sites), time-series analysis (use of data collected over time at a site), and several types of simulation models (site-specific models and more general/conceptual models). These and other data-intensive approaches discussed earlier in Chapter 5 can readily be applied to the challenging issues of assessing restoration successes in the diverse habitats of the GoM when appropriate datasets can be accessed.

Metrics and data availability for tracking long-term environmental background trends in the GoM have been discussed in Chapter 2. Of those, several are also key to assessing restoration efficacy at larger spatial scales, including the following:

- freshwater inflow (USGS flow gages) and water quality parameters, including nutrient concentrations (federal, state, or local water quality monitoring programs): these data are used to generate loading estimates to estuarine and coastal waters. Changes in nutrient loadings can be used to gage effectiveness of multiple nutrient reduction efforts within a watershed.
- land use and land cover (Landsat, MODIS, Sentinel-2 and -3, or other satellite imagery): these are used to track changes in coastal habitat areal extent; transformation of rural to urban land uses (for example); and connectivity between habitat types within watersheds and their receiving waterbodies. Changes in the areal extent of habitat types can be used to assess net changes from multiple and large-scale land conservation efforts.
- tide gage data and subsidence measurements: these are used to estimate local relative sea level rise.
- ocean and coastal acidification indicators, such as pH: these are essential for successful restoration and maintenance of commercially important shellfish.

Ambient water quality measurements: these are used to track light levels for key habitats, including submerged aquatic vegetation. Methods to assess chlorophyll *a* and total suspended solids using multi- and hyperspectral data have improved (Dou et al., 2018), allowing for their use in assessing stages of restoration in water quality condition (Boyer et al., 2009; Millette et al., 2019). Tracking chlorophyll *a* concentrations in estuarine and coastal waters over time has also been an effective method for assessing light availability for submerged aquatic and other benthic communities (Luo et al., 2020), and is associated with implementation of multiple nutrient reduction projects within larger watersheds (see Chapter 4).

Satellite data and models for assessing wetland change (Couvillion, 2021; Klemas, 2013; Taddeo et al., 2019): they may be useful tools for assessing restoration over significant spatial scales. Examples include mangroves (Lee and Yeh., 2009) and constructed wetlands (Li et al., 2020b).

Status and trends data for fish, birds, and marine mammals collected for treaty, regulatory, and management purposes in the GoM: these represent some of the most readily available long-term data and encompass significant spatial scales. These data can be useful metrics for integrating across broad political and ecological spatiotemporal scales when assessing ecosystem health and function beyond project scale. They include fish (Jordan et al., 2010; Schrandt et al., 2021); birds;[34] water birds (Ogden et al., 2014; Rajpar et al., 2018); and marine mammals (Bossart, 2010; Wells et al., 2004). These data, collected for multiple diverse objectives, can contribute to the development of various lines of evidence efforts to assess cumulative impacts of restoration across spatial scales.

---

[33] This point was also made in a presentation to the committee by Gregory Steyer of the U.S. Geological Survey (November 9, 2020).

[34] See https://gomamn.org.

TABLE 5.1. Measurement and Modeling Techniques to Support Cumulative Effects Assessments from Project-Level Restoration Efforts

Restoration Approaches and Anticipated Outcomes	Parameters and Indicators		Cumulative Effects (CE) Considerations		
Restoration Type	**Example Project-Level Parameters Useful for CE Assessment**	**Other Example Parameters and Indicators**	**Example Toolbox for CE Applications**	**Key Considerations Regarding Scale**	**Novel Techniques and Emerging Technologies**
Oysters	Density Mortality Areal dimensions Reef elevation (height) Shoreline change (living shorelines)	Reef connectivity and reefs across habitat and salinity gradients Rates of shell and reef accretion	Population models Conceptual models Indices	Reef system scale (elevation; connectivity of system) Gulf of Mexico scale (density, overall area, average increase in reef elevation over time) Fluctuations in ecosystem conditions critical to survival (e.g., salinity, dissolved oxygen) may underscore the importance of assessing reef systems across salinity and habitat gradients with a goal of maximizing survival under a variety of conditions Landings data can provide some indication of oysters' Gulf-wide distribution	Autonomous underwater vehicles to monitor essential variables, such as temperature, salinity, oxygen, and benthic cover Machine learning (ML) and deep learning (DL) models implemented on satellite data for water quality monitoring
Submerged Aquatic Vegetation (SAV)	Areal extent Density Percentage cover and composition Survival Biomass Water clarity	Proximity to other habitats Denitrification rates Biomass of benthos Light characteristics Changes in areal coverage Secondary production	Cumulative Net Ecosystem Improvement Conceptual models Monitored reference systems	For some SAV projects, a simple summation of SAV processes across the full GoM could be feasible. For example, using SAV coverage data (and composition and density data), apply local and, where needed, literature estimates of denitrification; sum across regions for an annual shallow water GoM estimate of nitrogen losses and compare with inputs to the GoM and to regions Time lags and measurable outcomes can vary depending on the scale of the project Large-scale efforts typically achieved via accompanying improvements in water quality and clarity	Autonomous underwater vehicle with imaging sensors to monitor benthic habitat cover or percentage cover of SAV SAV habitat maps using ML and DL models on remote sensing images from uncrewed aerial systems or satellites

TABLE 5.1. Measurement and Modeling Techniques to Support Cumulative Effects Assessments from Project-Level Restoration Efforts Continued

Wetlands (Salt Marshes and Mangroves)	Area of habitat restored, enhanced, created, or conserved Vegetation percentage cover and species composition Above-ground and below-ground biomass Gross primary production	Habitat connectivity and proximity to other habitats Changes in areal coverage Secondary production Additional process measurements, such as carbon sequestration and denitrification, are important in these systems and will need modeling and research community coordination and collaboration	Physical models Cumulative Net Ecosystem Improvement	Attributes, such as plant cover and use by fauna, can become well established within 1–5 years of restoration, while other features, such as below-ground biomass, can take one or more decades to approach levels of natural marshes	Crowdsourcing for reporting marsh dieback or other degradations CubeSats high temporal frequency useful for monitoring impacts of acute disturbance events Ground sensor networks with automated data collection and transmission (e.g., Phenocam or eddy covariance towers) Moderate resolution multispectral satellite data for frequent monitoring of wetland biophysical status
Water Quality, Including Nutrient Reduction	Sediment, nitrogen, phosphorus loads; ambient pollutant concentrations Bacteriological indicators	Likely will need quite a few process measurements (accretion rates, production rates, etc.) to develop useful models used to quantify CE benefits Delisting of waterbodies from 303(d) list for target impairments	Conceptual models Structured decision making Physical models	Lag times between field-scale management and response of relevant water quality metrics may take years of in-stream monitoring to detect Changes as a result of restoration action likely more difficult to detect as scale of assessment increases Comparison with long-term, system-scale data could help to separate restoration responses from other trends	Autonomous underwater vehicles or floating sensor network for water quality monitoring Satellite- or uncrewed aerial systems–based water quality modeling (bio-optical models, ML, or DL models) to monitor chlorophyll a, harmful algal blooms, colored dissolved organic matter, and total suspended solids
Birds	Abundance and community assemblage Habitat quality Area (for bird habitat projects)	Habitat fragmentation Habitat connectivity Nesting colony counts via air over time Change in nesting success	Population models Indices Structured decision making	Population status and trends best served by regional- or larger-scale assessment	Crowdsourcing and crowd sensing Uncrewed aerial systems

Blue Carbon and Primary Production as Integrative Assessment Metrics

In addition to the metrics and trends discussed in Chapter 2, coastal primary production and blue carbon storage are important indicators of ecosystem response and function (Macreadie et al., 2019; Mcleod et al., 2011; Sigman and Hain, 2012) and have the potential to be effective integrative metrics for evaluating the cumulative effect of restoration projects. Primary production in marine environments refers to organic carbon production by phytoplankton (Chassot et al., 2010; Sigman and Hain, 2012), while coastal blue carbon refers to the carbon produced and stored in coastal habitats—such as mangroves, marshes, or seagrasses—as biomass and soil organic carbon (Chmura et al., 2003; Laffoley and Grimsditch, 2009; Mcleod et al., 2011). Tropicalization (discussed in Chapter 2) affects blue carbon through the continuum of plant communities and replacement through climate change effects (Osland et al., 2022).

Primary production and blue carbon assimilation are considered integrative metrics because of their linkage to multiple ecosystem services; thus, they have been used as proxies for evaluating outcomes of coastal restoration projects (summarized in Windham-Myers et al., 2018). Some of these linkages are shown in Chapter 3 (Tables 3.1 and 3.2). Both metrics are positively linked to nutrient cycling, healthy and productive coastal wetland and submerged vegetation, and enhanced biodiversity. For example, sediment and nutrient from river runoff to wetlands improve wetland productivity and elevation, making them more resilient against relative sea level rise, and reducing the rate of habitat fragmentation. Nutrient uptake by submerged aquatic communities enhances their productivity and helps improve water clarity and fisheries (unless nutrient concentrations are too high, in which case algal blooms and eutrophication may occur). Healthy and productive coastal wetlands enhance bird counts, fisheries, tourism, and economic growth. Therefore, these metrics are integrative and can be effective toward cumulative effect assessment, with a caveat that the rate of change has to be evaluated along with a time-integrated production budget. Newly restored ecosystems will need time to produce the desired production budget due to the temporal lag of many years (summarized in Table 4.5, in Chapter 4) between ecosystem restoration activities and restoration of optimum ecosystem functions (Burden et al., 2019; Greiner et al., 2013; Osland et al., 2012). Therefore, budgets alone may not represent the effectiveness of restoration projects.

Both primary production and blue carbon have been measured in the field or can be estimated using models for a wide variety of ecosystems. They can be measured at a point scale using analysis of field-collected physical samples, moored sensors, or chamber-based data collection (Järveoja et al., 2018; Marra, 2002; UNESCO-IOC, 2012); at an area scale using tower-based (eddy flux tower) data collection and analysis (Forbrich and Giblin, 2015; Knox et al., 2018; Koopmans et al., 2020; Tokoro and Kuwae, 2018); or at a landscape scale using observations from satellites or airborne sensors (Behrenfeld et al., 2005, 2006; Platt and Sathyendranath, 1993; Yan et al., 2008). Recent advances in sensors and sending devices, airborne platforms, and increases in spatiotemporal coverage by earth observation satellites create new opportunities for restoration managers to deploy a combination of these techniques to estimate primary production and blue carbon, and to use them as integrative metrics for assessing the cumulative effect of multiple restoration projects.

Assessing Functionality

With the increasing availability of satellite-derived assessments of the areal extent of coastal habitats, it is possible to determine the total area of most of the restored salt marsh in the Gulf; it is more difficult to determine the areal extent of restored oyster reef and seagrass meadows. However, to the extent that satellite data are not readily available for oyster reefs and seagrasses, an estimate of the total area restored by *Deepwater Horizon*-associated restoration activities can be obtained by gathering information from all *Deepwater Horizon* funding entities on the size of areas restored and then summing them by habitat type. The *Deepwater Horizon* NRDA Trustees are in the process of developing their 5-year programmatic review (due by early 2022), which is expected to include an aggregate of information (e.g., areal extent of marsh restoration, linear miles of living shoreline) from NRDA projects under way to date.

Once the spatial extent of the areas restored is estimated, metrics that measure their functioning can be employed to compare their productivity to reference areas containing healthy examples of a similar coastal nursery habitat. One way of doing this is to calculate secondary production, a metric that estimates the accumulation of habitat-associated animal biomass through time and can serve as a valuable tool for measuring and assessing the outcome of restoration activities (Layman and Rypel, 2020; Sobocinski and Latour, 2015). This approach is an example of a line of evidence, also called cumulative net ecosystem improvement (CNEI) modeling (Diefenderfer et al., 2016, 2021; also see discussion in Chapter 3).

Because habitat-specific estimates of secondary production exist for reference areas of the major types of restored coastal habitats, a first-order assessment of the benefit a restored area could provide can be obtained by multiplying the area restored by literature-derived estimates of secondary production from nearby reference locations. However, this value will likely be a substantial overestimate of the true value of the restored area because, as much evidence indicates, it often takes many years before secondary production in restored habitats approaches that of reference areas (see Table 4.5, in Chapter 4) (Gray et al., 2002; Hering, 2009), and, in some cases, restored function is not equivalent even decades after restoration (Chellew, 2017). In addition, the potential synergistic and/or antagonistic effects of the proximity of restored nursery habitats (as discussed in Chapter 3) and effects of long-term environmental trends (as discussed in Chapter 2) could result in either positive or negative effects on their functioning. Despite these caveats, first-order estimates of restoration benefits could provide a baseline for further refinements as the project matures.

PROGRAM LEVEL ADAPTIVE MANAGEMENT STRATEGIES

As highlighted throughout this report, environmental background trends, especially those associated with climate change, are exhibiting higher variability over time. Restoration practices that have been successful in the past may no longer be adequate to compensate for the effects of anticipated changes in background trends. Adaptive management techniques can provide restoration program managers with the ability to revisit and update large-scale restoration strategies, based on periodic review of monitoring data and progress toward programmatic goals.

New Advances in Adaptive Management

As noted in Chapter 4, successful implementation of adaptive management in Gulf restoration has been limited; applying adaptive management across multiple projects and large spatial scales presents many challenges. However, this lack of success is not isolated to the GoM (Nagarkar and Raulund-Rasmussen, 2016; Zedler, 2017), despite successes in other coastal and estuarine restoration programs (Littles et al., 2022). Numerous authors have noted that adaptive management has not delivered its desired results (Nel and Roux, 2018; Westgate et al., 2013; Williams and Brown, 2014) and recommended ways to improve the efficacy of the process. McLoughlin et al. (2020) identified specific reasons for this situation: (1) adaptive management has been inserted into restoration policy without fully understanding what is needed to make it effective; (2) the theory of adaptive management is ahead of both practice and capabilities to support it; (3) a deficit of trust exists across institutions and organizations in structuring adaptive management; and (4) adaptive management remains an evolving process in which uncertainty is complicated by interactions between its social and ecological components. In addition, an essential aspect of adaptive management is that "monitoring has to be adequate to detect change resulting from management experiments" (Westgate et al., 2013, p. 129), although the ability to initiate and sustain long-term investigations is difficult (Lindenmayer and Likens, 2010). Like cumulative effects assessment and ecological synthesis in general (Kemp and Boynton, 2012), AM cannot take place without good-quality, accessible data that evaluate the causes of success or failure of the constructed projects, all of which are limited in the GoM as documented in this report.

Several recently developed ideas about employing AM at large scales may have utility in the Gulf. For example, multiple authors have noted that adaptive feedback learning loops may be more appropriately defined as double or triple loops involving multiple levels of decision making rather than a single loop of simply adjusting existing routines (McLoughlin et al., 2020; Pahl-Wostl, 2009). A spiral framework that recognizes a temporal element (Fernández-Giménez et al., 2019; Montambault et al., 2015) may more accurately capture long-term environmental trends that act on the Gulf and on projects attempting to restore parts of it, and may also represent the time lags, especially at large scales, in availability of key data. The AM process in the Gulf may therefore be represented in the form of a spiral moving forward in time, rather than as a static circle, as is more commonly depicted for AM.

Enhanced Recognition of the Need for Larger-Scale Adaptive Management

As described in Chapter 2, many external stressors impact GoM restoration. Three examples from that chapter are highlighted below to elevate the need for effective adaptive management:

- Where relative sea level rise is greater than 3 mm/yr, recent research suggests that marsh drowning will likely occur within a few centuries and that relative sea level rise in the range of 6–9 mm/yr will likely convert marshes into open water within a 50-year time frame (Törnqvist et al., 2020) without additional inputs of sediment. This means that even the relatively small sea level rise in Florida, which is above the 3 mm/yr threshold, will eventually cause adverse impacts on restored wetlands, while in some coastal areas of Texas and Louisiana, where rates are already in the range of 5–8 mm/yr (Argus et al., 2018), marsh drowning has been and will continue to be common and hugely important.
- Ocean acidification will become an increasing concern, specifically in relation to oyster reef formation and persistence. Based on NOAA data, subsurface waters in the northwestern GoM are acidifying at a rate greater than the global surface ocean rate, with the majority (59–70 percent) of the acidification being respiration driven (Hu et al., 2018).
- Based on a global literature review (He and Silliman, 2019), the effects of climate change and land use/land cover change on wetland ecosystem health could include impacts on species' range shifts and extinction rates, nutrient deposition, and habitat fragmentation.

Of relevance to this discussion of AM, the 2016 consent decree[35] with BP establishes provisions for additional payment of funds for unknown conditions and adaptive management, which can be used

> (1) to address injuries and/or losses to Natural Resources (including services provided by Natural Resources) resulting from the *Deepwater Horizon* Incident that are unknown to the Trustees as of July 2, 2015, including for any associated Natural Resource Damage assessment and planning activities, or (2) to adapt, enhance, supplement, or replace restoration projects or approaches initially selected by the Trustees.

These funds may be requested from BP beginning January 1, 2026. In order to maximize the use of these funds, and to prepare for the potential adaptation or enhancement of the restoration program, *Deepwater Horizon* funding entities have the opportunity to set up the infrastructure necessary to facilitate the evaluation of the restoration program and make adjustments now, if warranted, while preparing for the future. As discussed in Chapter 4, the *Deepwater Horizon Monitoring and Adaptive Management (MAM) Implementation Manual* outlines guidance for the implementation of project-level monitoring and adaptive management, with a recognition that future versions will address program level MAM. This work may be less effective if the *Deepwater Horizon* NRDA Trustees wait until 2026 and the establishment of the Unknown Conditions and Adaptive Management Trustee Implemention Group,[36] as billions of dollars will have already been expended by this date, and large-scale adjustments, if needed, may become cost prohibitive as restoration funds are depleted.

---

[35] See https://www.justice.gov/enrd/file/838066/download (p. 23, paragraph 21).

[36] See https://www.gulfspillrestoration.noaa.gov/co-trustees.

Increased recognition of the need for adaptive management beyond the project scale is becoming apparent. For example, in 2021, the Louisiana Trustee Implementation Group defined their MAM strategy to address larger-than-project scales and identified possible activities to address elements of the strategy. While all elements of this strategy may not be applicable for implementation in other states, they could provide an initial template when considering programmatic adaptive management guidance. Their objectives include that:

- *Deepwater Horizon* NRDA lessons learned are systematically captured and incorporated into future project selection, design, implementation, and evaluation;
- relative effectiveness of different restoration approaches are identified;
- the influence that *Deepwater Horizon* NRDA restoration has on ecosystem condition is understood at present and for comparison at 5-year intervals;
- access to and availability of collected data, monitoring, project documents, and lessons learned is increased; and
- communications about MAM within and across agencies, stakeholders, and the public are increased to support effective adaptive management of coastal restoration.

Synergistic and Antagonistic Considerations

Chapter 3 discusses synergistic and antagonistic effects of multiple stressors and restoration projects at watershed scales and beyond. Managers and restoration funders could enhance the efficacy of multiple restoration projects implemented over large spatial scales if they better understood how restoration projects may interact with one another in either positive or negative ways (Moreno-Mateos et al., 2020; Roy et al., 2016) and the potential for positive interactions among restoration projects going forward (Fitzsimons et al., 2020; Neeson et al., 2016; Ridlon et al., 2021; Wiesenburg et al., 2021). The prevalence of interactions among causal factors within ecosystems make ecosystem science, and the evaluation and management of ecosystem restoration, challenging because single explanations for causality are rare (Kemp and Boynton, 2012; Odum, 1971).

Including the implications of these synergistic or antagonistic interactions into an adaptive management process that involves multiple and diverse restoration projects across large spatial scales can be dauntingly complex (Diefenderfer et al., 2021; Maxwell et al., 2016; Walters, 1997), but not doing so can mean missed opportunities, or worse—negative interactions of multiple projects with the potential to negate anticipated ecosystem benefits. Although adaptive management has been more commonly applied to single projects, the complexity and large scale of restoration in the Gulf will need a more comprehensive approach that assesses both synergistic and antagonistic effects among projects.

The Importance of Synthesis for Adaptive Management and Cumulative Effects Assessments

Synthesis efforts are needed to determine how much the many localized restoration efforts, when taken together, have resulted in measurably improved coastal and estuarine ecosystems across the GoM region. In addition, such analyses provide a mechanism for adjusting efforts to produce better restoration outcomes.

To date, more than 30 percent of the *Deepwater Horizon* environmental restoration funds have been spent or committed, and a very substantial and diverse set of observations and measurements has been collected. Synthesis efforts, whether in the form of simulation modeling (Chapter 3), comparative analysis (Chapter 4), or other techniques described in this report, provide a quantitative framework for the analysis of these diverse datasets. A synthesis framework makes it possible to address difficult and exceedingly complex environmental questions and provide answers that lead to increased understanding of coastal and estuarine system dynamics and, ultimately, better management decisions.

In addition, synthesis work is often performed at spatial and temporal scales that can lead to actionable management. For example, measurements of denitrification rates in coastal sediments are of interest to biogeochemists but by themselves are of little use in helping to make management decisions. However, when used in a synthesis tool, such as a water quality model, the relative importance of nitrogen loss via denitrification becomes clear, as do the needs for management actions aimed at nutrient load reduction that will in turn help reduce the extent of deep-water hypoxia. Synthesis at scales relevant to management groups is also particularly needed in the Gulf because of strong and concerning trends in both chronic (e.g., sea level rise, tropicalization) and acute (e.g., hurricanes, floods) stressors that can directly and indirectly influence the success of restoration projects at all scales.

Existing synthesis efforts in the GoM—including the Texas Coast Ecosystem Health Report Card[37] (Harwell et al., 2019; McKinney et al., 2019), the Galveston Bay Report Card,[38] the Mississippi River Watershed Report Card,[39] and the Everglades Health Report Card[40]—provide a variety of models that make use of existing data and metrics to assess ecosystem health and inform resource management. A synthesis activity that assesses and informs GoM restoration at scales up to and including the entire Gulf region could benefit from lessons learned in these efforts, especially regarding how to work across different state and federal agencies and political boundaries.

Risks of Not Considering Large-Scale Restoration from a Cumulative Effects Approach

The scale of restoration investments (size and number of projects) will affect the ability to assess cumulative effects.[41] As the scale of restoration has grown to include multiple and diverse projects over broader geographies, so have the observations that positive interactions can be enhanced through careful planning, analysis, and adaptive management (Diefenderfer et al., 2021). While the scientific literature about positive cumulative effects in large-scale restoration is sparse, examples like the restoration of the multi-state Chesapeake Bay demonstrate that such benefits can occur (Gurbisz and Kemp, 2014; Testa et al., 2017).

Every restoration site has a particular hydrological and geomorphological context, as well as a set of biogeochemical processes that play important roles in sustaining ecological functions at both ecosystem and landscape scales (Brinson et al., 1993; Gosselink et al., 1989; NRC, 1992; Toth, 1995). Decisions made at the planning stage regarding the scale of modeling and evaluation may have critically important implications for success or failure to meet restoration objectives after implementation (Diefenderfer et al., 2005). The ability to enact changes to restoration approaches during implementation and to adaptively manage restoration sites when the context changes in unanticipated ways is also necessary to achieve project performance objectives (LoSchiavo et al., 2013).

Why Synthesis Is Often Recommended and Rarely Implemented

There are a number of factors that appear to inhibit or discourage synthesis activities even in cases in which synthesis is sorely needed to support management and assessment needs. Some of these factors have been discussed by Kemp and Boynton (2012) and Testa et al. (2017). First, synthesis is difficult. One part of this difficulty is the challenge of generating synthesis questions for which sufficient evidence exists, so that they can be successfully addressed. Another difficulty is that many scientists are more comfortable

[37] See https://www.harte.org/project/texas-coast-ecosystem-health-report-card.

[38] See https://www.galvbaygrade.org.

[39] See https://americaswatershed.org/reportcard.

[40] See https://www.saj.usace.army.mil/Missions/Environmental/Ecosystem-Restoration/RECOVER/2019-Everglades-Health-Report-Card.

[41] Presentation to the committee, Gregory Steyer, U.S. Geological Survey (November 9, 2020).

and skilled at traditional experimental hypothesis testing than the inductive approach mainly associated with synthesis (Kemp and Boynton, 2012). Additionally, synthesis is not a fast process; instead, it takes time to develop useful products. Finally, synthesis generally does not happen, especially at large spatial scales, without directed financial and institutional support. Such dedicated support is necessary because the issues being addressed often involve multiple states, and necessitates funding support that extends beyond normal project timelines and across jurisdictional boundaries.

The Importance of Coordination

Because most estuary- and watershed-scale programs in the GoM cross jurisdictional boundaries and include multiple federal, state, and local partners, effective and consistent cooperation is needed to assess cumulative effects of restoration efforts. For an example from another area, in the Chesapeake Bay states and agencies responsible for implementing the bay recovery effort found that one entity that represented all of the interested parties was necessary to coordinate the restoration effort and the assessment of the cumulative effects of the many restoration projects.[42] Similarly, Gross and Hagy (2017) evaluated 16 successful lake and estuary nutrient management programs and found that the attributes most associated with achieving restoration goals included leadership and coordination by a dedicated watershed management agency and governance through a bottom-up collaborative process. Examples of successful GoM restoration efforts at the estuary/watershed scales (discussed in Chapter 4) include coordination of restoration efforts by a dedicated watershed management entity.

Common Barriers to and Opportunities for Successful Synthesis

A comprehensive monitoring database or clearinghouse is obviously needed for synthesis activities. As indicated above, current data collection in the GoM is often thought to be inadequate for regional or GoM-wide synthesis.[43] There are also considerable inconsistencies in the types of variables and techniques used in monitoring programs across the Gulf (NASEM, 2017).

However, because there are 34 U.S. GoM estuaries, opportunities exist for preliminary, but potentially very useful, comparative analyses of both pre-*Deepwater Horizon* and contemporary data (see Box 5.1). Such comparative analyses could be aimed at assessing similarities and outliers among GoM estuaries on issues related to their sensitivity to watershed inputs, fisheries yields, and changes in water quality conditions (see Chapter 4 for details on comparative analyses). Comparative analyses could help managers assess aspects of landscape-scale processes that constrain possibilities for restoration of individual sites (as shown in Table 3.4, in Chapter 3). Furthermore, comparative analyses could inform the design of sampling and analysis at the watershed scale—which can differ based on watershed condition.

There is a lack of centralized data management, storage, and access across the many restoration projects. While there are several freely accessible data repositories operating in the Gulf region (e.g., GCOOS; NOAA's Data Integration Visualization Exploration and Reporting [DIVER] site), these need to be expanded to include data from the many new project-level restoration programs. Such an improved data system could provide incentive for larger-scale GoM synthesis efforts by researchers across academic, governmental, and nongovernmental institutions.

Funding for long-term monitoring efforts is often difficult to obtain. Commitment to long-term monitoring is especially important because Gulf restoration efforts are taking place in the context of changing long-term environmental trends (discussed in Chapter 2), which are likely to have significant influences on the success or failure of restoration projects. For example, long-term commitments to both monitoring

---

[42] See https://www.chesapeakebay.net/who/bay_program_history.

[43] See, for example, the August 4, 2020, Gulf Research Program webinar "Restoring the Gulf after Deepwater Horizon: Perspective from the Front Lines"; https://www.nationalacademies.org/event/08-04-2020/webinar-restoring-the-gulf-after-deepwater-horizon-perspective-from-the-front-lines.

BOX 5.1
Comparative Analysis—A GoM-wide Example

Throughout this report, environmental histories and restoration progress are summarized for several estuarine systems in the GoM, including the major approaches (e.g., monitoring, modeling, leadership) used in restoration efforts. While these efforts are extremely valuable, they fall short of approaching a GoM-wide assessment of environmental conditions and possible responses to restoration efforts.

One approach to starting GoM-wide assessments of environmental conditions and changes in those conditions is to use a comparative analysis approach. Kemp and Boynton (2012) note that comparative analysis involves using similar data from many different systems to develop a model, which quantifies how one or more key properties or processes vary in relation to differences in external drivers or other internal properties. Comparative analysis involves assembling data from diverse ecosystems (e.g., estuaries, watersheds, rivers) or studies (e.g., nutrient reduction projects, marsh restoration sites) and using these data to develop statistical approaches that can be used for comparison (Cloern and Jassby, 2010; Megrey et al., 2009; Kemp and Boynton, 2012). This method has been used effectively to address a range of questions, including how nutrient loading to coastal systems influences phytoplankton growth (Cloern and Jassby, 2010; Detenbeck et al., 2019; Monbet, 1992; Moorman et al., 2014;) or benthic invertebrate biomass and fisheries harvest (Deegan et al., 1986; DeWitt et al., 2011; MacKenzie et al., 2015).

There are several compelling reasons to consider a comparative analysis approach in the GoM: there are a substantial number of "estuarine replicates" to use in such analyses; many of the important features of these systems have been summarized; and data are readily available (e.g., Bianchi et al., 1998), although they likely need to be compiled and synthesized. As an example of possible analyses, the ratio of watershed area to estuarine surface area could be readily developed for these 34 estuarine systems. This simple ratio could provide a first cut at suggesting system sensitivity to inputs from drainage basins. To further improve this simple analysis, the water residence time could be added to the analysis, as limnologists have successfully done for decades in lake studies (e.g., Vollenweider, 1976). These data also appear to be available for many of the GoM estuaries (e.g., Bianchi et al., 1998). Other data, such as river flow rates, submerged aquatic vegetation coverage, and fisheries' harvests could also be used in a comparative analysis format. Outcomes from these analyses might suggest commonalities among estuaries, identify outliers, and suggest reasons for divergences. By using both pre-DWH and contemporary data, practitioners may start to see where and by how much restoration efforts are having an effect. A similar concept was employed by Allan et al. (2013) in the Great Lakes. This project produced a visual depiction of the entire Great Lakes system and demonstrated that joint spatial analysis of stressors and ecosystem services can provide a foundation to maximize benefits from restoration. These types of analyses, particularly those that are simple and largely descriptive, could be started now and serve as an initial effort at GoM-wide analyses, allowing practitioners to begin to see the GoM "as a whole."

and applied research are the central features of restoration success in Tampa Bay (Greening et al., 2014, Sherwood et al., 2016).

Finally, most monitoring indicators collected are "slice-in-time" measurements. Alone, they tell scientists and practitioners little about the underlying processes that control the state of these variables. Rate measurements, which capture changes in a variable as a function of time, are rare in monitoring programs for several reasons, including the increased cost per measurement and the greater time and expertise needed to make them. This is a challenge for ecological modeling of ecosystem restoration and species recovery, which relies on rate data (Buenau et al., 2014). Nevertheless, inclusion of rate measurements is warranted both for their useful monitoring metrics and, importantly, for their value in the calibration and verification of models and for advancing basic understanding of ecosystem performance. Without rate measurements, models cannot adequately predict rates of future change, a critically important tool for restoration managers. New sensors and methods discussed previously provide opportunities for reducing the cost of gathering rate measurements and expanding the geographic area to scales such as estuaries and watersheds, at which syntheses useful for cumulative effects assessment may be conducted.

There is an opportunity for research coupled with monitoring programs in areas related to threshold responses, tropicalization, and other "ecosystem unknowns," all of which could improve the understanding of restoration projects and their interactions, and thus, overall GoM restoration success:[44] this point was made in presentations to the committee by Kirk Rogers, USGS (September 29, 2020); Barry Keim, Louisiana State University (September 29, 2020); Michael Osland, U.S. Geological Survey (November 9, 2020); and Gregory Steyer, U.S. Geological Survey (November 10, 2020). New technology-driven monitoring, modeling, and visualization techniques represent opportunities that, coupled with traditional field-based monitoring, could generate a comprehensive long-term database for GoM, facilitating current and future GoM synthesis activities.

Lack of Entity(ies) Charged with Synthesizing, Assessing, and Reporting Progress on Larger Scales

Synthesis efforts have not been a common or high-priority activity in the Gulf region for several reasons. In many cases, data are just becoming available and thus synthesis has been limited by data availability.[45] Additionally, there is little incentive for synthesis by investigators working at the single-project level. Finally, cumulative effects evaluations for the entire near-shore Gulf system would involve multistate cooperation and funding support. Without concerted synthesis efforts, it seems unlikely that the degree to which Gulf habitats and ecosystems will be in better condition after very large amounts of restoration funding can be quantitatively determined. In the words of one presenter to the committee, states need to show that the Gulf-wide "restoration success needle" has moved in a positive direction.[46]

Effective Monitoring to Evaluate Ecological Restoration in the Gulf of Mexico (NASEM, 2017) notes that there is a clear need to create a collaborative multidisciplinary synthesis effort dedicated to Gulf-wide restoration issues involving cumulative effects quantification of restoration efforts, impacts of large-scale trends in system stressors on restoration projects, and better understanding the effects of both chronic and acute stressors on restoration success. The GoM is not without examples of synthesis activities that go far toward meeting some of the above goals. The water quality modeling work by Fennel and Laurent (2018) synthesized an enormous amount of hydrodynamic, nutrient loading, water quality, and other data to forecast the likely spatial response of the hypoxic zone in the northern Gulf to various nutrient load reduction schemes. USGS work with Mississippi River diversions aimed at wetland restoration is at large spatial scales and has shown cumulative effects gains in Louisiana.[47] Additional synthesis work as part of these studies could examine changes in food web and sediment nutrient dynamics as hypoxia is reduced or increased carbon sequestration and biogeochemical buffering of nutrient loads as waters flood over newly created wetland habitats. Estimation of a diverse set of cumulative effects is therefore possible and desirable at Gulf-wide scales.

Creation of a Gulf-wide synthesis effort and network for evaluation/forecasting and evaluating restoration scenarios has, to date, been a difficult task. One possible route to creating such an effort could involve a federal presence. For example, the EPA plays a central role in the large-scale, multistate Chesapeake Bay Program, which has increased coordination among states with air, land, and water monitoring; research; regulations; and synthesis efforts toward reaching restoration goals. In another effort, NSF has funded several synthesis centers that bring together and support working groups devoted to solving complex environmental problems, often using and analyzing diverse monitoring datasets. The creation of a synthesis center in the GoM would need to include incentives to initiate such an effort, and would need to include existing Gulf-wide entities (e.g., USGS, NOAA, EPA, RESTORE Centers of Excellence, Gulf of Mexico Alliance, the Gulf Research Program). A diagram showing the components needed to support data collation and synthesis at larger-than-project scales while utilizing an AM approach is shown in Figure 5.2.

---

[44] See https://www.nationalacademies.org/event/08-04-2020/webinar-restoring-the-gulf-after-deepwater-horizon-perspective-from-the-front-lines.

[45] Presentations to the committee from Jonathan Porthouse, National Fish and Wildlife Foundation, and Buck Sutter, Gulf Coast Ecosystem Restoration Council, in presentations to the committee (August 12, 2020).

[46] Presentations to the committee from Paul Mickle, Mississippi Department of Marine Resources (August 12, 2020).

[47] Presentation to the committee, Gregory Steyer, U.S. Geological Survey (November 10, 2020).

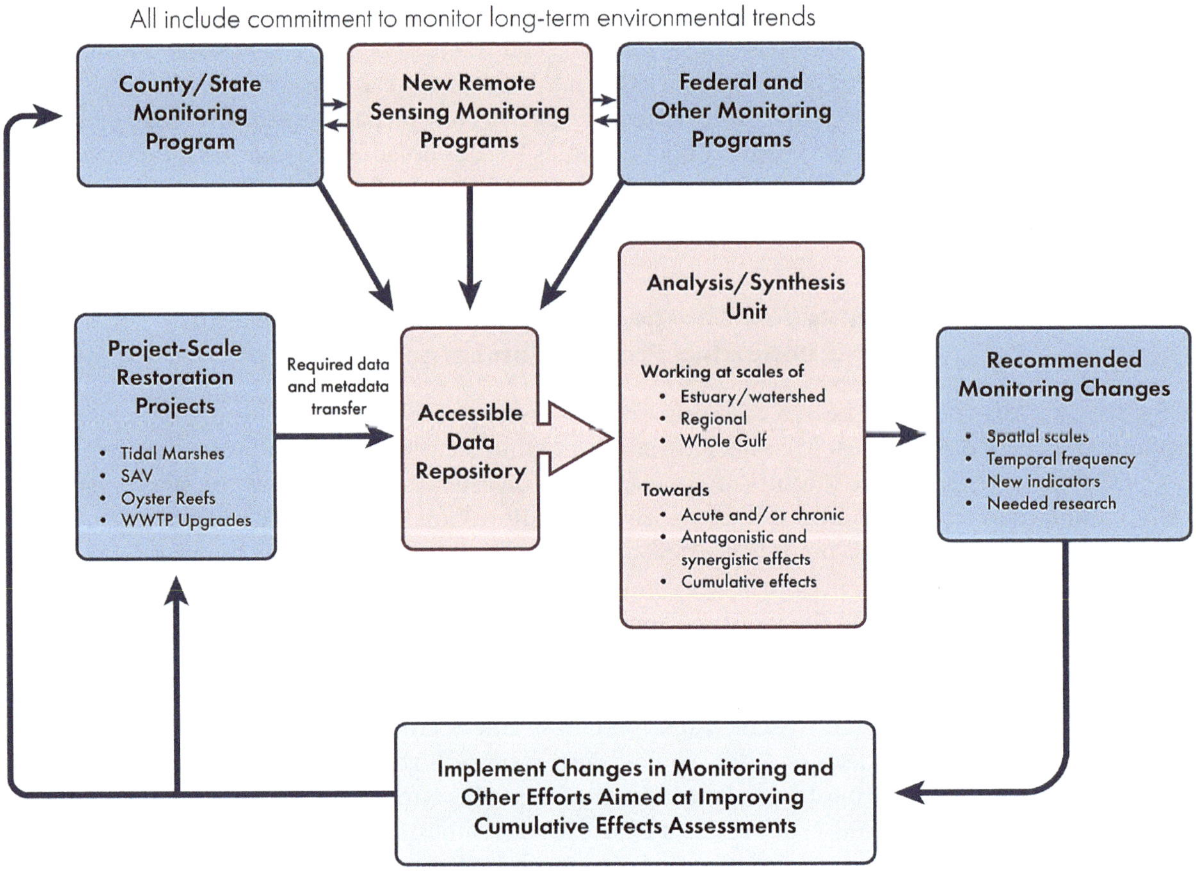

Figure 5.2. A schematic diagram of the components needed to "scale up" project-level restoration activities to regional and Gulf-wide spatial scales using an adaptive management approach. The boxes shown in red highlight key components that are currently weak or missing from the GoM, including: an accessible data repository—several exist, but data need to be added or linked; establishment of a GoM-wide restoration analysis/synthesis unit working at larger spatial scales to develop cumulative effects estimates and understanding; and better integration of county, state, federal, nongovernmental, non-traditional (including community scientists and crowdsourcing), and other monitoring/research program datasets with new remote sensing monitoring capabilities.

FINAL THOUGHTS

Significant funds have been expended or committed to date on *Deepwater Horizon*-funded projects. Progress has been made in advancing monitoring and modeling capabilities over the last 10 years (*Deepwater Horizon* NRDA Trustees 2017a, 2019; NOAA and USGS, 2019), and offer opportunities for further advancement. The emergence of new technologies that enhance monitoring and enable collection and processing of substantially greater amounts of data are occurring across disciplines, but have not yet typically been incorporated into restoration projects. Long-term environmental trends for some parameters and species are now available at local and regional scales, but data collection, analysis, and reporting are often inconsistent and existing efforts are not adequate to potentially detect Gulf-wide trends, and mechanisms to continue these efforts beyond the eventual sunset of *Deepwater Horizon* restoration programs have not been identified. Application of multiple lines of evidence to assess cumulative effects has been implemented in some long-term science-based management programs initiated prior to the *Deepwater Horizon* oil spill, but challenges remain in the development of the critical analysis and synthesis of the cumulative effects of *Deepwater Horizon* projects to date. Assessment of cumulative impacts—additive, synergistic, and possibly antagonistic effects—of multiple restoration projects of similar or diverse nature over spatial and temporal scales beyond that of an individual project is generally lacking. Development of synthesis capacity that can support an adaptive management process to integrate diverse restoration projects over significant spatial scales is recognized as needed by the *Deepwater Horizon* funding entities, but has not yet been initiated to a large degree. An underlying theme of this report is the need for integration of science and management of restoration activities. It is envisioned that each *Deepwater Horizon* funding entity, within its programmatic authority, can work cooperatively with others to realize this integration.

The committee recognizes the challenges faced by the Gulf Coast environmental restoration community, including recovery from the *Deepwater Horizon* oil spill, as well as the continued impacts of multiple hurricanes and other climatic events, and it applauds the progress made to date on recovery and restoration efforts. The assessment of the largest ecological restoration investment in history is an unprecedented challenge and opportunity. Learning achieved through the remainder of the settlement period will be the foundation for the next generation of managers, who inherit the responsibility for GoM ecosystems and communities. The following report conclusions and recommendations are provided to assist in supporting successful restoration efforts now and in the future.

CONCLUSIONS AND RECOMMENDATIONS

Conclusion 1: Adequate scientific evidence needed to evaluate cumulative effects of restoration on a regional scale in the Gulf of Mexico (GoM) is currently not available and, to date, no entity has been tasked to develop and implement a strategy to assess cumulative effects of environmental restoration efforts. Environmental benefits associated with multiple restoration projects have been observed within some GoM estuaries and watersheds, although not at larger scales. Without a focused effort and strategy, rigorous synthesis of the effects of multiple restoration projects at a regional or Gulf-wide scale cannot be conducted.

Conclusion 2: Because environmental changes can influence the success or failure of restoration efforts and can hinder the ability to detect potential cumulative effects of multiple restoration efforts, a thorough understanding of long-term environmental trends is essential for decision makers and restoration practitioners. Advanced monitoring techniques and approaches, including satellite remote sensing, connected sensor networks, and automation, can greatly assist in determining long-term environmental trends and assessing acute events. In addition, long-term environmental trends derived from targeted monitoring efforts can inform a range of analytical tools. The output from these tools can support the development of adaptive management actions that will subsequently improve restoration success and protect existing investments. Significant spatial and temporal gaps in monitoring GoM-wide environmental indicators and data collection and dissemination efforts limit development of this important and valuable capability. Specifically:

- Long-term environmental trends across the Gulf Coast states are monitored by a patchwork of government agencies, nonprofit organizations, and industries for a variety of reasons (e.g., regulatory, environmental tracking, performance evaluation). Study designs, data collection methods, analyses, and data availability vary and are often not comparable, making synthesis very difficult.
- One-time Gulf-wide monitoring studies are useful, but without periodic updates, do not generate enough information to determine long-term background trends needed for cumulative effects assessments.
- Key metrics to assess landscape-scale changes and support adaptive management include those necessary to estimate environmental trends associated with climate change; freshwater, nutrient, and sediment loading to coastal waters; land use/land cover; ambient water quality; status and trends of finfish and shellfish species, marine mammals, turtles, and birds; and primary and secondary production. For example, enhanced efforts and standardization of methods are needed for:
    - ambient water quality, the measurement of which can be enhanced by using high-spatio-temporal-resolution satellite data on chlorophyll a, suspended sediments, colored organic matter, and harmful algae;

- tide gage data and subsidence measurements to estimate local, relative sea level rise;
- estimating the extent and effects of ocean and coastal acidification, information essential for successful restoration and maintenance of commercially important shellfish and in the selection of materials for restoration efforts (e.g., oyster shells, limestone);
- tracking, targeting needed research, and managing the effects of tropicalization on fishery species, other species, and habitats; and
- estimating the ecological functioning of restored habitats, something not often measured even though the spatial extent of restored habitats is usually monitored and reported.

Recommendation A: Enhanced, consistent, and sustained long-term monitoring, analysis, synthesis, and reporting of environmental trends and indicators are urgently needed to enable the detection and tracking of cumulative effects of multiple restoration projects. Monitoring efforts should focus on developing the lines of evidence to support the assessment of cumulative effects at estuarine, regional, and larger scales. The *Deepwater Horizon* funding entities should immediately evaluate methods, identify funding mechanisms, and charge an entity to lead efforts to coordinate and enhance long-term priority monitoring efforts and promote consistent data collection, analysis, synthesis, and reporting among programs; support periodic assessments of collected data; assess the use of advanced techniques; and ensure data availability, with the goal of implementing these changes within 3–5 years.

Conclusion 3: The Gulf Coast environmental restoration community (federal agencies, states, nongovernmental organizations, and local public and private entities) has an opportunity to incorporate what has been learned from past and ongoing ecosystem restoration to inform future projects and programs supported by the remaining *Deepwater Horizon* funds. However, unless data and information from existing projects are made accessible and identification of information needed to assess cumulative effects of restoration efforts is undertaken more expeditiously, opportunities to improve the likelihood of success in the many projects remaining to be implemented will be greatly reduced or even permanently lost. Although it may be too early to fully assess cumulative effects of *Deepwater Horizon*-funded restoration efforts because of lag times between implementation and detection of effects, applying "lessons learned" from existing restoration efforts can help mitigate future risks of failure and ensure that *Deepwater Horizon* funds are invested wisely to increase the likelihood of meaningful and long-term Gulf of Mexico recovery and resilience.

Recommendation B: Restoration funding entities should adopt guidance to ensure that, as soon as they are available, all data, reports, and other project-specific information are deposited into freely accessible repositories that follow FAIR (Findable, Accessible, Interoperable, Reusable) principles. The *Deepwater Horizon* funding entities should identify and allocate resources to ensure that these data repositories remain functional throughout the life of each program, and additional support (as needed) should be sought to maintain data access in the future.

Recommendation C: The *Deepwater Horizon* funding entities should expedite the issuance of guidance for adaptive management and cumulative effects assessment at the programmatic scale for *Deepwater Horizon*-funded large-scale and multiple-restoration efforts. Guidance should include consistency in monitoring criteria that facilitate cumulative effects assessments.

Recommendation D: The *Deepwater Horizon* funding entities should immediately initiate a synthesis of available information from *Deepwater Horizon*-funded projects to assess characteristics of successful and unsuccessful restoration efforts. Results should be utilized in designing and implementing effective large-scale restoration projects within geographic areas of concern, and/or adjusting restoration approaches and techniques with the remaining funds from the *Deepwater Horizon* settlement.

Conclusion 4: Natural and anthropogenic drivers create multiple ecosystem pressures and stressors that act on restoration efforts over broad spatial scales, ranging from individual projects to entire ecosystems. The cumulative impacts of these pressures and stressors are often complex, resulting in synergistic and antagonistic effects of ecological significance. However, synergistic and antagonistic effects of large-scale restoration efforts in the Gulf of Mexico have not been assessed to date, and results from a limited number of assessments are mixed.

Recommendation E: The *Deepwater Horizon* funding entities should evaluate mechanisms that support cross-state and Gulf-wide collaboration among researchers, resource managers, and practitioners, with an objective to design and implement restoration efforts that allow assessment of antagonistic and synergistic effects.

Conclusion 5: The use of multiple lines of evidence to develop a framework to help assess cumulative effects for large-scale restoration efforts in the Gulf of Mexico has been proposed and, in some cases, applied. Assessment of cumulative effects of large-scale restoration is a recent research area and work on applying this research to restoration implementation is needed.

Conclusion 6: Opportunities exist now to prepare for the assessment of cumulative effects and restoration success from existing regional or large-scale restoration efforts in the Gulf of Mexico. These include:

- applying methods to assess functional equivalency between restored and natural sites;
- assessing the degree of environmental stress from natural and anthropogenic sources;
- applying a multiple-lines-of-evidence approach to assess cumulative effects at the estuary or watershed scale in preparation for Gulf-wide efforts;
- undertaking comparative analysis of estuaries or watersheds across the Gulf of Mexico to develop a greater understanding of similarities and differences among these systems; and
- evaluating expected benefits of a restoration effort as compared to a future condition without the effort.

These opportunities will involve consideration of changing environmental trends and a commitment to monitor, analyze, synthesize, and report results.

Recommendation F: To take advantage of the unprecedented opportunity to assess cumulative effects and inform restoration efforts ongoing and planned in the Gulf of Mexico, the *Deepwater Horizon* funding entities should evaluate and implement mechanisms necessary to address priority research needs and support efforts assess cumulative effects within the next 3–5 years. Mechanisms could include providing explicit responsibility to and support for existing Gulf-wide entities; development of an independent, regional, multidisciplinary, multiagency team; or a distribution of effort between existing entities.

Recommendation G: As additional monitoring data and scientific evidence become available, *Deepwater Horizon* program managers should continue to collaboratively develop and implement an adaptive management strategy for the Gulf of Mexico restoration effort, including the development of ecosystem conceptual models. Evaluation of priority issues should use the best available tools and methods, focus on progress of cumulative effects assessments and restoration objectives, and identify necessary changes to restoration approaches if needed. Mechanisms to continue these efforts beyond the eventual sunset of *Deepwater Horizon* restoration programs should be identified and implemented.

References

Achete, F., M. van der Wegen, J.A. Roelvink, and B. Jaffe. 2017. How can climate change and engineered water conveyance affect sediment dynamics in the San Francisco Bay-Delta system? *Climatic Change* 142(3):375-389. DOI: https://doi.org/10.1007/s10584-017-1954-8.

Adams, M.P., R.K. Hovey, M.R. Hipsey, L.C. Bruce, M. Ghisalberti, R.J. Lowe, R.K. Gruber, L. Ruiz-Montoya, P.S. Maxwell, D.P. Callaghan, G.A. Kendrick, and K.R. O'Brien. 2016. Feedback between sediment and light for seagrass: Where is it important? *Limnology and Oceanography* 61(6):1937-1955. DOI: https://doi.org/10.1002/lno.10319.

ADCNR, Alabama MRD, and NOAA (Alabama Department of Conservation and Natural Resources, Marine Resources Division, and the National Oceanic and Atmospheric Administration). 2021. *Coastal Alabama comprehensive oyster restoration strategy*. Deepwater Horizon Alabama Trustee Implementation Group. https://www.gulfspillrestoration.noaa.gov/sites/default/files/2021-12%20AL%20Final%20Coastal%20Alabama%20Comprehensive%20Oyster%20Restoration%20Strategy_508.pdf.

Alabama Gulf Coast Recovery Council. 2019. *Alabama state expenditure plan*. Submitted Pursuant to the Spill Impact Component of the RESTORE Act 33 U.S.C. § 1321(t)(3). https://restorethegulf.gov/sites/default/files/ALABAMA%20SEP%20-%20FINAL_508_4_1_19_0.pdf.

Allan, J.D. 2004. Landscapes and riverscapes: The influence of land use on stream ecosystems. *Annual Review of Ecology, Evolution, and Systematics* 35(1):257-284. DOI: https://doi.org/10.1146/annurev.ecolsys.35.120202.110122.

Allan, J.D., P.B. McIntyre, S.D. Smith, B.S. Halpern, G.L. Boyer, A. Buchsbaum, G.A. Burton, Jr., L.M. Campbell, W.L. Chadderton, J.J. Ciborowski, P.J. Doran, T. Eder, D.M. Infante, L.B. Johnson, C.A. Joseph, A.L. Marino, A. Prusevich, J.G. Read, J.B. Rose, E.S. Rutherford, S.P. Sowa, and A.D. Steinman. 2013. Joint analysis of stressors and ecosystem services to enhance restoration effectiveness. *Proceedings of the National Academy of Sciences of the United States of America* 110(1):372-377. DOI: https://doi.org/10.1073/pnas.1213841110.

Allison, M.A., C.R. Demas, B.A. Ebersole, B.A. Kleiss, C.D. Little, E.A. Meselhe, N.J. Powell, T.C. Pratt, and B.M. Vosburg. 2012. A water and sediment budget for the lower Mississippi-Atchafalaya River in flood years 2008-2010: Implications for sediment discharge to the oceans and coastal restoration in Louisiana. *Journal of Hydrology* 432-433:84-97. DOI: https://doi.org/10.1016/j.jhydrol.2012.02.020.

Allison, M.A., D.S. Biedenharn, T.A. Dahl, B. Kleiss, and C.D. Little. 2017. *Suspended sediment loads and tributary inputs in the Mississippi River below St. Louis, MO, 1990-2013 compared with earlier results.* Presented at American Geophysical Union, Fall Meeting, December 1, 2017.

Alpert, P., H. Messer, and N. David. 2016. Meteorology: Mobile networks aid weather monitoring. *Nature* 537(7622):617. DOI: https://doi.org/10.1038/537617e.

Altieri, A.H., and K.B. Gedan. 2014. Climate change and dead zones. *Global Change Biology* 21(4):1395-1406. DOI: https://doi.org/10.1111/gcb.12754.

Anderson, D.M., A.D. Cembella, and G.M. Hallegraeff. 2012. Progress in understanding harmful algal blooms: Paradigm shifts and new technologies for research, monitoring, and management. *Annual Review of Marine Science* 4:143-176. DOI: https://doi.org/10.1146/annurev-marine-120308-081121.

Anderson, J.B., D.J. Wallace, A.R. Simms, A.B. Rodriguez, R.W.R. Weight, and Z.P. Taha. 2016. Recycling sediments between source and sink during a eustatic cycle: Systems of late quaternary northwestern Gulf of Mexico Basin. *Earth-Science Reviews* 153:111-138. DOI: https://doi.org/10.1016/j.earscirev.2015.10.014.

Anderson, D.M., E. Fensin, C.J. Gobler, A.E. Hoeglund, K.A. Hubbard, D.M. Kulis, J.H. Landsberg, K.A. Lefebvre, P. Provoost, M.L. Richlen, J.L. Smith, A.R. Solow, and V.L. Trainer. 2021. Marine harmful algal blooms (HABs) in the United States: History, current status and future trends. *Harmful Algae* 102. DOI: https://doi.org/10.1016/j.hal.2021.101975.

Arfken, A., B. Song, J.S. Bowman, and M. Piehler. 2017. Denitrification potential of the eastern oyster microbiome using a 16S rRNA gene-based metabolic inference approach. *PLoS One* 12(9):e0185071. DOI: https://doi.org/10.1371/journal.pone.0185071

Argus, D.F., M. Shirzaei, and W.R. Peltier. 2018. *Subsidence along the Gulf and Atlantic coast of the United States exacerbates ocean inundation of the land produced by sea level rise.* Presented at American Geophysical Union Fall Meeting, December 1, 2018. https://ui.adsabs.harvard.edu/abs/2018AGUFM.G43B0711A

Armitage, A.R. 2021. Perspectives on maximizing coastal wetland restoration outcomes in anthropogenically altered ecosystems. *Estuaries and Coasts* 44:1699-1709. DOI: https://doi.org/10.1007/s12237-021-00907-4.

Armitage, A.R., W.E. Highfield, S.D. Brody, and P. Louchouarn. 2015. The contribution of mangrove expansion to salt marsh loss on the Texas Gulf Coast. *PLoS ONE* 10(5). DOI: https://doi.org/10.1371/journal.pone.0125404.

Aronson, R.B., and W.F. Precht. 2016. Physical and biological drivers of coral-reef dynamics. In *Coral reefs at the crossroads,* edited by D.K. Hubbard, C.S. Rogers, J.H. Lipps, and J.G.D. Stanley. Dordrecht: Springer Netherlands. Pp. 261-275.

Aronson, J., J.N. Blignaut, and T.B. Aronson. 2017. Conceptual frameworks and references for landscape-scale restoration: Reflecting back and looking forward. *Annals of the Missouri Botanical Garden* 102(2):188-200, 113.

Aronson, J., N. Goodwin, L. Orlando, C. Eisenberg, and A.T. Cross. 2020. A world of possibilities: Six restoration strategies to support the United Nation's decade on ecosystem restoration. *Restoration Ecology* 28(4):730-736. DOI: https://doi.org/10.1111/rec.13170.

ASM (American Society for Microbiology). 2020. Microbial Genomics of the Global Ocean System: Report on an American Academy of Microbiology (Academy), The American Geophysical Union (AGU), and The Gulf of Mexico Research Initiative (GoMRI) Colloquium held on 9 and 10 April 2019. Washington, DC: American Society for Microbiology. DOI: https://doi.org/10.1128/AAMCol.Apr.2019.

Babaeian, E., M. Sadeghi, S.B. Jones, C. Montzka, H. Vereecken, and M. Tuller. 2019. Ground, proximal, and satellite remote sensing of soil moisture. *Reviews of Geophysics* 57(2):530-616. DOI: https://doi.org/10.1029/2018rg000618.

Backer, L.C., J.H. Landsberg, M. Miller, K. Keel, and T.K. Taylor. 2013. Canine cyanotoxin poisonings in the United States (1920s–2012): Review of suspected and confirmed cases from three data sources. *Toxins* 5(9):1597-1628. DOI: https://doi.org/10.3390/toxins5091597.

Baggett, L.P., S.P. Powers, R.D. Brumbaugh, L.D. Coen, B.M. DeAngelis, J.K. Greene, B.T. Hancock, S.M. Morlock, B.L. Allen, D.L. Breitburg, D. Bushek, J.H. Grabowski, R.E. Grizzle, E.D. Grosholz, M.K. La Peyre, M.W. Luckenbach, K.A. McGraw, M.F. Piehler, S.R. Westby, and P.S.E. zu Ermgassen. 2015. Guidelines for evaluating performance of oyster habitat restoration. *Restoration Ecology* 23(6):737-745. DOI: https://doi.org/10.1111/rec.12262.

Bakker, D.C.E., B. Pfeil, C.S. Landa, N. Metzl, K.M. O'Brien, A. Olsen, K. Smith, C. Cosca, S. Harasawa, S.D. Jones, S.I. Nakaoka, Y. Nojiri, U. Schuster, T. Steinhoff, C. Sweeney, T. Takahashi, B. Tilbrook, C. Wada, R. Wanninkhof, S.R. Alin, C.F. Balestrini, L. Barbero, N.R. Bates, A.A. Bianchi, F. Bonou, J. Boutin, Y. Bozec, E.F. Burger, W.J. Cai, R.D. Castle, L. Chen, M. Chierici, K. Currie, W. Evans, C. Featherstone, R.A. Feely, A. Fransson, C. Goyet, N. Greenwood, L. Gregor, S. Hankin, N.J. Hardman-Mountford, J. Harlay, J. Hauck, M. Hoppema, M.P. Humphreys, C.W. Hunt, B. Huss, J.S.P. Ibánhez, T. Johannessen, R. Keeling, V. Kitidis, A. Körtzinger, A. Kozyr, E. Krasakopoulou, A. Kuwata, P. Landschützer, S.K. Lauvset, N. Lefèvre, C. Lo Monaco, A. Manke, J.T. Mathis, L. Merlivat, F.J. Millero, P.M.S. Monteiro, D.R. Munro, A. Murata, T. Newberger, A.M. Omar, T. Ono, K. Paterson, D. Pearce, D. Pierrot, L.L. Robbins, S. Saito, J. Salisbury, R. Schlitzer, B. Schneider, R. Schweitzer, R. Sieger, I. Skjelvan, K.F. Sullivan, S.C. Sutherland, A.J. Sutton, K. Tadokoro, M. Telszewski, M. Tuma, S.M.A.C. Van Heuven, D. Vandemark, B. Ward, A.J. Watson, and S. Xu. 2016. A multi-decade record of high-quality fCO_2 data in version 3 of the Surface Ocean CO_2 Atlas (SOCAT). *Earth System Science Data* 8(2):383-413. DOI: https://doi.org/10.5194/essd-8-383-2016.

Baldera, A., D.A. Hanson, and B. Kraft. 2018. Selecting indicators to monitor outcomes across projects and multiple restoration programs in the Gulf of Mexico. *Ecological Indicators* 89:559-571. DOI: https://doi.org/10.1016/j.ecolind.2018.01.025.

Barnett, K., and R.T. Belote. 2021. Modeling an aspirational connected network of protected areas across North America. *Ecological Applications* 31(6):e02387. DOI: https://doi.org/10.1002/eap.2387.

Barrett, N., J. Seiler, T. Anderson, S. Williams, S. Nichol, and S.N. Hill. 2010. *Autonomous Underwater Vehicle (AUV) for mapping marine biodiversity in coastal and shelf waters: Implications for marine management.* Presented at OCEAN 2010 IEEE-Sydney, Australia, May 24–27, 2010. DOI: https://doi.org/10.1109/OCEANSSYD.2010.5603860.

Bauer, J.E., W-J. Cai, P.A. Raymond, T.S. Bianchi, C.S. Hopkinson, P.A.G. Regnier. 2013. The changing carbon cycle of the coastal ocean. *Nature* 504: 61-70. DOI: https://doi.org/10.1038/nature12857.

Baumann, R.H., and R.E. Turner. 1990. Direct impacts of outer continental shelf activities on wetland loss in the central Gulf of Mexico. *Environmental Geology and Water Sciences* 15(3):189-198. DOI: https://doi.org/10.1007/BF01706410.

Baumann, M.S., G.F. Fricano, K. Fedeli, C.E. Schlemme, M.C. Christman, and M.V. Carle. 2020. Recovery of salt marsh invertebrates following habitat restoration: Implications for marsh restoration in the northern Gulf of Mexico. *Estuaries and Coasts* 43(7):1711-1721. DOI: https://doi.org/10.1007/s12237-018-0469-5.

Beck, M.W., R.D. Brumbaugh, L. Airoldi, A. Carranza, L.D. Coen, C. Crawford, O. Defeo, G.J. Edgar, B. Hancock, M.C. Kay, H.S. Lenihan, M.W. Luckenbach, C.L. Toropova, G. Zhang, and X. Guo. 2011. Oyster reefs at risk and recommendations for conservation, restoration, and management. *BioScience* 61(2):107-116. DOI: https://doi.org/10.1525/bio.2011.61.2.5.

Beck, M.W., E.T. Sherwood, J.R. Henkel, K. Dorans, K. Ireland, P. Varela, and SpringerLink. 2019. Assessment of the cumulative effects of restoration activities on water quality in Tampa Bay, Florida. *Estuaries and Coasts* 42:1774-1791. DOI: https://doi.org/10.1007/s12237-019-00619-w.

Beck, M.W., M. Burke, and G. Raulerson. 2020. *Tampa Bay water quality assessment.* Technical Report #01-20. St. Petersburg, Florida: Tampa Bay Estuary Program. https://drive.google.com/file/d/1Z_P3zahMFXSMyC7rW49MsjWP-jP_OqJF/view.

Beck, M., M. Burke, and E. Sherwood. 2021. *Tampa Bay estuary nutrient management compliance assessment.* Technical Report #06-21. St. Petersburg, Florida: Tampa Bay Estuary Program. https://tbep-tech.github.io/tbnmc-compliance-assessment-2020.

Beck, M.W., M. Burke, and G. Raulerson. 2022. *2021 Tampa Bay water quality assessment.* Technical Report #01-22. St. Petersburg, Florida: Tampa Bay Estuary Program.

Beerens, J., E. Noonburg, and D. Gawlik. 2015. Linking dynamic habitat selection with wading bird foraging distributions across resource gradients. *PLoS One* 10:e0128182. DOI: https://doi.org/10.1371/journal.pone.0128182.

Begam, M.M., T. Sutradhar, R. Chowdhury, C. Mukherjee, S.K. Basak, and K. Ray. 2017. Native salt-tolerant grass species for habitat restoration, their acclimation and contribution to improving edaphic conditions: A study from a degraded mangrove in the Indian Sundarbans. *Hydrobiologia* 803(1):373-387. DOI: https://doi.org/10.1007/s10750-017-3320-2.

Behrenfeld, M.J., E. Boss, D.A. Siegel, and D.M. Shea. 2005. Carbon-based ocean productivity and phytoplankton physiology from space. *Global Biogeochemical Cycles* 19(1). DOI: https://doi.org/10.1029/2004gb002299.

Behrenfeld, M.J., R.T. O'Malley, D.A. Siegel, C.R. McClain, J.L. Sarmiento, G.C. Feldman, A.J. Milligan, P.G. Falkowski, R.M. Letelier, and E.S. Boss. 2006. Climate-driven trends in contemporary ocean productivity. *Nature* 444(7120):752-755. DOI: https://doi.org/10.1038/nature05317.

Beijbom, O., P.J. Edmunds, C. Roelfsema, J. Smith, D.I. Kline, B.P. Neal, M.J. Dunlap, V. Moriarty, T.-Y. Fan, C.-J. Tan, S. Chan, T. Treibitz, A. Gamst, B.G. Mitchell, and D. Kriegman. 2015. Towards automated annotation of benthic survey images: Variability of human experts and operational modes of automation. *PLoS ONE* 10(7):e0130312. DOI: https://doi.org/10.1371/journal.pone.0130312.

Beisner, B.E., D.T. Haydon, and K. Cuddington. 2003. Alternative stable states in ecology. *Frontiers in Ecology and the Environment* 1(7):376-382. DOI: https://doi.org/10.1890/1540-9295(2003)001[0376:ASSIE]2.0.CO;2.

Berardelli, J. 2019. *How climate change is making hurricanes more dangerous*. Yale Climate Connections. https://yaleclimateconnections.org/2019/07/how-climate-change-is-making-hurricanes-more-dangerous/.

Bergstrom, E., J. Silva, C. Martins, and P. Horta. 2019. Seagrass can mitigate negative ocean acidification effects on calcifying algae. *Scientific Reports* 9(1):1932. DOI: https://doi.org/10.1038/s41598-018-35670-3.

Berkström, C., M. Gullström, R. Lindborg, A.W. Mwandya, S.A.S. Yahya, N. Kautsky, and M. Nyström. 2012. Exploring "knowns" and "unknowns" in tropical seascape connectivity with insights from East African coral reefs. *Estuarine, Coastal and Shelf Science* 107:1-21. DOI: https://doi.org/10.1016/j.ecss.2012.03.020.

Bernhardt, E.S., J.R. Blaszczak, C.D. Ficken, M.L. Fork, K.E. Kaiser, and E.C. Seybold. 2017. Control points in ecosystems: Moving beyond the hot spot hot moment concept. *Ecosystems* 20(4):665-682. DOI: https://doi.org/10.1007/s10021-016-0103-y.

Bertram, C., M. Quaas, T.B.H. Reusch, A.T. Vafeidis, C. Wolff, and W. Rickels. 2021. The blue carbon wealth of nations. *Nature Climate Change* 11(8):704-709. DOI: https://doi.org/10.1038/s41558-021-01089-4.

Bianchette, T.A., K.B. Liu, Y. Qiang, and N.S.N. Lam. 2016. Wetland accretion rates along coastal Louisiana: Spatial and temporal variability in light of hurricane Isaac's impacts. *Water (Switzerland)* 8(1). DOI: https://doi.org/10.3390/w8010001.

Bianchi, T.S., J.R. Pennock, and R.R. Twilley (Eds.). 1998. *Biogeochemistry of Gulf of Mexico estuaries*. Hoboken, NJ: Wiley Publishing Company. https://www.wiley.com/en-us/Biogeochemistry+of+Gulf+of+Mexico+Estuaries-p-9780471161745.

Biasutti, M., A.H. Sobel, S.J. Camargo, and T.T. Creyts. 2012. Projected changes in the physical climate of the Gulf Coast and Caribbean. *Climatic Change* 112(3-4):819-845. DOI: https://doi.org/10.1007/s10584-011-0254-y.

Biber, E. 2013. The challenge of collecting and using environmental monitoring data. *Ecology and Society* 18(4). DOI: https://doi.org/10.5751/ES-06117-180468.

Blake, E.S., C.W. Landsea, and E.J. Gibley. 2011. *The Deadliest, Costliest, and ost Intense United States Tropical Cyclones from 1851 to 2010 (and Other Frequently Requested Hurricane Facts). NOAA Technical Memorandum NWS NHC-6.* Miami, FL: U.S. Department of Commerce. 47 p.

Blum, M.D., and H.H. Roberts. 2009. Drowning of the Mississippi Delta due to insufficient sediment supply and global sea-level rise. *Nature Geoscience* 2(7):488-491. DOI: https://doi.org/10.1038/ngeo553.

Boddula, V., L. Ramaswamy, and D. Mishra. 2017. CyanoSense: A wireless remote sensor system using Raspberry-Pi and Arduino with application to algal bloom. In *2017 IEEE International Conference on AI & Mobile Services (AIMS)*: Honolulu, HI: IEEE. Pp. 85-88.

Boesch, D.F. 2020. *Envisioning the future of the Louisiana Gulf Coast.* Walton Family Foundation. https://8ce82b94a8c4fdc3ea6d-b1d233e3bc3cb10858bea65ff05e18f2.ssl.cf2.rackcdn.com/e0/46/6238d4b84d298ae415dc93a27b0e/envisioning-future-of-louisiana-coast-report-03-09-20.pdf.

Boothroyd, R.J., R.D. Williams, T.B. Hoey, B. Barrett, and O.A. Prasojo. 2020. Applications of Google Earth Engine in fluvial geomorphology for detecting river channel change. *WIREs Water* 8(1). DOI: https://doi.org/10.1002/wat2.1496.

Borja, A., I. Galparsoro, O. Solaun, I. Muxika, E.M. Tello, A . Uriarte, and V. Valencia. 2006. The European Water Framework Directive and the DPSIR, a methodological approach to assess the risk of failing to achieve good ecological status. *Estuarine Coastal and Shelf Science.* 66:84-96. DOI: https://doi.org/10.1016/j.csr.2005.05.004.

Bossart, G.D. 2010. Marine mammals as sentinel species for oceans and human health. *Veterinary Pathology* 48(3):676-690. DOI: https://doi.org/10.1177/0300985810388525.

Boström, C., S.J. Pittman, C. Simenstad, and R.T. Kneib. 2011. Seascape ecology of coastal biogenic habitats: advances, gaps, and challenges. *Marine Ecology Progress Series* 427:191-217. DOI: https://doi.org/10.3354/meps09051.

Boyer, J.N., C.R. Kelble, P.B. Ortner, and D.T. Rudnick. 2009. Phytoplankton bloom status: Chlorophyll a biomass as an indicator of water quality condition in the southern estuaries of Florida, USA. *Ecological Indicators* 9(6 SUPPL.):S56-S67. DOI: https://doi.org/10.1016/j.ecolind.2008.11.013.

Boynton, W.R., and W.M. Kemp. 2008. Estuaries. In *Nitrogen in the marine environment* (2nd ed.), edited by D.G. Capone, D.A. Bronk, M.R. Mulholland and E.J. Carpenter. San Diego, CA: Academic Press.

Boynton, W.R., G.H. Garber, R. Summers, and W.M. Kemp. 1995. Inputs, transformations, and transport of nitrogen and phosphorus in Chesapeake Bay and selected tributaries. *Estuaries* 18(18):285-314. DOI: https://doi.org/10.2307/1352640.

Bradley, P., and S. Yee. 2015. *Using the DPSIR framework to develop a conceptual model: Technical support document.* Washington, DC: U.S. Environmental Protection Agency.

Bradley, P., B. Jessup, S.J. Pittman, C.F.G. Jeffrey, and others. 2020. Development of a reef fish biological condition gradient model with quantitative decision rules for the protection and restoration of coral reef ecosystems. *Marine Pollution Bulletin.* 159. DOI: https://doi.org/10.1016/j.marpolbul.2020.111387.

Brady, D.C., J.M. Testa, D.M. Di Toro, W.R. Boynton, and W.M. Kemp. 2013. Sediment flux modeling: Calibration and application for coastal systems. *Estuarine, Coastal and Shelf Science* 117:107-124. DOI: https://doi.org/10.1016/j.ecss.2012.11.003.

Breitburg, D.L., and G.F. Riedel. 2005. Multiple stressors in marine systems. *Marine conservation biology: The science of maintaining the sea's biodiversity.* Marine Conservation Biology Institute. Pp. 167-182. https://repository.si.edu/bitstream/handle/10088/25003/serc_Breitburg_MarineConservationBiology_Chapter10.pdf?sequence=1&isAllowed=y.

Bricker, S.B., C.G. Clement, D.E. Pirhalla, S.P. Orlando, and D.R.G. Farrow. 1999. *National estuarine eutrophication assessment: Effects of nutrient enrichment in the nation's estuaries.* Silver Spring, MD: National Oceanic and Atmospheric Administration, National Ocean Service, Special Projects Office and the National Centers for Coastal Ocean Science. Pp. 1-328.

Bricker, S.B., B. Longstaff, W. Dennison, A. Jones, K. Boicourt, C. Wicks, and J. Woerner. 2008. Effects of nutrient enrichment in the nation's estuaries: A decade of change. *Harmful Algae* 8(1):21-32. DOI: https://doi.org/10.1016/j.hal.2008.08.028.

Brinson, M.M. 1993. *A hydrogeomorphic classification for wetlands.* Technical Report WRP-DE-4. Vicksburg, MS: U.S. Army Engineering Waterways Experiment Station.

Brown, C., K. Andrews, J. Brenner, J.W. Tunnell, C. Canfield, C. Dorsett, M. Driscoll, E. Johnson, and S. Kaderka. 2011. *Strategy for restoring the Gulf of Mexico (a cooperative NGO report).* Arlington, VA: The Nature Conservancy.

Brown, G.L., and K.C. Pevey. 2019. *Hydrodynamic, salinity, and morphological modeling study of a sediment diversion: An application of the adaptive hydraulics model/SEDLIB sediment transport library.* New Orleans, LA: U.S. Army Corps of Engineers.

Brudvig, L.A., and C.P. Catano. 2021. Prediction and uncertainty in restoration science. *Restoration Ecology* e13380. DOI: https://doi.org/10.1111/rec.13380.

Buenau, K.E., T.L. Hiller, and A.J. Tyre. 2014. Modelling the effects of river flow on population dynamics of piping plovers (Charadrius melodus) and least terns (Sternula antillarum) nesting on the Missouri River. *River Research and Applications* 30(8):964-975.

Burchard, H., K. Bolding, W. Kühn, A. Meister, T. Neumann, and L. Umlauf. 2006. Description of a flexible and extendable physical–biogeochemical model system for the water column. *Journal of Marine Systems* 61(3):180-211. DOI: https://doi.org/10.1016/j.jmarsys.2005.04.011.

Burden, A., A. Garbutt, and C.D. Evans. 2019. Effect of restoration on saltmarsh carbon accumulation in Eastern England. *Biology Letters* 15(1):20180773. DOI: https://doi.org/10.1098/rsbl.2018.0773.

Burger, J. 2017. Avian resources of the northern Gulf of Mexico. *Habitats and biota of the Gulf of Mexico: Before the Deepwater Horizon oil spill*, edited by C. Ward. New York: Springer. DOI: https://doi.org/10.1007/978-1-4939-3456-0_4.

Burke, K.L. 2018. Scientists in the wake of the hurricanes. *American Scientist* 106(2):69. DOI: https://doi.org/10.1511/2018.106.2.69.

Cahoon, D.R., D.J. Reed, J.W. Day, J.C. Lynch, A. Swales, and R.R. Lane. 2020. Applications and utility of the surface elevation table–marker horizon method for measuring wetland elevation and shallow soil subsidence-expansion. *Geo-Marine Letters* 40(5):809-815. DOI: https://doi.org/10.1007/s00367-020-00656-6.

Cai, W.J., X. Hu, W.J. Huang, M.C. Murrell, J.C. Lehrter, S.E. Lohrenz, W.C. Chou, W. Zhai, J.T. Hollibaugh, Y. Wang, P. Zhao, X. Guo, K. Gundersen, M. Dai, and G.C. Gong. 2011. Acidification of subsurface coastal waters enhanced by eutrophication. *Nature Geoscience* 4(11):766-770. DOI: https://doi.org/10.1038/ngeo1297.

Carl Kraft, B., and R. Crandall. 2019. A framework for considering climate change impacts in project selection for Deepwater Horizon restoration efforts. *Wetlands* 40(4):893-899. DOI: https://doi.org/10.1007/s13157-019-01226-y.

Campbell, A.D., and Y. Wang. 2020. Salt marsh monitoring along the mid-Atlantic coast by Google Earth Engine enabled time series. *PLoS One* 15(2):e0229605. DOI: https://doi.org/10.1371/journal.pone.0229605.

Carle, M.V., K.G. Benson, and J.F. Reinhardt. 2020. Quantifying the benefits of estuarine habitat restoration in the Gulf of Mexico: An introduction to the theme section. *Estuaries and Coasts* 43(7):1680-1691. DOI: https://doi.org/10.1007/s12237-020-00807-z.

Carlisle, D.M., D.M. Wolock, and M.R. Meador. 2011. Alteration of streamflow magnitudes and potential ecological consequences: A multiregional assessment. *Frontiers in Ecology and the Environment* 9(5):264-270. DOI: https://doi.org/10.1890/100053.

Carmichael, R.H., W.M. Graham, A. Aven, G. Worthy, S. Howden. 2012. Were multiple stressors a "perfect storm" for northern Gulf of Mexico bottlenose dolphins (*Tursiops truncatus*) in 2011? *PLoS ONE* 7(7): e41155. DOI: https://doi.org/10.1371/journal.pone.0041155.

Carpenter, S.R. 1993. Statistical analysis of the ecosystem experiments. In *The Trophic Cascade in Lakes (Cambridge Studies in Ecology)*, edited by S.R. Carpenter and J.F. Kitchell. Cambridge, UK: Cambridge University Press.

Carpenter, S.R., T.M. Frost, D. Heisey, and T.K. Kratz. 1989. Randomized intervention analysis and the interpretation of whole-ecosystem experiments. *Ecology* 70(4):1142-1152. DOI: https://doi.org/10.2307/1941382.

Carr, J., P. D'Odorico, K. McGlathery, and P. Wiberg. 2010. Stability and bistability of seagrass ecosystems in shallow coastal lagoons: Role of feedbacks with sediment resuspension and light attenuation. *Journal of Geophysical Research: Biogeosciences* 115(3). DOI: https://doi.org/10.1029/2009JG001103.

Carstensen, J., and C.M. Duarte. 2019. Drivers of pH variability in coastal ecosystems. *Environmental Science and Technology* 53(8):4020-4029. DOI: https://doi.org/10.1021/acs.est.8b03655.

Castañeda-Moya, E., V.H. Rivera-Monroy, R.M. Chambers, X. Zhao, L. Lamb-Wotton, A. Gorsky, E.E. Gaiser, T.G. Troxler, J.S. Kominoski, and M. Hiatt. 2020. Hurricanes fertilize mangrove forests in the Gulf of Mexico (Florida Everglades, USA). *Proceedings of the National Academy of Sciences of the United States of America* 117(9):4831-4841. DOI: 10.1073/pnas.1908597117.

Cerco, C., and T. Cole. 1993. Three-dimensional eutrophication model of Chesapeake Bay. *Journal of Environmental Engineering* 119:1006-1025. DOI: https://doi.org/10.1061/(ASCE)0733-9372(1993)119:6(1006).

Ceriani, S.A., P. Casale, M. Brost, E.H. Leone, and B.E. Witherington. 2019. Conservation implications of sea turtle nesting trends: Elusive recovery of a globally important loggerhead population. *Ecosphere* 10(11). DOI: https://doi.org/10.1002/ecs2.2936.

CERP (Comprehensive Everglades Restoration Plan). 2014. System status report. https://usace.contentdm.oclc.org/utils/getfile/collection/p16021coll7/id/8694.

Chagaris, D.D., W.F. Patterson, and M.S. Allen. 2020. Relative effects of multiple stressors on reef food webs in the northern Gulf of Mexico revealed via ecosystem modeling. *Frontiers in Marine Science* 7(513). DOI: https://doi.org/10.3389/fmars.2020.00513.

Chambers, R.M. 2021. Comparison of nutrient accrual in constructed living shoreline and natural fringing marshes. *Ocean & Coastal Management* 199:105401-102021. DOI: https://doi.org/10.1016/j.ocecoaman.2020.105401.

Chassot, E., S. Bonhommeau, N.K. Dulvy, F. Melin, R. Watson, D. Gascuel, and O. Le Pape. 2010. Global marine primary production constrains fisheries catches. *Ecology Letters* 13(4):495-505. DOI: https://doi.org/10.1111/j.1461-0248.2010.01443.x.

Chellew, M. 2007. *Spatial relationships for vegetation in restored and reference salt marshes in the Salmon River Estuary, Oregon*. Oregon State University. https://ir.library.oregonstate.edu/concern/graduate_thesis_or_dissertations/5425kg625.

Chen, S., and C. Hu. 2017. Estimating sea surface salinity in the northern Gulf of Mexico from satellite ocean color measurements. *Remote Sensing of Environment* 201:115-132. DOI: https://doi.org/10.1016/j.rse.2017.09.004.

Chmura, G.L., S.C. Anisfeld, D.R. Cahoon, and J.C. Lynch. 2003. Global carbon sequestration in tidal, saline wetland soils. *Global Biogeochemical Cycles* 17(4). DOI: https://doi.org/10.1029/2002gb001917.

Choice, Z.D., T.K. Frazer, and C.A. Jacoby. 2014. Light requirements of seagrasses determined from historical records of light attenuation along the Gulf coast of peninsular Florida. *Marine Pollution Bulletin* 81(1):94-102. DOI: https://doi.org/10.1016/j.marpolbul.2014.02.015.

Christensen, J.D., M.E. Monaco., and T.A. Lowery. 1997. An index to assess the sensitivity of Gulf of Mexico species to changes in estuarine salinity regimes. *Gulf and Caribbean Research* 9(4):219-229.

Cicchetti, G., and H. Greening. 2011. Estuarine biotope mosaics and habitat management goals: An application in Tampa Bay, FL, USA. *Estuaries and Coasts* 34(6):1278-1292.

Cicchetti, G., M. Pelletier, M. Pryor, S. Jackson, P. Bradley, S. Davies, C. Deacutis, K. Rocha, D. Santavy, and E. Shumchenia. 2017. *Implementing the biological condition gradient framework for management of estuaries and coasts*. Washington, DC: EPA Office of Research and Development. EPA/600/R-15/287.

Cigliano, J.A., R. Meyer, H.L. Ballard, A. Freitag, T.B. Phillips, and A. Wasser. 2015. Making marine and coastal citizen science matter. *Ocean & Coastal Management* 115:77-87. DOI: https://doi.org/10.1016/j.ocecoaman.2015.06.012.

Clark, J.S., and A.E. Gelfand. 2006. A future for models and data in environmental science. *Trends in Ecology & Evolution* 21(7):375-380. DOI: https://doi.org/10.1016/j.tree.2006.03.016.

Clewell, A.F., and J. Aronson. 2013. *Ecological restoration, second edition: Principles, values, and structure of an emerging profession (the science and practice of ecological restoration series)*. Washington, DC: Island Press.

Cloern, J.E., and A.D. Jassby. 2010. Patterns and scales of phytoplankton variability in estuarine–coastal ecosystems. *Estuaries and Coasts* 33(2):230-241. DOI: https://doi.org/10.1007/s12237-009-9195-3.

Clune, J.W., P.D. Capel, M.P. Miller, D.A. Burns, A.J. Sekellick, P.R. Claggett, R.H. Coupe, R.M. Fanelli, A.M. Garcia, J.P. Raffensperger, S. Terziotti, G. Bhatt, J.D. Blomquist, K.G. Hopkins, J.L. Keisman, L.C. Linker, G.W. Shenk, R.A. Smith, A.M. Soroka, J.S. Webber, D.M. Wolock, and Q. Zhang. 2021. *Nitrogen in the Chesapeake Bay watershed—A century of change, 1950–2050.* Reston, VA: U.S. Geological Survey. Pp. 168.

Coen, L.D., and M.W. Luckenbach. 2000. Developing success criteria and goals for evaluating oyster reef restoration: Ecological function or resource exploitation? *Ecological Engineering* 15(3-4):323-343. DOI: https://doi.org/10.1016/S0925-8574(00)00084-7.

Cole, A.M., M.J. Durako, and M.O. Hall. 2018. Multivariate analysis of water quality and benthic macrophyte communities in Florida Bay, USA reveals hurricane effects and susceptibility to seagrass die-off. *Frontiers in Plant Science* 9(630). DOI: https://doi.org/10.3389/fpls.2018.00630.

Comeaux, R.S., M.A. Allison, and T.S. Bianchi. 2012. Mangrove expansion in the Gulf of Mexico with climate change: Implications for wetland health and resistance to rising sea levels. *Estuarine, Coastal and Shelf Science* 96:81-95. DOI: https://doi.org/10.1016/j.ecss.2011.10.003.

Congdon, V.M., K.H. Dunton, J. Brenner, K.L. Goodin, and K.W. Ames. 2018. Ecological resilience indicators for seagrass ecosystems. In: *Ecological resilience indicators for five northern Gulf of Mexico ecosystems*. Arlington, VA: NatureServe. https://www.natureserve.org/sites/default/files/chapter_4_-_seagrass_indicators.pdf.

Conner, W.H., J.W. Day, Jr., R.H. Baumann, and J.M. Randall. 1989. Influence of hurricanes on coastal ecosystems along the northern Gulf of Mexico. *Wetlands Ecology and Management* 1(1):45-56. DOI: https://doi.org/10.1007/BF00177889.

Conrad, C.C., and K.G. Hilchey. 2011. A review of citizen science and community-based environmental monitoring: issues and opportunities. *Environmental Monitoring and Assessment* 176(1):273-291. DOI: https://doi.org/10.1007/s10661-010-1582-5.

Cooke, S.J., A.M. Rous, L.A. Donaldson, J.J. Taylor, T. Rytwinski, K.A. Prior, K.E. Smokorowski, and J.R. Bennett. 2018. Evidence-based restoration in the Anthropocene—From acting with purpose to acting for impact. *Restoration Ecology* 26(2):201-205. DOI: https://doi.org/10.1111/rec.12675.

Cooke, S.J., J.R. Bennett, and H.P. Jones. 2019. We have a long way to go if we want to realize the promise of the "Decade on Ecosystem Restoration." *Conservation Science and Practice* 1(12):e129. DOI: https://doi.org/10.1111/csp2.129.

Corbane, C., S. Lang, K. Pipkins, S. Alleaume, M. Deshayes, V.E. García Millán, T. Strasser, J. Vanden Borre, S. Toon, and F. Michael. 2015. Remote sensing for mapping natural habitats and their conservation status—New opportunities and challenges. *International Journal of Applied Earth Observations and Geoinformation* 37:7-16. DOI: https://doi.org/10.1016/j.jag.2014.11.005.

Corbellini, S., E. Di Francia, S. Grassini, L. Iannucci, L. Lombardo, and M. Parvis. 2018. Cloud based sensor network for environmental monitoring. *Measurement* 118:354-361. DOI: https://doi.org/10.1016/j.measurement.2017.09.049.

Côté, I.M., E.S. Darling, and C.J. Brown. 2016. Interactions among ecosystem stressors and their importance in conservation. *Proceedings of the Royal Society B: Biological Sciences* 283(1824):2015-2592. DOI: https://doi.org/0.1098/rspb.2015.2592.

CEQ (Council on Environmental Quality). 1997. Considering Cumulative Effects Under the National Environmental Policy Act. Washington, District of Columbia, USA. https://ceq.doe.gov/docs/ceq-publications/ccenepa/exec.pdf.

Couvillion, B.R. 2021. *Coastal wetland area change in the Gulf of Mexico, 1985-2020.* U.S. Geological Survey data release. DOI: https://doi.org/10.5066/P9ZQI7ZW.

Couvillion, B.R., H. Beck, D. Schoolmaster, and M. Fischer. 2017. Land area change in coastal Louisiana (1932 to 2016). *Scientific investigations map.* DOI: https://doi.org/10.3133/sim3381.

Couvillion, B.R., N.K. Ganju, and Z. Defne. 2021. *An unvegetated to vegetated ratio (UVVR) for coastal wetlands of the conterminous United States (2014-2018).* U.S. Geological Survey data release. DOI: https://doi.org/10.5066/P97DQXZP.

Cowardin, L.M., V. Carter, F.C. Golet, and E.T. LaRoe. 1979. *Classification of wetlands and deepwater habitats of the United States*. U.S. Department of the Interior, Fish and Wildlife Service. https://www.fws.gov/wetlands/documents/classification-of-wetlands-and-deepwater-habitats-of-the-united-states.pdf.

CPRA (Coastal Protection and Restoration Authority). 2017. *Louisiana's comprehensive master plan for a sustainable coast*. Baton Rouge, LA: Coastal Protection and Restoration Authority of Louisiana. https://coastal.la.gov/our-plan/2017-coastal-master-plan.

Crain, C.M., K. Kroeker, and B.S. Halpern. 2008. Interactive and cumulative effects of multiple human stressors in marine systems. *Ecology Letters* 11(12):1304-1315. DOI: https://doi.org/10.1111/j.1461-0248.2008.01253.x.

Crisci, C., B. Ghattas, and G. Perera. 2012. A review of supervised machine learning algorithms and their applications to ecological data. *Ecological Modelling* 240:113-122. DOI: https://doi.org/10.1016/j.ecolmodel.2012.03.001.

Dahl, T.E. 2011. *Status and trends of wetlands in the conterminous United States 2004 to 2009.* Washington, DC: U.S. Department of the Interior, Fish and Wildlife Service.

Dahl, T.E., and S.M. Stedman. 2013. *Status and trends of wetlands in the coastal watersheds of the conterminous United States 2004 to 2009.* Washington, DC: U.S. Department of the Interior, Fish and Wildlife Service; and National Oceanic and Atmospheric Administration, National Marine Fisheries Service.

Daniels, M.B., A. Sharpley, R.D. Harmel, and K. Anderson. 2018. The utilization of edge-of-field monitoring of agricultural runoff in addressing nonpoint source pollution. *Journal of Soil and Water Conservation* 73(1):1-8. DOI: https://doi.org/10.2489/jswc.73.1.1.

Daoust, R., T. Doss, M. Gorman, M. Harwell, and C. Ulrich. 2014. A 10-year ecosystem restoration community of practice tracks large-scale restoration trends. *Sapiens* 7.

Darling, E.S., and I.M. Côté. 2008. Quantifying the evidence for ecological synergies. *Ecology Letters* 11(12):1278-1286. DOI: https://doi.org/10.1111/j.1461-0248.2008.01243.x.

Davies, S.P., and S.K. Jackson. 2006. The biological condition gradient: A descriptive model for interpreting change in aquatic ecosystems. *Ecological Applications* 16(4):1251-1266. DOI: https://doi.org/10.1890/1051-0761(2006)016[1251:tbcgad]2.0.co;2.

Davis, R.A., Jr. 2016. *Barrier islands of the Florida Gulf Coast peninsula.* Lanham, MD: Rowman & Littlefield.

Davis, R.A. 2017. Sediments of the Gulf of Mexico. In *Habitats and biota of the Gulf of Mexico: Before the Deepwater Horizon Oil Spill* (Vol. 1), edited by C.H. Ward. New York: Springer New York. Pp. 165-215.

Day, J., J. Cable, J. Cowan, R. Delaune, K. de Mutsert, B. Fry, H. Mashriqui, D. Justic, P. Kemp, R. Lane, J. Rick, S. Rick, L.P. Rozas, G. Snedden, E. Swenson, R. Twilley, and B. Wissel. 2009. The impacts of pulsed reintroduction of river water on a Mississippi Delta Coastal Basin. *Journal of Coastal Research* 54:225-243. DOI: https://doi.org/10.2112/SI54-015.1.

Day, R H., S.T.A., J. Brenner, K. Goodin, D. Faber-Langendoen, K.W. Ames. 2018. *Ecological resilience indicators for mangrove ecosystems*. Arlington, VA: NatureServe.

Day, J.W., W.H. Conner, R.D. DeLaune, C.S. Hopkinson, R.G. Hunter, G.P. Shaffer, D. Kandalepas, R.F. Keim, G.P. Kemp, R.R. Lane, V.H. Rivera-Monroy, C.E. Sasser, J.R. White, and I.A. Vargas-Lopez. 2021. A review of 50 years of study of hydrology, wetland dynamics, aquatic metabolism, water quality and trophic status, and nutrient biogeochemistry in the barataria basin, Mississippi Delta-system functioning, human impacts and restoration approaches. *Water (Switzerland)* 13(5). DOI: https://doi.org/10.3390/w13050642.

de Groot, R.S., R. Alkemade, L. Braat, L. Hein, and L. Willemen. 2010. Challenges in integrating the concept of ecosystem services and values in landscape planning, management and decision making. *Ecological Complexity* 7(3):260-272. DOI: https://doi.org/10.1016/j.ecocom.2009.10.006.

de Mutsert, K., K.A. Lewis, E.D. White, and J. Buszowski. 2021. End-to-end modeling reveals species-specific effects of large-scale coastal restoration on living resources facing climate change. *Frontiers in Marine Science* 8(104). DOI: https://doi.org/10.3389/fmars.2021.624532.

de Sherbinin, A., A. Bowser, T.-R. Chuang, C. Cooper, F. Danielsen, R. Edmunds, P. Elias, E. Faustman, C. Hultquist, R. Mondardini, I. Popescu, A. Shonowo, and K. Sivakumar. 2021. The critical importance of citizen science data. *Frontiers in Climate* 3(20). DOI: https://doi.org/10.3389/fclim.2021.650760.

Dee, S.G., M.A. Torres, R.C. Martindale, A. Weiss, and K.L. DeLong. 2019. The future of reef ecosystems in the Gulf of Mexico: Insights from coupled climate model simulations and ancient hot-house reefs. *Frontiers in Marine Science* 6. DOI: https://doi.org/10.3389/fmars.2019.00691.

Deegan, L.A., J.W. Day, J.G. Gosselink, A. Yáñez-Arancibia, G.S. Chávez, and P. Sánchez-Gil. 1986. Relationships among physical characteristics, vegetation distribution and fisheries yield in Gulf of Mexico Estuaries. In *Estuarine variability*. Wolfe, DA: Academic Press.

Dell'Apa, A., J.P. Kilborn, and W.J. Harford. 2020. Advancing ecosystem management strategies for the Gulf of Mexico's fisheries resources: Implications for the development of a fishery ecosystem plan. *Bulletin of Marine Science* 96:617-640. https://doi.org/10.5343/bms.2019.0081.

D'Elia, C. F., W. R. Boynton, and J. G. Sanders. 2003. A watershed perspective on nutrient enrichment, science, and policy in the Patuxent River, Maryland: 1960-2000. *Estuaries* 26(2):171-185. DOI: https://doi.org/10.1007/bf02695960.

DeMaria, M., and J. Kaplan. 1994. Sea surface temperature and the maximum intensity of Atlantic tropical cyclones. *Journal of Climate* 7(9):1324-1334. DOI: https://doi.org/10.1175/1520-0442(1994)007<1324:Sstatm>2.0.Co;2.

DeMaso, S., M. Brasher, and J. Gleason. 2019. GoMAMN strategic bird monitoring guidelines: Waterfowl. In *Mississippi Agricultural and Forestry Experiment Station Research Bulletin*, edited by R.R. Wilson, A.M.V. Fournier, J.S. Gleason, J.E. Lyons, and M.S. Woodrey. Mississippi State University. Pp. 229-274

Denman, K.L. 2003. Modelling planktonic ecosystems: Parameterizing complexity. *Progress in Oceanography* 57(3):429-452. DOI: https://doi.org/10.1016/S0079-6611(03)00109-5.

Detenbeck, N.E., M. You, and D. Torre. 2019. Recent Changes in Nitrogen Sources and Load Components to Estuaries of the Contiguous United States. *Estuaries and Coasts* 42(8):2096-2113. DOI: https://doi.org/10.1007/s12237-019-00614-1.

DeWitt, T.H., S.R. Pacella, C. Folger, P.M. Eldridge. 2011. *Comparison of the structure of lower and upper estuary food webs for Yaquina Bay (OR).* Presented at Astoria, OR, March 3–5, 2011.

Diamond, J., T. Chan, J. Austin, C. Dalbom, and M. Davis. 2014. *Funding Deep Water Horizon restoration & recovery: How much, going where, for what?* White paper. New Orleans, LA: Environmental Law Institute & Tulane Institute on Water Resources Law & Policy.

Diaz, R.J., and R. Rosenberg. 2008. Spreading dead zones and consequences for marine ecosystems. *Science* 321(5891):926-929. DOI: https://doi.org/10.1126/science.1156401.

Diefenderfer, H.L, R.M. Thom, and J.E. Adkins. 2003. *Systematic approach to coastal ecosystem restoration.* Prepared for National Oceanic and Atmospheric Administration Coastal Services Center. Contract EA1330-02-RQ-0029. Battelle Contract 4418. Richland, WA: Pacific Northwest Division.

Diefenderfer, H.L., R.M. Thom, and K. Hofseth. 2005. A Framework for Risk Analysis for Ecological Restoration Projects. In *U.S. Army Corps of Engineers in Economics and Ecological Risk Assessment: Applications to Watershed Management, Environmental and Ecological Risk Assessment Series (Vol. 5)*, edited by R.J.F. Bruins and M.T. Heberling. Boca Raton, FL: CRC Press. Pp 57-105.

Diefenderfer, H.L., K.L. Sobocinski, R.M. Thom, C.W. May, A.B. Borde, S.L. Southard, J. Vavrinec, and N.K. Sather. 2009. Multiscale analysis of restoration priorities for marine shoreline planning. *Environmental Management* 44(4):712-731. DOI: https://doi.org/10.1007/s00267-009-9298-4.

Diefenderfer, H.L., R.M. Thom, G.E. Johnson, J.R. Skalski, K.A. Vogt, B.D. Ebberts, G.C. Roegner, and E.M. Dawley. 2011. Levels-of-evidence approach for assessing cumulative ecosystem response to estuary and river restoration programs. *Ecological Restoration* 29(1-2):111-132. DOI: https://doi.org/10.3368/er.29.1-2.111.

Diefenderfer, H., G.E. Johnson, J. Skalski, S. Breithaupt, and A. Coleman. 2012. Application of the diminishing returns concept in the hydroecologic restoration of riverscapes. *Landscape Ecology* 27. DOI: https://doi.org/10.1007/s10980-012-9713-8.

Diefenderfer, H.L., G.E. Johnson, R.M. Thom, K.E. Buenau, L.A. Weitkamp, C.M. Woodley, A.B. Borde, and R.K. Kropp. 2016. Evidence-based evaluation of the cumulative effects of ecosystem restoration. *Ecosphere* 7(3):e01242. DOI: https://doi.org/10.1002/ecs2.1242.

Diefenderfer, H.L., G.D. Steyer, M.C. Harwell, A.J. LoSchiavo, H.A. Neckles, D.M. Burdick, G.E. Johnson, K.E. Buenau, E. Trujillo, J.C. Callaway, R.M. Thom, N.K. Ganju, and R.R. Twilley. 2021. Applying cumulative effects to strategically advance large-scale ecosystem restoration. *Frontiers in Ecology and the Environment* 19(2):108-117. DOI: https://doi.org/10.1002/fee.2274.

DiGennaro, B., D. Reed, C. Swanson, L. Hastings, Z. Hymanson, M. Healey, S. Siegel, S. Cantrell, and B. Herbold. 2012. Using conceptual models in ecosystem restoration decision making: An example from the Sacramento-San Joaquin River Delta, California. *San Francisco Estuary and Watershed Science* 10(3). DOI: https://doi.org/10.15447/sfews.2012v10iss3art1.

Dillon, K., M. Peterson, and C. May. 2015. Functional equivalence of constructed and natural intertidal eastern oyster reef habitats in a northern Gulf of Mexico estuary. *Marine Ecology Progress Series* 528:187-203. DOI: https://doi.org/10.3354/meps11269.

DiMarco, S.F., J. Strauss, N. May, R.L. Mullins-Perry, E.L. Grossman, and D. Shormann. 2012. Texas coastal hypoxia linked to Brazos River discharge as revealed by oxygen isotopes. *Aquatic Geochemistry* 18(2):159-181. DOI: https://doi.org/10.1007/s10498-011-9156-x.

Dixon, L.K. 2000. Establishing light requirements for the seagrass Thalassia testudinum: An example from Tampa Bay, Florida. In *Seagrasses: Monitoring, Ecology, Physiology, and Management*, edited by S.A. Bortone. Boca Raton, FL: CRC Press. Pp. 9-31.

Dokken, Q. 2011. IXTOC I versus Macondo well blowout: Anatomy of an oil spill event then and now. *International Oil Spill Conference Proceedings* 2011:abs420. DOI: https://doi.org/10.7901/2169-3358-2011-1-420.

Doney, S.C., D.S. Busch, S.R. Cooley, and K.J. Kroeker. 2020. The impacts of ocean acidification on marine ecosystems and reliant human communities. *Annual Review of Environment and Resources* 45:83-112.

Dorado, S., T. Booe, J. Steichen, A.S. McInnes, R. Windham, A. Shepard, A.E.B. Lucchese, H. Preischel, J.L. Pinckney, S.E. Davis, D.L. Roelke, and A. Quigg. 2015. Towards an understanding of the interactions between freshwater inflows and phytoplankton communities in a subtropical estuary in the Gulf of Mexico. *PLoS ONE* 10(7):e0130931. DOI: https://doi.org/10.1371/journal.pone.0130931.

Doren, R.F., J.H. Richards, and J.C. Volin. 2009. A conceptual ecological model to facilitate understanding the role of invasive species in large-scale ecosystem restoration. *Ecological Indicators* 9(6 SUPPL.):S150-S160. DOI: https://doi.org/10.1016/j.ecolind.2008.06.007.

Dorgham, M.M. 2014. Effects of Eutrophication. In *Eutrophication: Causes, Consequences and Control*, edited by A.A. Ansari, and S.S. Gill. Dordrecht: Springer Netherlands. DOI: https://doi.org/10.1007/978-94-007-7814-6_3

Dorward-King, E.J., G.W. II Suter, L.A. Kapustka, D.R. Mount, D.K. Reed-Judkins, S.M. Cormier, S.D. Dyer, M.G. Luxon, R. Parrish, and G.A. Jr. Burton. 2001. Distinguishing among factors that influence ecosystems. In *Ecological variability: Separating natural from anthropogenic causes of ecosystem impairment*, edited by D.J. Baird and G.A. Burton Jr. Pensacola, Florida: SETAC Press. Pp. 1–26.

Dou, Z., L. Cui, J. Li, Y. Zhu, C. Gao, X. Pan, Y. Lei, M. Zhang, X. Zhao, and W. Li. 2018. Hyperspectral estimation of the chlorophyll content in short-term and long-term restorations of mangrove in Quanzhou Bay estuary, China. *Sustainability* 10(4):1127. DOI: https://doi.org/10.3390/su10041127.

Dougherty, A., C. Harpold, and J. Clark. 2010. *Ecosystems and Fisheries-Oceanography Coordinated Investigations (EcoFOCI) field manual. AFSC Processed Rep. 2010-02.* Seattle, WA: Alaska Fisheries Science Center, National Oceanic and Atmospheric Administration, National Marine Fisheries Service.

Downes, B.J., L.A. Barmuta, P.G. Fairweather, D.P. Faith, M.J. Keough, P.S. Lake, B.D. Mapstone, and G.P. Quinn. 2002. *Monitoring ecological impacts: Concepts and practice in flowing waters.* Cambridge, UK: Cambridge University Press.

D'Sa, E.J., M. Korobkin, and D.S. Ko. 2011. Effects of Hurricane Ike on the Louisiana-Texas coast from satellite and model data. *Remote Sensing Letters* 2(1):11-19. DOI: https://doi.org/10.1080/01431161.2010.489057.

D'souza, N.A., A. Subramaniam, A.R. Juhl, M. Hafez, A. Chekalyuk, S. Phan, B. Yan, I.R. MacDonald, S.C. Weber, and J.P. Montoya. 2016. Elevated surface chlorophyll associated with natural oil seeps in the Gulf of Mexico. *Nature Geoscience* 9(3):215-218. DOI: https://doi.org/10.1038/ngeo2631.

Dunn, D.E. 1996. *Trends in nutrient inflows to the Gulf of Mexico from streams draining the conterminous United States.* 96:4113. Washington, DC: U.S. Department of the Interior, U.S. Geological Survey.

Duke, T., and W.L. Kruczynski. 1992. *Status and trends of emergent and submerged vegetated habitats, Gulf of Mexico, U.S.A.* Washington, DC: Environmental Protection Agency, EPA-800-R-92-003. Office of Water, Gulf of Mexico Program. Pp. 161.

DWH NRDA LA TIG (Deepwater Horizon Natural Resource Damage Assessment and Restoration Louisiana Trustee Implementation Group). 2021. *Monitoring and adaptive management strategy.* Baton Rouge, LA: Louisiana Trustee Implementation Group. https://la-dwh.com/wp-content/uploads/2021/09/MAM-strategy.pdf.

DWH NRDA Trustees. 2016. *Deepwater Horizon Oil Spill: Final programmatic damage assessment and restoration plan and final programmatic environmental impact statement.* Washington, DC: National Oceanic and Atmospheric Administration. https://www.gulfspillrestoration.noaa.gov/restoration-planning/gulf-plan.

DWH NRDA Trustees. 2017a. Monitoring and adaptive management strategy procedures and guidelines manual version 1.0. In *Trustee council standard operating procedures for implementation of the natural resource restoration for the DWH oil spill.* https://www.gulfspillrestoration.noaa.gov/sites/default/files/2018_01_TC_MAM_Procedures_Guidelines_Manual_12-2017_508_c.pdf.

DWH NRDA Trustees. 2017b. *Deepwater Horizon Oil Spill natural resource damage assessment: Strategic framework for oyster restoration activities.* http://www.gulfspillrestoration.noaa.gov/restoration-planning/gulf-plan.

DWH NRDA Trustees. 2017c. *Deepwater Horizon Oil Spill natural resource damage assessment: Strategic framework for marine mammal restoration activities.* http://www.gulfspillrestoration.noaa.gov/restoration-planning/gulf-plan.

DWH NRDA Trustees. 2017d. *Deepwater Horizon Oil Spill natural resource damage assessment: Strategic framework for sea turtle restoration activities.* http://www.gulfspillrestoration.noaa.gov/restoration-planning/gulf-plan.

DWH NRDA Trustees. 2019. *Monitoring and adaptive management procedures and guidelines manual, version 1.0. Originally released December 2017, updated August 2019.* https://www.gulfspillrestoration.noaa.gov/sites/default/files/2019-08%20MAM_Manual_FULL_Updated%202019.pdf

DWH NRDA Trustees. 2021. *Deepwater Horizon natural resource damage assessment 2021 programmatic review.* https://www.gulfspillrestoration.noaa.gov/sites/default/files/2021-11_Deepwater_Horizon_TC_Final_2021_Programmatic_Review.pdf.

DWH Open Ocean Trustee Implementation Group. 2020. *Open Ocean Trustee Implementation Group monitoring and adaptive management strategy.* http://www.gulfspillrestoration.noaa.gov/.

Dzwonkowski, B., S. Fournier, J.T. Reager, S. Milroy, K. Park, A.M. Shiller, A.T. Greer, I. Soto, S.L. Dykstra, and V. Sanial. 2018. Tracking sea surface salinity and dissolved oxygen on a river-influenced, seasonally stratified shelf, Mississippi Bight, northern Gulf of Mexico. *Continental Shelf Research* 169:25-33. DOI: https://doi.org/10.1016/j.csr.2018.09.009.

Earle, P.S., D.C. Bowden, and M.R. Guy. 2012. Twitter earthquake detection: Earthquake monitoring in a social world. *Annals of Geophysics* 54(6):708-715. DOI: https://doi.org/10.4401/ag-5364. DOI: 10.4401/ag-5364.

Ebberts, B.D., B.D. Zelinsky, J.P. Karnezis, C.A. Studebaker, S. Lopez-Johnston, A.M. Creason, L. Krasnow, G.E. Johnson, and R.M. Thom. 2018. Estuary ecosystem restoration: Implementing and institutionalizing adaptive management. *Restoration Ecology* 26(2):360-369.

Ebbets, A.L., D.R. Lane, and P. Dixon. 2020. Using meta-analysis to develop evidence-based recovery trajectories of vegetation and soils in restored wetlands in the northern Gulf of Mexico. *Estuaries and Coasts* 43:1692-1710. DOI: https://doi.org/10.1007/s12237-019-00536-y.

Edwards, A.M. 2001. Adding detritus to a nutrient–phytoplankto–zooplankton model: A dynamical-systems approach. *Journal of Plankton Research* 23:389-413. DOI: https://doi.org/10.1093/plankt/23.4.389.

Eger, A.M., E. Marzinelli, P. Gribben, C.R. Johnson, C. Layton, P.D. Steinberg, G. Wood, B.R. Silliman, and A. Vergés. 2020. Playing to the positives: Using synergies to enhance kelp forest restoration. *Frontiers in Marine Science* 7(544). DOI: https://doi.org/10.3389/fmars.2020.00544.

ELI (Environmental Law Institute). 2020. *Gulf Coast Recovery and Restoration: 10-Year Review*. http://eli-ocean.org/wp-content/blogs.dir/2/files/Gulf-Restoration-Recovery-10-Year-Review.pdf.

Ellison, A.M., A.J. Felson, and D.A. Friess. 2020. Mangrove rehabilitation and restoration as experimental adaptive management. *Frontiers in Marine Science* 7. DOI: https://doi.org/10.3389/fmars.2020.00327.

Elsey-Quirk, T., S.A. Graham, I.A. Mendelssohn, G. Snedden, J.W. Day, R.R. Twilley, G. Shaffer, L.A. Sharp, J. Pahl, and R.R. Lane. 2019. Mississippi River sediment diversions and coastal wetland sustainability: Synthesis of responses to freshwater, sediment, and nutrient inputs. *Estuarine, Coastal and Shelf Science* 221:170-183. DOI: https://doi.org/10.1016/j.ecss.2019.03.002.

Emilien, A.-V., C. Thomas, and H. Thomas. 2021. UAV & satellite synergies for optical remote sensing applications: A literature review. *Science of Remote Sensing* 3:100019. DOI: https://doi.org/10.1016/j.srs.2021.100019.

Engle, M.A., and B. Brunner. 2019. Considerations in the application of machine learning to aqueous geochemistry: Origin of produced waters in the northern U.S. Gulf Coast Basin. *Applied Computing and Geosciences* 3-4:100012. DOI: https://doi.org/10.1016/j.acags.2019.100012.

Engström-Öst, J., O. Glippa, R.A. Feely, M. Kanerva, J.E. Keister, S.R. Alin, B.R. Carter, A.K. McLaskey, K.A. Vuori, and N. Bednaršek. 2019. Eco-physiological responses of copepods and pteropods to ocean warming and acidification. *Scientific Reports* 9(1). DOI: https://doi.org/10.1038/s41598-019-41213-1.

Enwright, N.M., W.M. SooHoo, J.L. Dugas ,C.P. Conzelmann, C. Laurenzano, D.M. Lee, K. Mouton, and S.J. Stelly. 2020. *Louisiana Barrier Island Comprehensive Monitoring Program: Mapping habitats in beach, dune, and intertidal environments along the Louisiana Gulf of Mexico shoreline, 2008 and 2015–16.* Reston, VA: U.S. Geological Survey Open-File Report 2020–1030, 57 p., DOI: https://doi.org/10.3133/ofr20201030.

EPA (Environmental Protection Agency). 1999. *Ecological condition of estuaries in the Gulf of Mexico* (EPA 620-R-98- 004). Washington, DC: Environmental Protection Agency, Office of Research and Development, National Health and Environmental Effects Research Laboratory, Gulf Ecology Division, Gulf Breeze, Florida.

EPA. 2008. *Integrated modeling for integrated environmental decision making* (EPA 100/R-08/010). https://www.epa.gov/sites/default/files/2015-02/documents/im4iedm_white_paper_final_epa100r08010.pdf.

EPA. 2015. *National coastal condition assessment 2010* (EPA 841-R-15-006). Washington, DC: Office of Water and Office of Research and Development. Pp. 1-129.

EPA. 2016. *A practitioner's guide to the biological condition gradient: A framework to describe incremental change in aquatic ecosystems.* https://www.epa.gov/sites/default/files/2016-02/documents/bcg-practioners-guide-report.pdf.

EPA. 2021. *National coastal condition assessment: A collaborative survey of the nation's estuaries and Great Lakes nearshore waters* (EPA 841-R-21-00). Washington, DC: Environmental Protection Agency. Pp 1-87.

Erickson, K.A., J. West, M.A. Dance, T.M. Farmer, J.C. Ballenger, and S.R. Midway. 2021. Changing climate associated with the range-wide decline of an estuarine finfish. *Global Change Biology* 27(11):2520-2536. DOI: https://doi.org/10.1111/gcb.15568.

Eshleman, K.N., and R.D. Sabo. 2016. Declining nitrate-N yields in the Upper Potomac River Basin: What is really driving progress under the Chesapeake Bay restoration? *Atmospheric Environment* 146:280-289. DOI: https://doi.org/10.1016/j.atmosenv.2016.07.004.

Evans Ogden, L. 2020. Drones help bridge the gaps in assessing global change. *Eos* 101. DOI: https://doi.org/10.1029/2020EO148306.

Fagherazzi, S., D.M. FitzGerald, R.W. Fulweiler, Z. Hughes, P.L. Wiberg, K.J. McGlathery, J.T. Morris, T.J. Tolhurst, L.A. Deegan, and D.S. Johnson. 2013. Ecogeomorphology of salt marshes. In *Treatise on Geomorphology*, edited by J.F. Shroder. San Diego: Academic Press. DOI: https://doi.org/10.1016/B978-0-12-374739-6.00329-8.

Falcone, J.A., J.C. Murphy, and L.A. Sprague. 2019. Regional patterns of anthropogenic influences on streams and rivers in the conterminous United States, from the early 1970s to 2012. *Journal of Land Use Science* 13(6):585-614. DOI: https://doi.org/10.1080/1747423X.2019.1590473.

Farley, S.S., A. Dawson, S.J. Goring, and J.W. Williams. 2018. Situating ecology as a big-data science: Current advances, challenges, and solutions. *BioScience* 68(8):563-576. DOI: https://doi.org/10.1093/biosci/biy068.

Feher, L.C., M.J. Osland, K.T. Griffith, J.B. Grace, R.J. Howard, C.L. Stagg, N.M. Enwright, K.W. Krauss, C.A. Gabler, R.H. Day, and K. Rogers. 2017. Linear and nonlinear effects of temperature and precipitation on ecosystem properties in tidal saline wetlands. *Ecosphere* 8(10). DOI: https://doi.org/10.1002/ecs2.1956.

Felder, D.L., D.K. Camp, and J.W. Tunnell, Jr. 2009. An introduction to Gulf of Mexico biodiversity assessment. *Gulf of Mexico Origin, Waters, and Biota* 1:1-14.

Feller, I.C., D.A. Friess, K.W. Krauss, and R.R. Lewis III. 2017. The state of the world's mangroves in the 21st century under climate change. *Hydrobiologia* 803(1). DOI: https://doi.org/10.1007/s10750-017-3331-z.

Fennel, K., and A. Laurent. 2018. *N* and P as ultimate and proximate limiting nutrients in the northern Gulf of Mexico: Implications for hypoxia reduction strategies. *Biogeosciences,* 15(10):3121-3131. DOI: https://doi.org/10.5194/bg-15-3121-2018.

Fernández-Giménez, M.E., D.J. Augustine, L.M. Porensky, H. Wilmer, J.D. Derner, D.D. Briske, and M.O. Stewart. 2019. Complexity fosters learning in collaborative adaptive management. *Ecology and Society* 24(2). DOI: https://doi.org/10.5751/ES-10963-240229.

Filbee-Dexter, K., J. Pittman, H.A. Haig, S.M. Alexander, C.C. Symons, and M.J. Burke. 2017. Ecological surprise: Concept, synthesis, and social dimensions. *Ecosphere* 8(12):e02005. DOI: https://doi.org/10.1002/ecs2.2005.

Finn, C., P. Abbeel, and S. Levine. 2017. Model-agnostic meta-learning for fast adaptation of deep networks. *Proceedings of the 34th International Conference on Machine Learning*. Sydney, Australia. DOI: https://doi.org/10.48550/arXiv.1703.03400.

Fischenich, J.C. 2008. *The application of conceptual models to ecosystem restoration* (RDC/EBA TN-08-1). Vicksberg, MS: U.S. Army Corps of Engineers.

Fischenich, J.C., K.E. Buenau, J.L. Bonneau, C. Fleming, D.R. Marmorek, M. Nelitz, C.L. Murray, B. Ma, G. Long, and C.J. Schwarz. 2018. *Science and adaptive management plan: Missouri River recovery program.* Washington, DC: U.S. Army Corps of Engineers.

Fischer, J., M. Riechers, J. Loos, B. Martin-Lopez, and V.M. Temperton. 2021. Making the UN decade on ecosystem restoration a social-ecological endeavour. *Trends in Ecology & Evolution* 36(1):20-28. DOI: https://doi.org/10.1016/j.tree.2020.08.018.

Fitch, W.A., K.E. Kirby, J.J. Dragna, D.D. Kuchler, D.K. Haycraft, R.C. Godfrey, J.A. Langan, B.E. Fields, H. Karis, M.T. Regan, and R.C. Brock. 2013. BP and Anadarko's Phase 2 Pre-Trial Memorandum Quantification Segment. Document Submitted in the U.S. District Court for the Eastern District of Louisiana MDL No. 2179 Section J. In Re: Oil Spill by the Oil Rig "Deepwater Horizon" in the Gulf of Mexico, on April 20, 2010. Document 11266. Filed September 5, 2013. 14 p.

Fitzsimons, J.A., S. Branigan, C.L. Gillies, R.D. Brumbaugh, J. Cheng, B.M. DeAngelis, L. Geselbracht, B. Hancock, A. Jeffs, T. McDonald, I.M. McLeod, B. Pogoda, S.J. Theuerkauf, M. Thomas, S. Westby, and P.S.E. zu Ermgassen. 2020. *Restoring shellfish reefs: Global guidelines for practitioners and scientists.* Hoboken, NJ: Wiley.

Flynn, K.J. 2001. A mechanistic model for describing dynamic multi-nutrient, light, temperature interactions in phytoplankton. *Journal of Plankton Research* 23(9):977-997. DOI: https://doi.org/10.1093/plankt/23.9.977.

Fodrie, F.J., K.L. Heck, Jr., S.P. Powers, W. Graham, and K. Robinson. 2010. Climate-related, decadal-scale assemblage changes of seagrass-associated fishes in the northern Gulf of Mexico. *Global Change Biology* 16(1):48-59. DOI: https://doi.org/10.1111/j.1365-2486.2009.01889.x.

Foley, M.M., J.A. Warrick, A. Ritchie, A.W. Stevens, P.B. Shafroth, J.J. Duda, M.M. Beirne, R. Paradis, G. Gelfenbaum, R. McCoy, and E.S. Cubley. 2017. Coastal habitat and biological community response to dam removal on the Elwha River. *Ecological Monographs* 87(4):552-577. DOI: https://doi.org/10.1002/ecm.1268.

Forbrich, I., and A.E. Giblin. 2015. Marsh–atmosphere CO exchange in a New England salt marsh. *Journal of Geophysical Research: Biogeosciences* 120(9):1825-1838. DOI: https://doi.org/10.1002/2015jg003044.

Frasier, K.E., A. Solsona-Berga, L. Stokes, and J.A. Hildebrand. 2020. Impacts of the Deepwater Horizon oil spill on marine mammals and sea turtles. In *Deep oil spills: Facts, fate and effects*, edited by S.A. Murawski, C.H. Ainsworth, S. Gilbert, D.J. Hollander, C.B. Paris, M. Schlüter, and D.L. Wetzel. Berlin: Springer Nature. Pp. 431-462. DOI: https://doi.org/10.1007/978-3-030-11605-7.

Frederick, P., and C. Green. 2019. GoMAMN strategic bird monitoring guidelines: Wading birds. In *Strategic bird monitoring guidelines for the northern Gulf of Mexico*, edited by R.R. Wilson, A.M.V. Fournier, J.S. Gleason, J.E. Lyons, and M.S. Woodrey. Mississippi State University. Pp. 203-228. https://gomamn.org/wp-content/uploads/2020/02/chapter8-1.pdf.

Freedman, B. 2015. Ecological effects of environmental stressors. *Oxford research encyclopedia of environmental science*. DOI: https://doi.org/10.1093/acrefore/9780199389414.013.1.

Fry, E.L., E.S. Pilgrim, J.R.B. Tallowin, R.S. Smith, S.R. Mortimer, D.A. Beaumont, J. Simkin, S.J. Harris, R.S. Shiel, H. Quirk, K.A. Harrison, C.S. Lawson, P.J. Hobbs, and R.D. Bardgett. 2017. Plant, soil and microbial controls on grassland diversity restoration: A long-term, multi-site mesocosm experiment. *Journal of Applied Ecology* 54(5):1320-1330. DOI: https://doi.org/10.1111/1365-2664.12869.

Fujiwara, M., F. Martinez-Andrade, R.J.D. Wells, M. Fisher, M. Pawluk, and M.C. Livernois. 2019. Climate-related factors cause changes in the diversity of fish and invertebrates in subtropical coast of the Gulf of Mexico. *Communications Biology* 2(1). DOI: https://doi.org/10.1038/s42003-019-0650-9.

Fulton, E.A. 2010. Approaches to end-to-end ecosystem models. *Journal of Marine Systems* 81(1):171-183. DOI: https://doi.org/10.1016/j.jmarsys.2009.12.012.

GAO (U.S. Government Accountability Office). 2021. *Offshore oil and gas. Updated regulations needed to improve pipeline oversight and decommissioning.* GAO-21-293. https://www.gao.gov/assets/gao-21-293.pdf.

Ganju, N.K., Z. Defne, M.L. Kirwan, S. Fagherazzi, A. D'Alpaos, and L. Carniello. 2017. Spatially integrative metrics reveal hidden vulnerability of microtidal salt marshes. *Nature Communications* 8. DOI: https://doi.org/10.1038/ncomms14156.

Gann, G.D., T. McDonald, B. Walder, J. Aronson, C.R. Nelson, J. Jonson, J.G. Hallett, C. Eisenberg, M.R. Guariguata, J. Liu, F. Hua, C. Echeverría, E. Gonzales, N. Shaw, K. Decleer, and K.W. Dixon. 2019. International principles and standards for the practice of ecological restoration (2nd ed.). *Restoration Ecology* 27(S1):S1-S46. DOI: https://doi.org/10.1111/rec.13035.

GBEP (Galveston Bay Estuary Program). 2009. *Charting the course to 2015: Galveston Bay Strategic Action Plan.* https://www.tceq.texas.gov/assets/public/comm_exec/pubs/gi/gi-385.pdf.

GBEP. 2018. *The Galveston Bay Plan, 2nd edition: Comprehensive conservation and management plan for the galveston bay estuary.* https://gbep.texas.gov/wp-content/uploads/2019/08/CCMP_2ndEdition_FINAL-TCEQ-Approved-DRAFT.pdf.

GBNEP (Galveston Bay National Estuary Program). 1995. *The Galveston Bay Plan.* https://www.tceq.texas.gov/assets/public/comm_exec/pubs/gbnep/gbnep-49/index.html.

Geary, W.L., M. Bode, T.S. Doherty, E.A. Fulton, D.G. Nimmo, A.I.T. Tulloch, V.J.D. Tulloch, and E.G. Ritchie. 2020. A guide to ecosystem models and their environmental applications. *Nature Ecology & Evolution* 4(11):1459-1471. DOI: https://doi.org/10.1038/s41559-020-01298-8.

Geller, G., P.N. Halpin, B. Helmuth, A.K. Skidmore, M. Abrams, N. Aguirre, M. Blair, E. Botha, M. Colloff, T. Dawson, J. Franklin, N. Horning, C. James, W. Magnusson, M.J. Santos, S.R. Schill, and K. Williams. 2017. Remote sensing for biodiversity. In *The GEO Handbook on Biodiversity Observation Networks*, edited by M. Walters and R.J. Scholes. New York: Springer. Pp. 19-38.

Gentile, J.H., M.A. Harwell, W. Cropper, Jr., C.C. Harwell, D. DeAngelis, S. Davis, J.C. Ogden, and D. Lirman. 2001. Ecological conceptual models: A framework and case study on ecosystem management for South Florida sustainability. *Science of the Total Environment* 274(1):231-253. DOI: https://doi.org/10.1016/S0048-9697(01)00746-X.

Geraldi, N.R., S.P. Powers, K.L. Heck, and J. Cebrian. 2009. Can habitat restoration be redundant? Response of mobile fishes and crustaceans to oyster reef restoration in marsh tidal creeks. *Marine Ecology Progress Series* 389:171-180. DOI: https://doi.org/10.3354/meps08224.

Ghannam, R.B., and S.M. Techtmann. 2021. Machine learning applications in microbial ecology, human microbiome studies, and environmental monitoring. *Computational and Structural Biotechnology Journal* 19:1092-1107. DOI: https://doi.org/10.1016/j.csbj.2021.01.028.

Ghermandi, A., and M. Sinclair. 2019. Passive crowdsourcing of social media in environmental research: A systematic map. *Global Environmental Change* 55:36-47. DOI: https://doi.org/10.1016/j.gloenvcha.2019.02.003.

Ghosh, S., D.R. Mishra, and A.A. Gitelson. 2016. Long-term monitoring of biophysical characteristics of tidal wetlands in the northern Gulf of Mexico—A methodological approach using MODIS. *Remote Sensing of Environment* 173:39-58. DOI: https://doi.org/10.1016/j.rse.2015.11.015.

Gibbens, S. 2020. How Powerful hurricanes hasten the disappearance of Louisiana's wetlands. *National Geographic*. https://www.nationalgeographic.com/science/article/how-hurricane-laura-hastens-louisiana-wetland-loss.

Gil-Agudelo, D.L., C.E. Cintra-Buenrostro, J. Brenner, P. González-Díaz, W. Kiene, C. Lustic, and H. Pérez-España. 2020. Coral reefs in the Gulf of Mexico large marine ecosystem: Conservation status, challenges, and opportunities. *Frontiers in Marine Science* 6. DOI: https://doi.org/10.3389/fmars.2019.00807.

Gilbert, D., N.N. Rabalais, R.J. Díaz, and J. Zhang. 2010. Evidence for greater oxygen decline rates in the coastal ocean than in the open ocean. *Biogeosciences* 7(7):2283-2296. DOI: https://doi.org/10.5194/bg-7-2283-2010.

Gilby, B.L., A.D. Olds, C.H. Peterson, R.M. Connolly, C.M. Voss, M.J. Bishop, M. Elliott, J.H. Grabowski, N.L. Ortodossi, and T.A. Schlacher. 2018. Maximizing the benefits of oyster reef restoration for finfish and their fisheries. *Fish and Fisheries* 19(5):931-947. DOI: https://doi.org/10.1111/faf.12301.

Gilby, B.L., M.P. Weinstein, R. Baker, J. Cebrian, S.B. Alford, A. Chelsky, D. Colombano, R.M. Connolly, C.A. Currin, I.C. Feller, A. Frank, J.A. Goeke, L.A. Goodridge Gaines, F.E. Hardcastle, C.J. Henderson, C.W. Martin, A.E. McDonald, B.H. Morrison, A.D. Olds, J.S. Rehage, N.J. Waltham, and S.L. Ziegler. 2021. Human actions alter tidal marsh seascapes and the provision of ecosystem services. *Estuaries and Coasts* 44(6):1628-1636. DOI: https://doi.org/10.1007/s12237-020-00830-0.

Gittman, R.K., S.B. Scyphers, C.S. Smith, I.P. Neylan, J.H. Grabowski. 2016. Ecological consequences of shoreline hardening: A meta-analysis. *BioScience* 66(9):763-773. DOI: https://doi.org/10.1093/biosci/biw091.

Gledhill, D.K., R. Wanninkhof, F.J. Millero, and M. Eakin. 2008. Ocean acidification of the Greater Caribbean Region 1996-2006. *Journal of Geophysical Research: Oceans* 113(10). DOI: https://doi.org/10.1029/2007JC004629.

Gledhill, J.H., A.F. Barnett, M. Slattery, K.L. Willett, G.L. Easson, S.S. Otts, and D.J. Gochfeld. 2020. Mass mortality of the eastern oyster *Crassostrea virginica* in the western Mississippi Sound following unprecedented Mississippi River flooding in 2019. *Journal of Shellfish Research* 39(2):235-244. DOI: https://doi.org/10.2983/035.039.0205.

Glibert, P.M., J. Icarus Allen, Y. Artioli, A. Beusen, L. Bouwman, J. Harle, R. Holmes, and J. Holt. 2014. Vulnerability of coastal ecosystems to changes in harmful algal bloom distribution in response to climate change: Projections based on model analysis. *Global Change Biology* 20(12):3845-3858. DOI: https://doi.org/10.1111/gcb.12662.

Glick, P., J. Clough, A. Polaczyk, B. Couvillion, and B. Nunley. 2013. Potential effects of sea-level rise on coastal wetlands in Southeastern Louisiana. *Journal of Coastal Research* 63(SPL.ISSUE):211-233. DOI: https://doi.org/10.2112/SI63-0017.1.

Goldstein, E.B., and G. Coco. 2015. Machine learning components in deterministic models: Hybrid synergy in the age of data. *Frontiers in Environmental Science* 3(33). DOI: https://doi.org/10.3389/fenvs.2015.00033.

GOMA (Gulf of Mexico Alliance) Water Quality Team. 2013. *Gulf of Mexico water-quality monitoring: Providing water-quality information to support informed resource management and public knowledge.* https://sarasota.wateratlas.usf.edu/upload/documents/GOMA-Gulf-monitoring-white-paper-Final.pdf.

Gonzalez, V.M., F.A. Garcia-Moreno, J.A. Melby, N.C. Nadal-Caraballo, and E.S. Godsey. 2020. *Alabama barrier island restoration assessment life-cycle structure response modeling.* Coastal and Hydraulics Laboratory. ERDC/CHL TR-20-5. Vicksburg, MS: U.S. Army Engineer Research and Development Center.

Goodin, K.L., D. Faber-Langendoen, J. Brenner, S.T. Allen, R.H. Day, V.M. Congdon, C. Shepard, K.E. Cummings, C.L. Stagg, C.A. Gabler, M. Osland, K.H. Dunton, R.R. Ruzicka, K. Semon-Lunz, D. Reed, M. Love. 2018. Ecological resilience indicators for five northern Gulf of Mexico ecosystems. *NatureServe*. https://www.natureserve.org/publications/ecological-resilience-indicators-five-northern-gulf-mexico-ecosystems.

Gosselink, J.G., L.C. Lee, and Society of Wetland Scientists. 1989. *Cumulative impact assessment in bottomland hardwood forests*. Wilmington, NC: Society of Wetland Scientists.

Grabowski, J.H., A.R. Hughes, D.L. Kimbro, and M.A. Dolan. 2005. How habitat setting influences restored oyster reef communities. *Ecology* 86(7):1926-1935. DOI: https://doi.org/10.1890/04-0690.

Gray, A., C.A. Simenstad, D.L. Bottom, and T.J. Cornwell. 2002. Contrasting functional performance of juvenile salmon habitat in recovering wetlands of the Salmon River estuary, Oregon, U.S.A. *Restoration Ecology* 10(3):514-526. DOI: https://doi.org/10.1046/j.1526-100X.2002.01039.x.

Gredzens, C., and D.J. Shaver. 2020. Satellite tracking can inform population-level dispersal to foraging grounds of post-nesting Kemp's ridley sea turtles. *Frontiers in Marine Science* 7(559). DOI: https://doi.org/10.3389/fmars.2020.00559.

Greening, H., A. Janicki, E.T. Sherwood, R. Pribble, and J.O.R. Johansson. 2014. Ecosystem responses to long-term nutrient management in an urban estuary: Tampa Bay, Florida, USA. *Estuarine, Coastal and Shelf Science* 151:A1-A16. DOI: https://doi.org/10.1016/j.ecss.2014.10.003.

Greet, J., J. Angus Webb, and R.D. Cousen. 2011. The importance of seasonal flow timing for riparian vegetation dynamics: A systematic review using causal criteria analysis. *Freshwater Biology* 56(7):1231-1247. DOI: https://doi.org/10.1111/j.1365-2427.2011.02564.x.

Greiner, J.T., K.J. McGlathery, J. Gunnell, and B.A. McKee. 2013. Seagrass restoration enhances "blue carbon" sequestration in coastal waters. *PLoS ONE* 8(8):e72469. DOI: https://doi.org/10.1371/journal.pone.0072469.

Griffith, A.W., and C.J. Gobler. 2020. Harmful algal blooms: A climate change co-stressor in marine and freshwater ecosystems. *Harmful Algae* 91:101590. DOI: https://doi.org/10.1016/j.hal.2019.03.008.

Groffman, P.M., C.T. Driscoll, G.E. Likens, T.J. Fahey, R.T. Holmes, C. Eagar, and J.D. Aber. 2004. Nor gloom of night: A new conceptual model for the Hubbard Brook Ecosystem Study. *BioScience* 54(2):139-148. DOI: https://doi.org/10.1641/0006-3568(2004)054[0139:Ngonan]2.0.Co;2.

Groffman, P.M., K. Butterbach-Bahl, R.W. Fulweiler, A.J. Gold, J.L. Morse, E.K. Stander, C. Tague, C. Tonitto, and P. Vidon. 2009. Challenges to incorporating spatially and temporally explicit phenomena (hotspots and hot moments) in denitrification models. *Biogeochemistry* 93(1/2):49-77.

Gross, C., and J.D. Hagy, III. 2017. Attributes of successful actions to restore lakes and estuaries degraded by nutrient pollution. *Journal of Environmental Management* 187:122-136. DOI: https://doi.org/10.1016/j.jenvman.2016.11.018.

Grüss, A., H.A. Perryman, E.A. Babcock, S.R. Sagarese, J.T. Thorson, C.H. Ainsworth, E.J. Anderson, K. Brennan, M.D. Campbell, M.C. Christman, S. Cross, M.D. Drexler, J. Marcus Drymon, C.L. Gardner, D.S. Hanisko, J. Hendon, C.C. Koenig, M. Love, F. Martinez-Andrade, J. Morris, B.T. Noble, M.A. Nuttall, J. Osborne, C. Pattengill-Semmens, A.G. Pollack, T.T. Sutton, and T.S. Switzer. 2018. Monitoring programs of the U.S. Gulf of Mexico: Inventory, development and use of a large monitoring database to map fish and invertebrate spatial distributions. *Reviews in Fish Biology and Fisheries* 28(4):667-691. DOI: https://doi.org/10.1007/s11160-018-9525-2.

GSMFC (Gulf States Marine Fisheries Commission). 2012. *The oyster fishery of the Gulf of Mexico, United States: A regional management plan.* Ocean Springs, MS: Gulf States Marine Fisheries Commission.

Gulf Coast Ecosystem Restoration Task Force. 2011. *Gulf of Mexico Regional Ecosystem Restoration Strategy.* https://www.energy.gov/sites/prod/files/GulfCoastReport_Full_12-02_508.pdf.

Gurbisz, C., and W.M. Kemp. 2014. Unexpected resurgence of a large submersed plant bed in Chesapeake Bay Analysis of time series data. *Limnology and Oceanography* 59(2):482-494.

Guthrie, A.G., D.M. Bilkovic, M. Mitchell, R. Chambers, J.S. Thompson, and R.E. Isdell. 2022. Ecological equivalency of living shorelines and natural marshes for fish and crustacean communities. *Ecological Engineering* 176:106511. DOI: https://doi.org/10.1016/j.ecoleng.2021.106511.

Halanych, K.M., C.H. Ainsworth, E.E. Cordes, R.E. Dodge, M. Huettel, I.A. Mendelssohn, S.A. Murawski, C.B. Paris-Limouzy, P.T. Schwing, R.F. Shaw, and T. Sutton. 2021. Effects of petroleum by-products and dispersants on ecosystems. *Oceanography* 34(1):152-163.

Halpern, B.S., B.R. Silliman, J.D. Olden, J. Bruno, and M.D. Bertness. 2007. Incorporating positive interactions in aquatic restoration and conservation. *Frontiers in Ecology and the Environment* 5:153-160.

Halpern, B.S., M. Frazier, J. Potapenko, K.S. Casey, K. Koenig, C. Longo, J.S. Lowndes, R.C. Rockwood, E.R. Selig, K.A. Selkoe, and S. Walbridge. 2015. Spatial and temporal changes in cumulative human impacts on the world's ocean. *Nature Communications* 6(1):7615. DOI: https://doi.org/10.1038/ncomms8615.

Hampton, S.E., C.A. Strasser, J.J. Tewksbury, W.K. Gram, A.E. Budden, A.L. Batcheller, C.S. Duke, and J.H. Porter. 2013. Big data and the future of ecology. *Frontiers in Ecology and the Environment* 11(3):156-162. DOI: https://doi.org/10.1890/120103.

Handley, L.R., and C.M. Lockwood. 2020. *Seagrass status and trends update for the northern Gulf of Mexico: 2002–2017.* Springs, MS: Gulf of Mexico Alliance.

Handley, L., D. Altsman, and R. DeMay. 2007. *Seagrass status and trends in the northern Gulf of Mexico, 1940-2002.* U.S. Geological Survey Scientific Investigations Report 2006-5287 and U.S. Environmental Protection Agency 855-R-04-003.

Handley, L.R., K.A. Spear, R. Baumstark, R. Moyer, and C.A. Thatcher. 2012. Introduction to emergent wetlands. In *Emergent Wetlands Status and Trends in the Northern Gulf of Mexico: 1950-2010.* Washington, DC: U.S. Geological Survey and Environmental Protection Agency.

Handley, L., Lockwood C., Kenworthy, J., and M. Finkbeiner. 2018 (Revised 2020). *A seagrass monitoring approach for the Gulf of Mexico.* Springs, MS: Gulf of Mexico Alliance. https://gulfofmexicoalliance.org/tools-and-resources/publications/.

HARC (Houston Advanced Research Center). 2020. *A characterization of the Galveston Bay ecosystem.* https://www.stateofgalvbay.org/welcome.

Harned, D., J. Atkins, and J. Harvill. 2004. Nutrient mass balance and trends, Mobile River Basin, Alabama, Georgia, and Mississippi. *Journal of the American Water Resources Association* 40:765-793. DOI: https://doi.org/10.1111/j.1752-1688.2004.tb04458.x.

Harned, D.A., E.L. Staub, K.L. Peak, K.M. Tighe, and S. Terziotti. 2009. Trends in water quality in the Southeastern United States, 1973–2005: U.S. Geological Survey Scientific Investigations Report 2009–5268, 25 p. DOI: https://doi.org/10.3133/sir20095268.

Hart, J.K., and K. Martinez. 2015. Toward an environmental internet of things. *Earth and Space Science* 2(5):194-200. DOI: https://doi.org/10.1002/2014EA000044.

Hart, K.M., M.M. Lamont, A.R. Iverson, and B.J. Smith. 2020. The importance of the northeastern Gulf of Mexico to foraging loggerhead sea turtles. *Frontiers in Marine Science* 7. DOI: https://doi.org/10.3389/fmars.2020.00330.

Harwell, M.A., J.H. Gentile, L.D. McKinney, J.W. Tunnell, W.C. Dennison, R.H. Kelsey, K.M. Stanzel, G.W. Stunz, K. Withers, and J. Tunnell. 2019. Conceptual framework for assessing ecosystem health. *Integrated Environmental Assessment and Management* 15(4):544-564. DOI: https://doi.org/10.1002/ieam.4152.

He, Q., and B.R. Silliman. 2019. Climate change, human impacts, and coastal ecosystems in the anthropocene. *Current Biology* 29(19):R1021-R1035. DOI: https://doi.org/10.1016/j.cub.2019.08.042.

Heck, K.L., F.J. Fodrie, S. Madsen, C.J. Baillie, and D.A. Byron. 2015. Seagrass consumption by native and a tropically associated fish species: Potential impacts of the tropicalization of the northern Gulf of Mexico. *Marine Ecology Progress Series* 520:165-173. DOI: https://doi.org/10.3354/meps11104.

Heimann, D.C., L.A. Sprague, and D.W. Blevins, 2011. *Trends in suspended-sediment loads and concentrations in the Mississippi River Basin, 1950–2009.* Washington, DC: U.S. Department of the Interior and U.S. Geological Survey. Pp. 1-33.

Heinle, D., C. D'Elia, J. Taft, J. Wilson, and M. Cole-Jones. 1980. *Historical review of water quality and climate data from Chesapeake Bay with emphasis on effects of enrichment.* Washington, DC: Environmental Protection Agency.

Hendriks, I.E., Y.S. Olsen, L. Ramajo, L. Basso, A. Steckbauer, T.S. Moore, J. Howard, and C.M. Duarte. 2014. Photosynthetic activity buffers ocean acidification in seagrass meadows. *Biogeosciences* 11(2):333-346. DOI: https://doi.org/10.5194/bg-11-333-2014.

Herbert, D.A., W.B. Perry, B.J. Cosby, and J.W. Fourqurean. 2011. Projected reorganization of Florida Bay Seagrass communities in response to the increased freshwater inflow of everglades restoration. *Estuaries and Coasts* 34(5):973. DOI: https://doi.org/10.1007/s12237-011-9388-4.

Hering, D.K. 2009. *Growth, residence, and movement of juvenile Chinook salmon within restored and reference estuarine marsh channels in Salmon River, Oregon*. Corvallis, OR: Oregon State University.

Hestir, E.L., D.H. Schoellhamer, J. Greenberg, T. Morgan-King, and S.L. Ustin. 2016. The effect of submerged aquatic vegetation expansion on a declining turbidity trend in the Sacramento-San Joaquin River delta. *Estuaries and Coasts* 39(4):1100-1112. DOI: https://doi.org/10.1007/s12237-015-0055-z.

Heuer, R.M., and M. Grosell. 2014. Physiological impacts of elevated carbon dioxide and ocean acidification on fish. *American Journal of Physiology—Regulatory Integrative and Comparative Physiology* 307(9):R1061-R1084. DOI: https://doi.org/10.1152/ajpregu.00064.2014.

Hiatt, M., E. Castañeda-Moya, R. Twilley, B. R. Hodges, and P. Passalacqua. 2018. Channel-island connectivity affects water exposure time distributions in a coastal river delta. *Water Resources Research* 54(3):2212-2232. DOI: https://doi.org/10.1002/2017WR021289.

Hijuelos, A.C. and S.A. Hemmerling. (2015). *Coastwide and barataria basin monitoring plans for Louisiana's system-wide assessment and monitoring program (SWAMP).* Baton Rouge, LA: The Water Institute of the Gulf.

Hill, A.B. 1965. The environment and disease: Association or causation. *Proceedings of the Royal Society of Medicine* 58(5):295-300.

Hine, A.C., G.R. Brooks, R.A. Davis, Jr., L.J. Doyle, G. Gelfenbaum, S.D. Locker, and D.C. Twichell. 2001. *A summary of findings of the West-Central Florida coastal studies project USGS open file report 01-303.* Reston, VA: U.S. Geological Survey.

Hobbie, J.E. 2000. *Estuarine science: A synthetic approach to research and practice.* Washington, DC: Island Press.

Hobbs, R.J., and D.A. Norton. 1996. Towards a conceptual framework for restoration ecology. *Restoration Ecology* 4(2):93-110. DOI: https://doi.org/10.1111/j.1526-100X.1996.tb00112.x.

Hodges, B.R. 2014. Hydrodynamical modeling. In *Reference module in earth systems and environmental sciences.* Amsterdam, Netherlands: Elsevier. DOI: https://doi.org/10.1016/B978-0-12-409548-9.09123-5.

Hodgson, E.E., and B.S. Halpern. 2018. Investigating cumulative effects across ecological scales. *Conservation Biology* 33(1):22-32. DOI: https://doi.org/10.1111/cobi.13125.
Hodgson, E.E., B.S. Halpern, and T.E. Essington. 2019. Moving beyond silos in cumulative effects assessment. *Frontiers in Ecology and Evolution* 7(211). DOI: https://doi.org/10.3389/fevo.2019.00211.
Holland, G., and C.L. Bruyère. 2014. Recent intense hurricane response to global climate change. *Climate Dynamics: Observational, Theoretical and Computational Research on the Climate System* 42(3-4):617-627. DOI: https://doi.org/10.1007/s00382-013-1713-0.
Holling, C.S. (Ed.) 1978. *Adaptive environmental assessment and management*. Chichester: John Wiley & Sons.
Hollweg, T.A., M.C. Christman, J. Lipton, B.P. Wallace, M.T. Huisenga, D.R. Lane, and K.G. Benson. 2020. Meta-analysis of nekton recovery following marsh restoration in the northern Gulf of Mexico. *Estuaries and Coasts* 43(7):1746-1763. DOI: https://doi.org/10.1007/s12237-019-00630-1.
Hood, R.R., G.W. Shenk, R.L. Dixon, S.M.C. Smith, W.P. Ball, J.O. Bash, R. Batiuk, K. Boomer, D.C. Brady, C. Cerco, P. Claggett, K. de Mutsert, Z.M. Easton, A.J. Elmore, M.A.M. Friedrichs, L.A. Harris, T.F. Ihde, L. Lacher, L. Li, L.C. Linker, A. Miller, J. Moriarty, G.B. Noe, G.E. Onyullo, K. Rose, K. Skalak, R. Tian, T.L. Veith, L. Wainger, D. Weller, and Y.J. Zhang. 2021. The Chesapeake Bay program modeling system: Overview and recommendations for future development. *Ecological Modelling* 456:109635. DOI: https://doi.org/10.1016/j.ecolmodel.2021.109635.
Hoos, A.B., and V.L. Roland II. 2019. *Spatially referenced models of streamflow and nitrogen, phosphorus, and suspended-sediment loads in the southeastern United States*. Reston, VA: U.S. Geological Survey. DOI: https://doi.org/10.3133/sir20195135
Howe, H.F. and C. Martínez-Garza. 2014. Restoration as experiment. *Botanical Sciences* 92(4). DOI: https://doi.org/10.17129/botsci.146.
Hu, C., R.H. Weisberg, Y. Liu, L. Zheng, K.L. Daly, D.C. English, J. Zhao, and G.A. Vargo. 2011. Did the northeastern Gulf of Mexico become greener after the Deepwater Horizon oil spill? *Geophysical Research Letters* 38(9). DOI: https://doi.org/10.1029/2011gl047184.
Hu, X., J.B. Pollack, M.R. McCutcheon, P.A. Montagna, and Z. Ouyang. 2015. Long-term alkalinity decrease and acidification of estuaries in northwestern gulf of Mexico. *Environmental Science and Technology* 49(6):3401-3409. DOI: https://doi.org/10.1021/es505945p.
Hu, X., M.F. Nuttall, H. Wang, H. Yao, C.J. Staryk, M.R. McCutcheon, R.J. Eckert, J.A. Embesi, M.A. Johnston, E.L. Hickerson, G.P. Schmahl, D. Manzello, I.C. Enochs, S. DiMarco, and L. Barbero. 2018. Seasonal variability of carbonate chemistry and decadal changes in waters of a marine sanctuary in the Northwestern Gulf of Mexico. *Marine Chemistry* 205:16-28. DOI: https://doi.org/10.1016/j.marchem.2018.07.006.
Huang, B., V.F. Banzon, E. Freeman, J. Lawrimore, W. Liu, T.C. Peterson, T.M. Smith, P.W. Thorne, S.D. Woodruff, and H.M. Zhang. 2015. Part I: Upgrades and intercomparisons. In Extended reconstructed sea surface temperature version 4 (ERSST.v4). *Journal of Climate* 28(3):911-930. DOI: https://doi.org/10.1175/JCLI-D-14-00006.1.
Huang, Y., D. Nicholson, B. Huang, and N. Cassar. 2021. Global estimates of marine gross primary production based on machine learning upscaling of field observations. *Global Biogeochemical Cycles* 35(3):e2020GB006718. DOI: https://doi.org/10.1029/2020GB006718.
Humphries, G., Magness, D.R., and Huettmann, F. (Eds.) 2018. *Machine learning for ecology and sustainable resource management.* Switzerland: Springer Nature.
Hydroqual, I. 1981. *Water quality analysis of the Patuxent River.* Washington, DC: Environmental Protection Agency.
Hyndes, G.A., K.L. Heck, Jr., A. Vergés, E.S. Harvey, G.A. Kendrick, P.S. Lavery, K. McMahon, R.J. Orth, A. Pearce, M. Vanderklift, T. Wernberg, S. Whiting, and S. Wilson. 2016. Accelerating tropicalization and the transformation of temperate seagrass meadows. *BioScience* 66(11):938-945. DOI: https://doi.org/10.1093/biosci/biw111.

INRS (Iowa Nutrient Reduction Strategy). 2017. *A science and technology-based framework to assess and reduce nutrients of Iowa waters and the Gulf of Mexico.* Des Moines, IA: Iowa Department of Agriculture and Land Stewardship, Iowa Department of Natural Resources, and Iowa State University College of Agriculture and Life Sciences. https://www.nutrientstrategy.iastate.edu/sites/default/files/documents/2017%20INRS%20Complete_Revised%202017_12_11.pdf.

IPCC (Intergovernmental Panel on Climate Change). 2019. *IPCC Special Report on the ocean and cryosphere in a changing climate,* edited by H.-O. Pörtner, D.C. Roberts, V. Masson-Delmotte, P. Zhai, M. Tignor, E. Poloczanska, K. Mintenbeck, A. Alegría, M. Nicolai, A. Okem, J. Petzold, B. Rama, N.M. Weyer. Cambridge, UK and New York, NY: Cambridge University Press, 755 pp. DOI: https://doi.org/10.1017/9781009157964.

IPCC. 2021. Climate change 2021: The physical science basis. Contribution of working group I to the *Sixth Assessment Report of the Intergovernmental Panel on Climate Change.* Cambridge, UK: Cambridge University Press.

Irlandi, E.A., and M.K. Crawford. 1997. Habitat linkages: The effect of intertidal saltmarshes and adjacent subtidal habitats on abundance, movement, and growth of an estuarine fish. *Oecologia* 110(2):222-230. DOI: https://doi.org/10.1007/s004420050154.

Isdell, R.E., D.M. Bilkovic, A.G. Guthrie, M.M. Mitchell, R.M. Chambers, M. Leu, and C. Hershner. 2021. Living shorelines achieve functional equivalence to natural fringe marshes across multiple ecological metrics. *PeerJ* 9:e11815. DOI: https://doi.org/10.7717/peerj.11815.

Jackson, K., G.R. Brooks, and R.A. Larson. 2021. Of marsh and mangrove: Coupled biophysical and anthropogenic drivers of 20th century wetland conversion in Tampa Bay Estuary, Florida (USA). *Anthropocene* 34. DOI: https://doi.org/10.1016/j.ancene.2021.100295.

Jaiswal, R.K., S. Ali, and B. Bharti. 2020. Comparative evaluation of conceptual and physical rainfall–runoff models. *Applied Water Science* 10(1):48. DOI: https://doi.org/10.1007/s13201-019-1122-6.

Jankowski, K.L., T.E. Törnqvist, and A.M. Fernandes. 2017. Vulnerability of Louisiana's coastal wetlands to present-day rates of relative sea-level rise. *Nature Communications* 8(1):14792. DOI: https://doi.org/10.1038/ncomms14792.

Järveoja, J., M.B. Nilsson, M. Gažovič, P.M. Crill, and M. Peichl. 2018. Partitioning of the net CO2 exchange using an automated chamber system reveals plant phenology as key control of production and respiration fluxes in a boreal peatland. *Global Change Biology* 24(8):3436-3451. DOI: https://doi.org/10.1111/gcb.14292.

Jin, S., L. Yang, P. Danielson, C. Homer, J. Fry, and G. Xian. 2013. A comprehensive change detection method for updating the National Land Cover Database to circa 2011. *Remote Sensing of Environment* 132:159-175. DOI: https://doi.org/10.1016/j.rse.2013.01.012.

Johnston, J.M., M.C. Barber, K. Wolfe, M. Galvin, M. Cyterski, and R. Parmar. 2017. An integrated ecological modeling system for assessing impacts of multiple stressors on stream and riverine ecosystem services within river basins. *Ecological Modelling* 354:104-114. DOI: https://doi.org/10.1016/j.ecolmodel.2017.03.021.

Johnston, J.M., J.H. Novak, and S.R. Kraemer. 2000. Multimedia integrated modeling for environmental protection: Introduction to a collaborative framework. *Environmental Monitoring and Assessment* 63(1):253-263. DOI: https://doi.org/10.1023/A:1006464407117.

Joint Nature Conservation Committee. 2018. *Autonomous underwater vehicles for use in marine benthic monitoring.* Marine Monitoring Platform Guidelines No. 2. Peterborough, UK: JNCC. https://data.jncc.gov.uk/data/f52a772a-1d81-4cab-b850-7a9e32d0fef6/JNCC-MMPG-002-FINAL-WEB.pdf.

Jones, A.R., Z.A. Doubleday, T.A.A. Prowse, K.H. Wiltshire, M.R. Deveney, T. Ward, S.L. Scrivens, P. Cassey, L.G. O'Connell, and B.M. Gillanders. 2018. Capturing expert uncertainty in spatial cumulative impact assessments. *Scientific Reports* 8(1):1469. DOI: https://doi.org/10.1038/s41598-018-19354-6.

Jones, D.O.B., A.R. Gates, V.A.I. Huvenne, A.B. Phillips, and B.J. Bett. 2019. Autonomous marine environmental monitoring: Application in decommissioned oil fields. *Science of the Total Environment* 668:835-853. DOI: https://doi.org/10.1016/j.scitotenv.2019.02.310.

Jordan, S.J., L.M. Smith, and J.A. Nestlerode. 2008. Cumulative effects of coastal habitat alterations on fishery resources: Toward prediction at regional scales. *Ecology and Society* 14(1): 16. http://www.ecologyandsociety.org/vol14/iss1/art16.

Jordan, S.J., M.A. Lewis, L.M. Harwell, and L.R. Goodman. 2010. Summer fish communities in northern Gulf of Mexico estuaries: Indices of ecological condition. *Ecological Indicators* 10(2):504-515. DOI: https://doi.org/10.1016/j.ecolind.2009.09.003.

Joye, S.B. 2015. Deepwater Horizon, 5 years on. *Science* 349(6248):592-593. DOI: https://doi.org/10.1126/science.aab4133.

Justić, D., V. Kourafalou, G. Mariotti, S. He, R. Weisberg, Y. Androulidakis, C. Barker, A. Bracco, B. Dzwonkowski, C. Hu, H. Huang, G. Jacobs, M. Le Hénaff, Y. Liu, S. Morey, J. Nittrouer, E. Overton, C.B. Paris, B.J. Roberts, K. Rose, A. Valle-Levinson, and J. Wiggert. 2021. Transport processes in the Gulf of Mexico along the river-estuary-shelf-ocean continuum: A review of research from the Gulf of Mexico research initiative. *Estuaries and Coasts*. DOI: https://doi.org/10.1007/s12237-021-01005-1.

Kaiser, M.J., and S. Narra. 2019. U.S. Gulf of Mexico pipeline activity statistics, trends and correlations. *Ships and Offshore Structures* 14(1):1-22. DOI: https://doi.org/10.1080/17445302.2018.1472517.

Kakogeorgiou, I., and K. Karantzalos. 2021. Evaluating explainable artificial intelligence methods for multi-label deep learning classification tasks in remote sensing. *International Journal of Applied Earth Observation and Geoinformation* 103:102520. DOI: https://doi.org/10.1016/j.jag.2021.102520.

Kannenberg, S.A., D.R. Bowling, and W.R.L. Anderegg. 2020. Hot moments in ecosystem fluxes: High GPP anomalies exert outsized influence on the carbon cycle and are differentially driven by moisture availability across biomes. *Environmental Research Letters* 15(5):054004. DOI: https://doi.org/10.1088/1748-9326/ab7b97.

Karnauskas, M., C.R. Kelble, S. Regan, C. Quenée, R. Allee, M. Jepson, A. Freitag, J.K. Craig, C. Carollo, L. Barbero, N. Trifonova, D. Hanisko, and G. Zapfe. 2017. *2017 Ecosystem Status Report Update for the Gulf of Mexico. NOAA Technical Memorandum NMFS-SEFSC-706.* Miami, FL: U.S. Department of Commerce. Pp. 1-56.

Karr, J.R., E.R. Larson, and E.W. Chu. 2022. Ecological integrity is both real and valuable. *Conservation Science and Practice* 4(2):e583. DOI: https://doi.org/10.1111/csp2.583.

Kealoha, A.K., K.E.F. Shamberger, S.F. DiMarco, K.M. Thyng, R.D. Hetland, D.P. Manzello, N.C. Slowey, and I.C. Enochs. 2020. Surface water CO_2 variability in the Gulf of Mexico (1996–2017). *Scientific Reports* 10(1). DOI: https://doi.org/10.1038/s41598-020-68924-0.

Keim, B.D., R.A. Muller, and G.W. Stone. 2007. Spatiotemporal patterns and return periods of tropical storm and hurricane strikes from Texas to Maine. *Journal of Climate* 20(14):3498-3509. DOI: https://doi.org/10.1175/JCLI4187.1.

Keiser, D.A., C.L. Kling, and J.S. Shapiro. 2018. The low but uncertain measured benefits of US water quality policy. *Proceedings of the National Academy of Sciences* 116(12):5262-5269. DOI: https://doi.org/10.1073/pnas.1802870115.

Keithly, W.R., Jr., and K.J. Roberts. 2017. Commercial and recreational fisheries of the Gulf of Mexico. In *Habitats and biota of the Gulf of Mexico: Before the Deepwater Horizon oil spill.* New York: Springer. DOI: https://doi.org/10.1007/978-1-4939-3456-0_2.

Kellar, N.M., T.R. Speakman, C.R. Smith, S.M. Lane, and others. 2017. Low reproductive success rates of common bottlenose dolphins *Tursiops truncatus* in the northern Gulf of Mexico following the *Deepwater Horizon* disaster (2010-2015). Endangered Species Research 33:143-158. DOI: https://doi.org/10.3354/esr00775.

Kelly, R., A. Fleming, G.T. Pecl, J. von Gönner, and A. Bonn. 2020. Citizen science and marine conservation: A global review. *Philosophical Transactions of the Royal Society B: Biological Sciences* 375(1814):20190461. DOI: https://doi.org/10.1098/rstb.2019.0461.

Kemp, W.M., and W.R. Boynton. 2012. Synthesis in estuarine and coastal ecological research: What is it, why is it important, and how do we teach it? *Estuaries and Coasts* 35(1):1-22. DOI: https://doi.org/10.1007/s12237-011-9464-9.

Kemp, G.P., J.W. Day, Jr., A. Yáñez-Arancibia, and N.S. Peyronnin. 2016. Can continental shelf river plumes in the northern and southern Gulf of Mexico promote ecological resilience in a time of climate change? *Water (Switzerland)* 8(3). DOI: https://doi.org/10.3390/w8030083.

Kennicutt, M.C. 2017. Water Quality of the Gulf of Mexico. In *Habitats and biota of the Gulf of Mexico: Before the Deepwater Horizon oil spill* (Vol. 1), edited by C.H. Ward. New York: Springer.

Kentula, M. 2000. Perspectives on setting success criteria for wetland restoration. *Ecological Engineering* 15:199-209. DOI: https://doi.org/10.1016/S0925-8574(00)00076-8.

Keogh, M.E., and T.E. Törnqvist. 2019. Measuring rates of present-day relative sea-level rise in low-elevation coastal zones: A critical evaluation. *Ocean Science* 15(1):61-73. DOI: https://doi.org/10.5194/os-15-61-2019.

Keogh, M.E., A.S. Kolker, G.A. Snedden, and A.A. Renfro. 2019. Hydrodynamic controls on sediment retention in an emerging diversion-fed delta. *Geomorphology* 332:100-111. DOI: https://doi.org/10.1016/j.geomorph.2019.02.008.

Keown, M.P., E.A. Dardeau, Jr., and E.M. Causey. 1986. Historic trends in the sediment flow regime of the Mississippi River. *Water Resources Research* 22(11):1555-1564. DOI: https://doi.org/10.1029/WR022i011p01555.

Kersting, K. 2018. Machine learning and artificial intelligence: Two Fellow travelers on the quest for intelligent behavior in machines. *Frontiers in Big Data* 1. DOI: https://doi.org/10.3389/fdata.2018.00006.

Kesel, R.H. 1988. The decline in the suspended load of the Lower Mississippi River and its influence on adjacent wetlands. *Environmental Geology and Water Sciences* 11(3):271-281. DOI: https://doi.org/10.1007/BF02574816.

Kleiss, B.A., J.C. Murphy, C.M. Mayne, J.P. Allgeier, A.B. Edmondson, K.C. Ginsberg, K.E. Jones, T.J. Lauth, E.L. Moe, J.W. Murphy, and M.A. Allison. 2021. Incorporating water quality analysis into navigation assessments as demonstrated in the Mississippi River Basin. *Journal of Waterway, Port, Coastal and Ocean Engineering* 147(5). DOI: https://doi.org/10.1061/(ASCE)WW.1943-5460.0000651.

Klemas, V. 2013. Remote sensing of coastal wetland biomass: An overview. *Journal of Coastal Research* 29(5):1016-1028. DOI: https://doi.org/10.2112/JCOASTRES-D-12-00237.1.

Knox, S.H., L. Windham-Myers, F. Anderson, C. Sturtevant, and B. Bergamaschi. 2018. Direct and indirect effects of tides on ecosystem-scale CO_2 exchange in a brackish tidal marsh in northern California. *Journal of Geophysical Research: Biogeosciences* 123(3):787-806. DOI: https://doi.org/10.1002/2017jg004048.

Konisky, R.A., D.M. Burdick, M. Dionne, and H.A. Neckles. 2006. A regional assessment of salt marsh restoration and monitoring in the Gulf of Maine. *Restoration Ecology* 14:516-525. DOI: https://doi.org/10.1111/j.1526-100X.2006.00163.x

Koopmans, D., M. Holtappels, A. Chennu, M. Weber, and D. de Beer. 2020. High net primary production of Mediterranean seagrass (*Posidonia oceanica*) meadows determined with aquatic eddy covariance. *Frontiers in Marine Science* 7. DOI: https://doi.org/10.3389/fmars.2020.00118.

Kozlov, M. 2021. Hurricane Ida forces Louisiana researchers to rethink their future. *Nature* 597(7876):313-314. DOI: https://doi.org/10.1038/d41586-021-02456-z.

Kucklick, J., L. Schwacke, R. Wells, A. Hohn, A. Guichard, J. Yordy, L. Hansen, E. Zolman, R. Wilson, J. Litz, D. Nowacek, T. Rowles, R. Pugh, B. Balmer, C. Sinclair and P. Rosel. 2011. Bottlenose dolphins as indicators of persistent organic pollutants in the western North Atlantic Ocean and northern Gulf of Mexico. *Environmental Science and Technology* 45: 4270-4277.

Kuhn, N.L., I.A. Mendelssohn, and D.J. Reed. 1999. Altered hydrology effects on Louisiana salt marsh function. *Wetlands* 19:617-626.

La Peyre, M.K., D. Aguilar Marshall, L.S. Miller, and A.T. Humphries. 2019. Oyster reefs in northern Gulf of Mexico estuaries harbor diverse fish and decapod crustacean assemblages: A meta-synthesis. *Frontiers in Marine Science* 6. DOI: https://doi.org/10.3389/fmars.2019.00666.

Ladouceur, E., and N. Shackelford. 2021. The power of data synthesis to shape the future of the restoration community and capacity. *Restoration Ecology* 29(1):e13251. DOI: https://doi.org/10.1111/rec.13251.

Laffoley, D. and G. Grimsditch. 2009. *The management of natural coastal carbon sinks*. Gland, Switzerland: International Union for Conservation of Nature.

Larkum, A.W.D., R.J. Orth, and C.M. Duarte. 2006. *Seagrasses: Biology, Ecology and Conservation*. Dordrecht: Springer Netherlands. 676 pp. DOI: https://doi.org/10.1007/978-1-4020-2983-7

Lamba, A., P. Cassey, R.R. Segaran, and L.P. Koh. 2019. Deep learning for environmental conservation. *Current Biology* 29(19):R977-R982. DOI: https://doi.org/10.1016/j.cub.2019.08.016.

Landsea, C.W., G.A. Vecchi, L. Bengtsson, and T.R. Knutson. 2010. Impact of duration thresholds on Atlantic tropical cyclone counts. *Journal of Climate* 23(10):2508-2519. DOI: https://doi.org/10.1175/2009JCLI3034.1.

Lane, R.R., J.W. Day, Jr., and J.N. Day. 2006. Wetland surface elevation, vertical accretion, and subsidence at three Louisiana estuaries receiving diverted Mississippi River water. *Wetlands* 26(4):1130-1142. DOI: https://doi.org/10.1672/0277-5212(2006)26[1130:WSEVAA]2.0.CO;2.

Laniak, G.F., G. Olchin, J. Goodall, A. Voinov, M. Hill, P. Glynn, G. Whelan, G. Geller, N. Quinn, M. Blind, S. Peckham, S. Reaney, N. Gaber, R. Kennedy, and A. Hughes. 2013. Integrated environmental modeling: A vision and roadmap for the future. *Environmental Modeling & Software* 39:3-23. DOI: https://doi.org/10.1016/j.envsoft.2012.09.006.

Lapenna, V., Soldovieri, F. 2021. Integration of space and in-situ techniques: A new paradigm for the monitoring and surveillance. *Remote Sensing of Environment* 253:112192. DOI: https://doi.org/10.1016/j.rse.2020.112192.

Laurent, A., K. Fennel, W.J. Cai, W.J. Huang, L. Barbero, and R. Wanninkhof. 2017. Eutrophication-induced acidification of coastal waters in the northern Gulf of Mexico: Insights into origin and processes from a coupled physical-biogeochemical model. *Geophysical Research Letters* 44(2):946-956. DOI: https://doi.org/10.1002/2016GL071881.

Layman, C.A., and A.L. Rypel. 2020. Secondary production is an underutilized metric to assess restoration initiatives. *Food Webs* 25:e00174. DOI: https://doi.org/10.1016/j.fooweb.2020.e00174.

Leach, C., T. Coulthard, A. Barkwith, D.R. Parsons, and S. Manson. 2021. The Coastline Evolution Model 2D (CEM2D) V1.1. *Geoscientific Model Development* 14(9):5507-5523. DOI: https://doi.org/10.5194/gmd-14-5507-2021.

LeCun, Y., Y. Bengio, and G. Hinton. 2015. Deep learning. *Nature* 521(7553):436-444. DOI: https://doi.org/10.1038/nature14539.

Leduc, A.O., P.L. Munday, G.E. Brown, and M.C. Ferrari. 2013. Effects of acidification on olfactory-mediated behaviour in freshwater and marine ecosystems: A synthesis. *Philosophical Transactions of the Royal Society B* 368(1627):20120447. DOI: https://doi.org/10.1098/rstb.2012.0447.

Lee, T.-M., and H.-C. Yeh. 2009. Applying remote sensing techniques to monitor shifting wetland vegetation: A case study of Danshui River estuary mangrove communities, Taiwan. *Ecological Engineering* 35(4):487-496. DOI: https://doi.org/10.1016/j.ecoleng.2008.01.007.

Lee, Y.J., W.R. Boynton, M. Li, and Y. Li. 2013. Role of late Winter-Spring wind influencing summer hypoxia in Chesapeake Bay. *Estuaries and Coasts* 36(4):683-696. DOI: https://doi.org/10.1007/s12237-013-9592-5.

Lefcheck, J.S., R.J. Orth, W.C. Dennison, D.J. Wilcox, R.R. Murphy, J. Keisman, C. Gurbisz, M. Hannam, J.B. Landry, K.A. Moore, C.J. Patrick, J. Testa, D.E. Weller, and R.A. Batiuk. 2018. Long-term nutrient reductions lead to the unprecedented recovery of a temperate coastal region. *Proceedings of the National Academy of Sciences* 115(14):3658-3662. DOI: https://doi.org/10.1073/pnas.1715798115.

Lehr, B., S. Bristol, and A. Possolo. 2010. *Oil budget calculator (OBC): Deepwater Horizon.* Technical documentation: A report to the National Incident Command. Washington, DC: The Federal Interagency Solutions Group.

Lester, R.E., and P.G. Fairweather. 2011. Ecosystem states: Creating a data-derived, ecosystem-scale ecological response model that is explicit in space and time. *Ecological Modelling* 222(15):2690-2703. DOI: https://doi.org/10.1016/j.ecolmodel.2011.05.009.

Lester, S.E., A.K. Dubel, G. Hernán, J. McHenry, and A. Rassweiler. 2020. Spatial Planning Principles for Marine Ecosystem Restoration. *Frontiers in Marine Science* 7. DOI: https://doi.org/10.3389/fmars.2020.00328.

Lewis, R.R., III. 2005. Ecological engineering for successful management and restoration of mangrove forests. *Ecological Engineering* 24(4 SPEC. ISS.):403-418. DOI: https://doi.org/10.1016/j.ecoleng.2004.10.003.

Li, S., A. Christensen, and R.R. Twilley. 2020a. Benthic fluxes of dissolved oxygen and nutrients across hydrogeomorphic zones in a coastal deltaic floodplain within the Mississippi River Delta plain. *Biogeochemistry* 149(2):115-140. DOI: https://doi.org/10.1007/s10533-020-00665-8.

Li, W., Z. Dou, L. Cui, R. Wang, Z. Zhao, S. Cui, Y. Lei, J. Li, X. Zhao, and X. Zhai. 2020b. Suitability of hyperspectral data for monitoring nitrogen and phosphorus content in constructed wetlands. *Remote Sensing Letters* 11(5):495-504. DOI: https://doi.org/10.1080/2150704X.2020.1734247.

Lin, J.P. and B.A. Kleiss. 2007. *A Wetland Restoration Spatial Decision Support System for the Mississippi Gulf Coast* (ERDC/EL TR-07-12). Vicksburg, MS: U.S. Army Engineer Research and Development Center. https://apps.dtic.mil/sti/pdfs/ADA471738.pdf.

Lindenmayer, D.B., and G.E. Likens. 2010. The science and application of ecological monitoring. *Biological Conservation* 143(6):1317-1328. DOI: https://doi.org/10.1016/j.biocon.2010.02.013.

Linker, L.C., G.W. Shenk, P. Wang, K.J. Hopkins, and S. Pokharel. 2002. A short history of Chesapeake Bay modeling and the next generation of watershed and estuarine models. *Proceedings of the Water Environment Federation* 2002(2):569-582. DOI: https://doi.org/10.2175/193864702785665021.

Linker, Lewis C., Richard A. Batiuk, Gary W. Shenk, and Carl F. Cerco. 2013. Development of the Chesapeake Bay watershed total maximum daily load allocation. *Journal of the American Water Resources Association.* 49(5):986-1006. DOI: https://doi.org/10.1111/jawr.12105.

Littles, C., J. Karnezis, K. Blauvelt, A. Creason, H. Diefenderfer, G. Johnson, L. Krasnow, and P. Trask. 2022. Adaptive management of large-scale ecosystem restoration: Increasing certainty of habitat outcomes in the Columbia River Estuary, USA. *Restoration Ecology* e13634. DOI: https://doi.org/10.1111/rec.13634.

Liu, H., J. Gilmartin, C. Li, and K. Li. 2021a. Detection of time-varying pulsed event effects on estuarine pelagic communities with ecological indicators after catastrophic hurricanes. *Ecological Indicators* 123:107327. DOI: https://doi.org/10.1016/j.ecolind.2020.107327.

Liu, L., T. Shi, H. Gao, X. Zhang, Q. Han, and X. Hu. 2021b. Long-term cross calibration of HJ-1A CCD1 and Terra MODIS reflective solar bands. *Scientific Reports* 11(1):7386. DOI: https://doi.org/10.1038/s41598-021-86619-y.

Livingston, R.J., R.L. Howell, X. Niu, G.F. Lewis, and G.C. Woodsum. 1999. Recovery of Oyster Reefs (Crassostrea Virginica) in a Gulf Estuary Following Disturbance by Two Hurricanes. *Bulletin of Marine Science* 64(3):465-483.

LoSchiavo, A.J., J.W. Vearil, R.G. Best, R.E. Burns, S. Gray, M.C. Harwell, E.B. Hines, S. Traxler, A.R. McLean, and T.S. Clair. 2013. Lessons learned from the first decade of adaptive management in comprehensive everglades restoration. *Ecology and Society* 18(4). DOI: https://doi.org/10.5751/ES-06065-180470.

Lotze, H.K., H.S. Lenihan, B.J. Bourque, R.H. Bradbury, R.G. Cooke, M.C. Kay, S.M. Kidwell, M.X. Kirby, C.H. Peterson, and J.B.C. Jackson. 2006. Depletion, degradation, and recovery potential of estuaries and coastal seas. *Science* 312(5781):1806-1809. DOI: https://doi.org/10.1126/science.1128035.

Love, M., A. Baldera, C. Robbins, and R. Spies. 2015. *Charting the Gulf: Analyzing the gaps in long-term monitoring of the Gulf of Mexico*. New Orleans, LA: Ocean Conservancy.

Love, M., C. Robbins, A. Baldera, S. Eastman, A. Bolten, R. Hardy, R. Herren, T. Metz, H. Vander Zanden, B. Wallace, M. Robbins, C. Baldera, A. Herren, and V. Zanden. 2017. *Restoration without borders: An assessment of cumulative stressors to guide large-scale, integrated restoration of sea turtles in the Gulf of Mexico*. New Orleans, LA: Ocean Conservancy. DOI: https://doi.org/10.13140/RG.2.2.19832.55049.

Ludwig, D.F., J. Iannuzzi, T.J. Iannuzzi, and J.K. Shisler. 2010. Spatial and temporal habitat use patterns by water birds in an urban estuarine ecosystem: Implications for ecosystem management and restoration. *Human and Ecological Risk Assessment: An International Journal* 16(1):163-184. DOI: https://doi.org/10.1080/10807030903459106.

Lugo, A.E., and S.C. Snedaker. 1974. The Ecology of Mangroves. *Annual Review of Ecology and Systematics* 5(1):39-64. DOI: https://doi.org/10.1146/annurev.es.05.110174.000351.

Luo, J., R. Pu, H. Duan, R. Ma, Z. Mao, Y. Zeng, L. Huang, and Q. Xiao. 2020. Evaluating the influences of harvesting activity and eutrophication on loss of aquatic vegetations in Taihu Lake, China. *International Journal of Applied Earth Observations and Geoinformation* 87. DOI: https://doi.org/10.1016/j.jag.2019.102038.

Luoma S.N., W.C., J. Gerritsen, A. Hatch, P. Jepson, T. Reynoldson, and R.M. Thom. 2001. Separating stressor influences from environmental variability: Eight case studies from aquatic and terrestrial ecosystems. In *Ecological variability: Separating natural from anthropogenic causes of ecosystem impairment*, edited by D.J. Baird and G. Allen Burton. Pensacola, FL: SETAC Press. PNWD-SA-5435.

Lusk, M.G., G.S. Toor, Y.Y. Yang, S. Mechtensimer, M. De, and T.A. Obreza. 2017. A review of the fate and transport of nitrogen, phosphorus, pathogens, and trace organic chemicals in septic systems. *Critical Reviews in Environmental Science and Technology* 47(7):455-541. DOI: https://doi.org/10.1080/10643389.2017.1327787.

Mabus, R. 2010. *America's Gulf Coast: A long-term recovery plan after the Deepwater Horizon oil spill ("Mabus Plan")*. New Orleans, LA: Gulf Coast Ecosystem Restoration Council.

MacKenzie, R.A., M. Dionne, J. Miller, M. Haas, and P.A. Morgan. 2015. Community structure and abundance of benthic infaunal invertebrates in maine fringing marsh ecosystems. *Estuaries and Coasts* 38(4):1317-1334. DOI: https://doi.org/10.1007/s12237-015-9977-8.

Macreadie, P.I., A. Anton, J.A. Raven, N. Beaumont, R.M. Connolly, D.A. Friess, J.J. Kelleway, H. Kennedy, T. Kuwae, P.S. Lavery, C.E. Lovelock, D.A. Smale, E.T. Apostolaki, T.B. Atwood, J. Baldock, T.S. Bianchi, G.L. Chmura, B.D. Eyre, J.W. Fourqurean, J.M. Hall-Spencer, M. Huxham, I.E. Hendriks, D. Krause-Jensen, D. Laffoley, T. Luisetti, N. Marba, P. Masque, K.J. McGlathery, J.P. Megonigal, D. Murdiyarso, B.D. Russell, R. Santos, O. Serrano, B.R. Silliman, K. Watanabe, and C.M. Duarte. 2019. The future of blue carbon science. *Nature Communications* 10(1):3998. DOI: https://doi.org/10.1038/s41467-019-11693-w.

Magliocca, N.R., T.K. Rudel, P.H. Verburg, W.J. McConnell, O. Mertz, K. Gerstner, A. Heinimann, and E.C. Ellis. 2015. Synthesis in land change science: Methodological patterns, challenges, and guidelines. *Regional Environmental Change* 15(2):211-226. DOI: https://doi.org/10.1007/s10113-014-0626-8.

Mallin, M.A., and C.A. Corbett. 2006. How hurricane attributes determine the extent of environmental effects: Multiple hurricanes and different coastal systems. *Estuaries and Coasts* 29(6):1046-1061. DOI: https://doi.org/0.1007/BF02798667.

Manning, A.D., D.B. Lindenmayer, and J. Fischer. 2006. Stretch goals and backcasting: Approaches for overcoming barriers to large-scale ecological restoration. *Restoration Ecology* 14(4):487-492. DOI: https://doi.org/10.1111/j.1526-100X.2006.00159.x.

Mao, F., K. Khamis, S. Krause, J. Clark, and D.M. Hannah. 2019. Low-cost environmental sensor networks: Recent advances and future directions. *Frontiers in Earth Science* 7. DOI: https://doi.org/10.3389/feart.2019.00221.

Mason, A.L., J.C. Taylor, and I.R. MacDonald. 2019. *An integrated assessment of oil and gas release into the marine environment at the former Taylor Energy MC20 Site*. Technical Memorandum 260. Silver Spring, MD: National Oceanic and Atmospheric Administration, National Ocean Service, National Centers for Coastal Ocean Science. DOI: https://doi.org/10.25923/kykm-sn39.

Marra, J. 2002. Approaches to the measurement of plankton production. In *Phytoplankton productivity*. Oxford, UK: Blackwell Science.

Marshak, A.R., and K.L. Heck, Jr. 2017. Interactions between range-expanding tropical fishes and the northern Gulf of Mexico red snapper Lutjanus campechanus. *Journal of Fish Biology* 91(4):1139-1165. DOI: https://doi.org/10.1111/jfb.13406.

Martin, S., E.L. Sparks, A.J. Constantin, J. Cebrian, and J.A. Cherry. 2021. Restoring fringing tidal marshes for ecological function and ecosystem resilience to moderate sea-level rise in the northern Gulf of Mexico. *Environmental Management* 67(2):384-397. DOI: https://doi.org/10.1007/s00267-020-01410-5.

Martinez-Andrade, F. 2018. *Trends in relative abundance and size of selected finfishes and shellfishes along the Texas coast: November 1975–December 2016.* Management Data Series No. 293. Austin, TX: Texas Parks and Wildlife, Coastal Fisheries Division.

Matear, R.J. 1995. Parameter optimization and analysis of ecosystem models using simulated annealing: A case study at Station P. *Journal of Marine Research* 53(4):571-607. DOI: https://doi.org/10.1357/0022240953213098.

Maxim, L., J.H. Spangenberg, and M. O'Connor. 2009. An analysis of risks for biodiversity under the DPSIR framework. *Ecological Economics* 69(1):12-23. DOI: https://doi.org/10.1016/j.ecolecon.2009.03.017.

Maxwell, P.S., J.S. Eklöf, M.M. van Katwijk, K.R. O'Brien, M. de la Torre-Castro, C. Boström, T.J. Bouma, D. Krause-Jensen, R.K.F. Unsworth, B.I. van Tussenbroek, and T. van der Heide. 2016. *The fundamental role of ecological feedback mechanisms for the adaptive management of seagrass ecosystems: A review*. Cambridge, UK: Cambridge Philosophical Society.

Maxwell, P.S., J.S. Eklöf, M.M. van Katwijk, K.R. O'Brien, M. de la Torre-Castro, C. Boström, T.J. Bouma, D. Krause-Jensen, R.K.F. Unsworth, B.I. van Tussenbroek, and T. van der Heide. 2017. The fundamental role of ecological feedback mechanisms for the adaptive management of seagrass ecosystems—A review. *Biological Reviews* 92(3):1521-1538. DOI: https://doi.org/10.1111/brv.12294.

Mayton, B., G. Dublon, S. Russell, E.F. Lynch, D.D. Haddad, V. Ramasubramanian, C. Duhart, G. Davenport, and J.A. Paradiso. 2017. The networked sensory landscape: Capturing and experiencing ecological change across scales. *Presence: Teleoperators and Virtual Environments* 26(2):182-209. DOI: https://doi.org/10.1162/PRES_a_00292.

MBNEP CCMP (Mobile Bay National Estuary Program Management Conference Comprehensive Conservation and Management Plan). *Update year one work plan.* 2018. Mobile, AL: Mobile Bay National Estuary Program. https://www.mobilebaynep.com/assets/pdf/U-1-2018-2019WorkplanFINAL2.pdf.

McBryan, T.L., K. Anttila, T.M. Healy, and P.M. Schulte. 2013. Responses to temperature and hypoxia as interacting stressors in fish: Implications for adaptation to environmental change. *Integrative and Comparative Biology* 53(4):648-659. DOI: https://doi.org/10.1093/icb/ict066.

McCarthy, M.J., F.E. Muller-Karger, D.B. Otis, and P. Méndez-Lâzaro. 2018a. Impacts of 40 years of land cover change on water quality in Tampa Bay, Florida. *Cogent Geoscience* 4:1422956. DOI: https://doi.org/10.1080/23312041.2017.1422956.

McCarthy, M.J., D.B. Otis, P. Méndez-Lázaro, and F.E. Muller-Karger. 2018b. Water quality drivers in 11 Gulf of Mexico estuaries. *Remote Sensing* 10(2):255. DOI: https://doi.org/10.3390/rs10020255.

McClain, M.E., E.W. Boyer, C.L. Dent, S.E. Gergel, N.B. Grimm, P.M. Groffman, S.C. Hart, J.W. Harvey, C.A. Johnston, E. Mayorga, W.H. McDowell, and G. Pinay. 2003. Biogeochemical hot spots and hot moments at the interface of terrestrial and aquatic ecosystems. *Ecosystems* 6(4):301-312. DOI: https://doi.org/10.1007/s10021-003-0161-9.

McCombs, J.W., N.D. Herold, S. Burkhalter, and C.J. Robinson. 2016. Accuracy assessment of NOAA coastal change analysis program 2006-2010 land cover and land cover change data. *Photogrammetric Engineering and Remote Sensing* 82:711-718.

McDonald, R.B., R.M. Moody, K.L. Heck, and J. Cebrian. 2016. Fish, macroinvertebrate and epifaunal communities in shallow coastal lagoons with varying seagrass cover of the northern Gulf of Mexico. *Estuaries and Coasts* 39(3):718-730. DOI: https://doi.org/10.1007/s12237-015-0031-7.

McKinley, D.C., A.J. Miller-Rushing, H.L. Ballard, R. Bonney, H. Brown, S.C. Cook-Patton, D.M. Evans, R.A. French, J.K. Parrish, T.B. Phillips, S.F. Ryan, L.A. Shanley, J.L. Shirk, K.F. Stepenuck, J.F. Weltzin, A. Wiggins, O.D. Boyle, R.D. Briggs, S.F. Chapin, D.A. Hewitt, P.W. Preuss, and M.A. Soukup. 2017. Citizen science can improve conservation science, natural resource management, and environmental protection. *Biological Conservation* 208:15-28. DOI: https://doi.org/10.1016/j.biocon.2016.05.015.

McKinney, L.D., J.W. Tunnell, J. Beseres Pollack, W. Dennison, J. Francis, H. Kelsey, C. Onuf, G. Stunz, M. Wetz, K. Withers, M.A. Harwell, and J.H. Gentile. 2019. *Texas Coast Ecosystem Health Report Card.* Corpus Christi, TX: Harte Research Institute.

McLellan, E., D. Robertson, K. Schilling, M. Tomer, J. Kostel, D. Smith, and K. King. 2015. Reducing nitrogen export from the corn belt to the Gulf of Mexico: Agricultural strategies for remediating hypoxia. *Journal of the American Water Resources Association* 51(1):263-289. DOI: https://doi.org/10.1111/jawr.12246.

McLeod, E., G.L. Chmura, S. Bouillon, R. Salm, M. Björk, C.M. Duarte, C.E. Lovelock, W.H. Schlesinger, and B.R. Silliman. 2011. A blueprint for blue carbon: Toward an improved understanding of the role of vegetated coastal habitats in sequestering CO_2. *Frontiers in Ecology and the Environment* 9(10):552-560. DOI: https://doi.org/10.1890/110004.

McLoughlin, C.A., M.C. Thoms, and M. Parsons. 2020. Reflexive learning in adaptive management: A case study of environmental water management in the Murray Darling Basin, Australia. *River Research and Applications* 36(4):681-694. DOI: https://doi.org/10.1002/rra.3607.

McNutt, M.K., R. Camilli, T.J. Crone, G.D. Guthrie, P.A. Hsieh, T.B. Ryerson, O. Savas, and F. Shaffer. 2012. Review of flow rate estimates of the Deepwater Horizon oil spill. *Proceedings of the National Academy of Sciences* 109(50):20260-20267. DOI: https://doi.org/10.1073/pnas.1112139108.

Meade, R.H., and J.A. Moody. 2010. Causes for the decline of suspended-sediment discharge in the Mississippi River system, 1940–2007. *Hydrological Processes* 24(1):35-49. DOI: https://doi.org/10.1002/hyp.7477.

Megrey, B.A., J.S. Link, G.L. Hunt, and E. Moksness. 2009. Comparative marine ecosystem analysis: Applications, opportunities, and lessons learned. *Progress in Oceanography* 81(1):2-9. DOI: https://doi.org/10.1016/j.pocean.2009.04.002.

Meiman J. and M. Segura. 2019. *Monitoring seagrass in national parks of the Gulf Coast network: Protocol narrative.* Natural Resource Report. NPS/GULN/NRR—2019/1973. Fort Collins, CO: National Park Service.

Meli, P., J.M. Rey Benayas, P. Balvanera, and M. Martínez Ramos. 2014. Restoration enhances wetland biodiversity and ecosystem service supply, but results are context-dependent: A meta-analysis. *PLoS ONE* 9(4):e93507. DOI: https://doi.org/10.1371/journal.pone.0093507.

Mendelssohn, I.A., and K.L. McKee. 2000. Saltmarshes and mangroves. In *North American terrestrial vegetation* (2nd ed.), edited by M.G. Barbour and W.D. Billings. New York: Cambridge University Press.

Mendoza R., N. Arreaga, J. Hernández, V. Segovia, I. Jasso, and D. Pérez. 2011. Aquatic Invasive Species in the Rio Bravo / Laguna Madre Ecological Region. Background Paper 2011-2. Montreal, Canada: Commission for Environmental Cooperation. 146 pp. http://www3.cec.org/islandora/en/item/10259-aquatic-invasive-species-in-rio-bravolaguna-madre-ecological-region-en.pdf

Merritt, W.S., R.A. Letcher, and A.J. Jakeman. 2003. A review of erosion and sediment transport models. *Environmental Modeling & Software* 18(8):761-799. DOI: https://doi.org/10.1016/S1364-8152(03)00078-1.

Meselhe, E., A.M. Khalifa, K. Hu, J. Lewis, and A.A. Tavakoly. 2022. Influence of key environmental drivers on the performance of sediment diversions. *Water* 14(1):24.

Meyer, C.A., and R. Pu. 2012. Seagrass resource assessment using remote sensing methods in St. Joseph Sound and Clearwater Harbor, Florida, USA. *Environmental Monitoring and Assessment* 184(2):1131-1143. DOI: https://doi.org/10.1007/s10661-011-2028-4.

Middleton, B.A. 2009. Regeneration of coastal marsh vegetation impacted by hurricanes Katrina and Rita. *Wetlands* 29(1):54-65. DOI: https://doi.org/10.1672/08-18.1.

Millette, N.C., C. Kelble, A. Linhoss, S. Ashby, L. Visser, and SpringerLink. 2019. Using spatial variability in the rate of change of chlorophyll a to improve water quality management in a subtropical oligotrophic estuary. *Estuaries and Coasts* 42:1792-1803. DOI: https://doi.org/10.1007/s12237-019-00610-5.

Mishra, D.R., A. Kumar, L. Ramaswamy, V.K. Boddula, M.C. Das, B.P. Page, and S.J. Weber. 2020. CyanoTRACKER: A cloud-based integrated multi-platform architecture for global observation of cyanobacterial harmful algal blooms. *Harmful Algae* 96:101828. DOI: https://doi.org/10.1016/j.hal.2020.101828.

Mize, S., J.C. Murphy, T.H. Diehl, and D.K. Demcheck. 2018. Suspended-sediment concentrations and loads in the lower Mississippi and Atchafalaya rivers decreased by half between 1980 and 2015. *Journal of Hydrology* 564:1-11. DOI: https://doi.org/10.1016/j.jhydrol.2018.05.068.

Mo, Y., M.S. Kearney, and R.E. Turner. 2020. The resilience of coastal marshes to hurricanes: The potential impact of excess nutrients. *Environment International* 138. DOI: https://doi.org/10.1016/j.envint.2019.105409.

Monbet, Y. 1992. Control of phytoplankton biomass in estuaries: A comparative analysis of microtidal and macrotidal estuaries. *Estuaries* 15(4):563-571. DOI: https://doi.org/10.2307/1352398.

Montagna, P.A., X. Hu, T.A. Palmer, and M. Wetz. 2018. Effect of hydrological variability on the biogeochemistry of estuaries across a regional climatic gradient. *Limnology and Oceanography* 63(6):2465-2478. DOI: https://doi.org/10.1002/lno.10953.

Montalvo, A.J., C.K. Faulk, and G.J. Holt. 2012. Sex determination in southern flounder, Paralichthys lethostigma, from the Texas Gulf Coast. *Journal of Experimental Marine Biology and Ecology* 432-433:186-190. DOI: https://doi.org/10.1016/j.jembe.2012.07.017.

Montambault, J.R., S. Wongbusarakum, T. Leberer, E. Joseph, W. Andrew, F. Castro, B. Nevitt, Y. Golbuu, N.W. Oldiais, C.R. Groves, W. Kostka, and P. Houk. 2015. Use of monitoring data to support conservation management and policy decisions in Micronesia. *Conservation Biology* 29(5):1279-1289. DOI: https://doi.org/10.1111/cobi.12542.

Moore, K., B. Neikirk, E.C. Shields, J. Jarvis, and D. Parrish. 2008. Water quality conditions and restoration of submerged aquatic vegetation (SAV) in the tidal freshwater James River 2007. *Special Reports in Applied Marine Science and Ocean Engineering* No. 401. Gloucester Point, VA: Virginia Institute of Marine Science. DOI: https://doi.org/10.21220/V51169.

Moore, C., A. Barnard, P. Fietzek, M.R. Lewis, H.M. Sosik, S. White, and O. Zielinski. 2009. Optical tools for ocean monitoring and research. *Ocean Science* 5(4):661-684. DOI: https://doi.org/10.5194/os-5-661-2009.

Moore, J.F., W.E. Pine, III, P.C. Frederick, S. Beck, M. Moreno, M.J. Dodrill, M. Boone, L. Sturmer, and S. Yurek. 2020. Trends in oyster populations in the northeastern Gulf of Mexico: An assessment of river discharge and fishing effects over time and space. *Marine and Coastal Fisheries* 12(3):191-204. DOI: https://doi.org/10.1002/mcf2.10117.

Moorman, M.C., A.B. Hoos, S.B. Bricker, R.B. Moore, A.M. García, and S.W. Ator. 2014. *Nutrient load summaries for major lakes and estuaries of the eastern United States, 2002.* Reston, VA: U.S. Geological Survey.

Moreno-Madrinan, M.J., M.Z. Al-Hamdan, D.L. Rickman, and F.E. Muller-Karger. 2010. Using the surface reflectance MODIS terra product to estimate turbidity in Tampa Bay, Florida. *Remote Sensing* 2(12):2713-2728. DOI: https://doi.org/10.3390/rs2122713.

Moreno-Mateos, D., and F.A. Comin. 2010. Integrating objectives and scales for planning and implementing wetland restoration and creation in agricultural landscapes. *Journal of Environmental Management* 91(11):2087-2095. DOI: https://doi.org/10.1016/j.jenvman.2010.06.002.

Moreno-Mateos, D., A. Alberdi, E. Morriën, W.H. van der Putten, A. Rodríguez-Uña, and D. Montoya. 2020. The long-term restoration of ecosystem complexity. *Nature Ecology and Evolution* 4(5):676-685.

Moriasi, N., M. Daniel, W. Gitau, N. Pai, and P. Daggupati. 2015. Hydrologic and water quality models: Performance measures and evaluation criteria. *Transactions of the ASABE* 58(6):1763-1785. DOI: https://doi.org/10.13031/trans.58.10715.

Morton, R.A., and J.A. Barras. 2011. Hurricane impacts on coastal wetlands: A half-century record of storm-generated features from Southern Louisiana. *Journal of Coastal Research* 27(6 A):27-43. DOI: https://doi.org/10.2112/JCOASTRES-D-10-00185.1.

Moser, S.C., M.A. Davidson, P. Kirshen, P. Mulvaney, J. F. Murley, J.E. Neumann, L. Petes, and D. Reed. 2014. Coastal zone development and ecosystems. In *Climate change impacts in the United States: The third national climate assessment*, edited by J.M. Melillo, T.C. Terese Richmond, and G.W. Yohe. U.S. Global Change Research Program. Pp. 579-618. DOI: https://doi.org/10.7930/J0MS3QNW.

Murawski, S.A., J.W. Fleeger, W.F. Patterson, III, C. Hu, K. Daly, I. Romero, and G.A. Toro-Farmer. 2016. How did the Deepwater Horizon oil spill affect coastal and continental shelf ecosystems of the Gulf of Mexico? *Oceanography* 29(3):160-173. DOI: https://doi.org/10.5670/oceanog.2016.80.

Murawski, S.A., J.P. Kilborn, A.C. Bejarano, D. Chagaris, D. Donaldson, F.J. Hernandez, T.C. MacDonald, C. Newton, E. Peebles, and K.L. Robinson. 2021. A synthesis of Deepwater Horizon impacts on coastal and nearshore living marine resources. *Frontiers in Marine Science* 7(1212). DOI: https://doi.org/10.3389/fmars.2020.594862.

Murphy, J.C., R.M. Hirsch, and L.A. Sprague. 2013. *Nitrate in the Mississippi River and its tributaries, 1980–2010—An update*. U.S. Geological Survey Scientific Investigations Report 2013–5169, 31 p.

Nagarkar, M., and K. Raulund-Rasmussen. 2016. An appraisal of adaptive management planning and implementation in ecological restoration: Case studies from the San Francisco Bay Delta, USA. *Ecology and Society* 21. DOI: https://doi.org/10.5751/ES-08521-210243.

Nagelkerken, I., M. Sheaves, R. Baker, and R.M. Connolly. 2015. The seascape nursery: a novel spatial approach to identify and manage nurseries for coastal marine fauna. *Fish and Fisheries* 16(2):362-371. DOI: https://doi.org/10.1111/faf.12057.

NASEM (National Academies of Sciences, Engineering, and Medicine). 2015. *Continuity of NASA Earth observations from space: A value framework*. Washington, DC: The National Academies Press.

NASEM. 2016. *Achieving science with cubesats: Thinking inside the box*. Washington, DC: The National Academies Press.

NASEM. 2017. *Effective monitoring to evaluate ecological restoration in the Gulf of Mexico*. Washington, DC: The National Academies Press.

NASEM. 2018a. *Understanding the long-term evolution of the coupled natural-human coastal system: The future of the U.S. Gulf Coast*. Washington, DC: The National Academies Press.

NASEM. 2018b. *Progress toward restoring the Everglades: The seventh biennial review—2018*. Washington, DC: The National Academies Press.

NASEM. 2018c. *Thriving on our changing planet: A decadal strategy for earth observation from space*. Washington, DC: The National Academies Press. DOI: https://doi.org/10.17226/24938.

NASEM. 2019. *The Gulf research program annual report 2018*. Washington, DC: The National Academies Press.

NASEM. 2020. *Advancing aerial mobility: A national blueprint*. Washington, DC: The National Academies Press.

NASEM. 2021. *Progress toward restoring the everglades: The eighth biennial review—2020*. Washington, DC: The National Academies Press.

National Infrastructure Commission (UK). 2021. Natural Capital and Environmental Net Gain: A Discussion Paper. https://nic.org.uk/app/uploads/Updated-Natural-Capital-Paper-Web-Version-Feb-2021.pdf

Needham, H.F., and B.D. Keim. 2012. A storm surge database for the US Gulf Coast. *International Journal of Climatology* 32(14):2108-2123. DOI: https://doi.org/10.1002/joc.2425.

Neeson, T.M., M.C. Ferris, M.W. Diebel, P.J. Doran, J.R. O'Hanley, and P.B. McIntyre. 2015. Enhancing ecosystem restoration efficiency through spatial and temporal coordination. *Proceedings of the National Academy of Sciences* 112(19):6236-6241. DOI: https://doi.org/10.1073/pnas.1423812112.

Neeson, T.M., S.D. Smith, J.D. Allan, and P.B. McIntyre. 2016. Prioritizing ecological restoration among sites in multi-stressor landscapes. *Ecological Applications: A Publication of the Ecological Society of America* 26(6):1785-1796. DOI: https://doi.org/10.1890/15-0948.1.

Nehir, M., M. Esposito, C. Begler, C. Frank, O. Zielinski, and E.P. Achterberg. 2021. Improved calibration and data processing procedures of OPUS optical sensor for high-resolution in situ monitoring of nitrate in seawater. *Frontiers in Marine Science* 8. DOI: https://doi.org/10.3389/fmars.2021.663800.

Nel, J.L., and D.J. Roux. 2018. *Planning for the protection and management of freshwater ecosystems inside and outside protected areas*. London: Routledge.

Nelson, D. 2015. *Estuarine salinity zones in Gulf of Mexico Atlas*. Stennis Space Center (MS): National Centers for Environmental Information. https://www.ncei.noaa.gov/maps/gulf-data-atlas/atlas.htm?plate=Salinity-Zones.

NFWF (National Fish and Wildlife Foundation). 2020. *Gulf environmental benefit fund, five-year report, 2013–2018.* Lafayette, LA: National Fish and Wildlife Foundation.

Nienhuis, J.H., T.E. Törnqvist, K.L. Jankowski, A.M. Fernandes, and M.E. Keogh. 2017. A new subsidence map for coastal Louisiana. *GSA Today* 27(9):58-59. DOI: https://doi.org/10.1130/GSATG337GW.1.

Niven, D.K., and G.S. Butcher. 2011. Status and trends of wintering coastal species along the northern Gulf of Mexico, 1965-2011. *American Birds* 65:12-19.

Nixon, S.W. 1987. Chesapeake Bay nutrient budgets—A reassessment. *Biogeochemistry* 4(1):77-90. DOI: https://doi.org/10.1007/BF02187363.

Nixon, S.W. 1995. Coastal marine eutrophication: A definition, social causes, and future concerns. *Ophelia* 41(1):199-219. DOI: https://doi.org/10.1080/00785236.1995.10422044.

Nixon, Z., S. Zengel, M. Baker, M. Steinhoff, G. Fricano, S. Rouhani, and J. Michel. 2016. Shoreline oiling from the Deepwater Horizon oil spill. *Marine Pollution Bulletin* 107(1):170-178. DOI: https://doi.org/10.1016/j.marpolbul.2016.04.003.

NMFS (National Marine Fisheries Service). 2021. *Fisheries of the United States, 2019.* https://media.fisheries.noaa.gov/2021-05/FUS2019-FINAL-webready-2.3.pdf?null.

NOAA (National Oceanic and Atmospheric Administration). 2015a. *Guidance for considering the use of living shorelines.* Silver Spring, MD: National Oceanic and Atmospheric Administration. https://www.habitatblueprint.noaa.gov/wp-content/uploads/2018/01/NOAA-Guidance-for-Considering-the-Use-of-Living-Shorelines_2015.pdf.

NOAA. 2015b. *A strategy for a healthy Gulf of Mexico*. Silver Spring, MD: National Oceanic and Atmospheric Administration.

NOAA and USGS (National Oceanic and Atmospheric Administration and U.S. Geological Survey). 2019. *Council Monitoring and Assessment Program (CMAP): Compilation of existing habitat and water quality monitoring, and mapping assessments for the Gulf of Mexico region*. NOAA NOS NCCOS Technical Memorandum 269. Silver Spring, MD: National Oceanic and Atmospheric Administration and U.S. Geological Survey. DOI: https://doi.org/10.25923/bpj6-z187.

Nolte, S., E.C. Koppenaal, P. Esselink, K.S. Dijkema, M. Schuerch, A.V. De Groot, J.P. Bakker, and S. Temmerman. 2013. Measuring sedimentation in tidal marshes: A review on methods and their applicability in biogeomorphological studies. *Journal of Coastal Conservation* 17(3):301-325. DOI: https://doi.org/10.1007/s11852-013-0238-3.

Norouzzadeh, M.S., A. Nguyen, M. Kosmala, A. Swanson, M.S. Palmer, C. Packer, and J. Clune. 2018. Automatically identifying, counting, and describing wild animals in camera-trap images with deep learning. *Proceedings of the National Academy of Sciences* 115(25):E5716-E5725. DOI: https://doi.org/10.1073/pnas.1719367115.

Norris, R.H., J.A. Webb, S.J. Nichols, M.J. Stewardson, and E.T. Harrison. 2012. Analyzing cause and effect in environmental assessments: Using weighted evidence from the literature. *Freshwater Science* 31(1):5-21. DOI: https://doi.org/10.1899/11-027.1.

Nowlin, W., A. Jochens, B. Kirkpatrick. 2015. *Harmful Algal Bloom Integrated Observing System (HABIOS) Plan.* https://gcoos.org/wp-content/uploads/2020/01/HABIOS-Plan_final_9_3_15.pdf.

NRC (National Research Council). 1992. *Restoration of aquatic ecosystems: Science, technology, and public policy.* Washington, DC: The National Academies Press.

NRC. 2000. *Improving the collection, management, and use of marine fisheries data.* Washington, DC: The National Academies Press.

NRC. 2004. *Adaptive management for water resources project planning.* Washington, DC: The National Academies Press.

NRC. 2010. *Assessment of sea-turtle status and trends: Integrating demography and abundance.* Washington, DC: The National Academies Press.

NRC. 2012. *Improving water quality in the Mississippi River basin and Northern Gulf of Mexico: Strategies and priorities.* Washington, DC: The National Academies Press.

NRC. 2013. *An ecosystem services approach to assessing the impacts of the Deepwater Horizon Oil Spill in the Gulf of Mexico.* Washington, DC: The National Academies Press.

Nundloll, V., B. Porter, G.S. Blair, B. Emmett, J. Cosby, D.L. Jones, D. Chadwick, B. Winterbourn, P. Beattie, G. Dean, R. Shaw, W. Shelley, M. Brown, and I. Ullah. 2019. The Design and Deployment of an End-To-End IoT Infrastructure for the Natural Environment. *Future Internet* 11(6):129. DOI: https://doi.org/10.3390/fi11060129.

Nyman, J.A., and R.H. Chabreck. 1995a. *Fire in coastal marshes: History and recent concerns.* Presented at Proceedings of the Tall Timbers Fire Ecology Conference, No. 19, Tallahassee, FL.

Nyman, J.A., C.R. Crozier, and R.D. DeLaune. 1995b. Roles and patterns of hurricane sedimentation in an estuarine marsh landscape. *Estuarine, Coastal and Shelf Science* 40(6):665-679. DOI: https://doi.org/10.1006/ecss.1995.0045.

Nyman, J.A., R.J. Walters, R.D. Delaune, and W.H. Patrick Jr. 2006. Marsh vertical accretion via vegetative growth. *Estuarine, Coastal and Shelf Science* 69(3-4):370-380. DOI: https://doi.org/10.1016/j.ecss.2006.05.041.

O'Connell, J.L., and M. Alber. 2016. A smart classifier for extracting environmental data from digital image time-series: Applications for PhenoCam data in a tidal salt marsh. *Environmental Modelling & Software* 84:134-139. DOI: https://doi.org/10.1016/j.envsoft.2016.06.025.

O'Connell, J.L., D.R. Mishra, M. Alber, and K.B. Byrd. 2021. BERM: A belowground ecosystem resiliency model for estimating Spartina alterniflora belowground biomass. *New Phytologist* 232(1):425-439. DOI: https://doi.org/10.1111/nph.17607.

O'Connor, B.S., F.E. Muller-Karger, R.W. Nero, C. Hu, and E.B. Peebles. 2016. The role of Mississippi River discharge in offshore phytoplankton blooming in the northeastern Gulf of Mexico during August 2010. *Remote Sensing of Environment* 173:133-144. DOI: https://doi.org/10.1016/j.rse.2015.11.004.

Odum, H.T. 1971. *Environment, Power, and Society.* New York: Wiley Interscience.

Odum, H.T., and B.J. Copeland. 1972. Functional classification of coastal ecological systems of the United States. *Memoir of the Geological Society of America* 133:9-28. DOI: https://doi.org/10.1130/MEM133-p9.

Oelsner, G.P., and E.G. Stets. 2019. Recent trends in nutrient and sediment loading to coastal areas of the conterminous U.S.: Insights and global context. *Science of the Total Environment* 654:1225-1240. DOI: https://doi.org/10.1016/j.scitotenv.2018.10.437.

Oesterwind, D., A. Rau, and A. Zaiko. 2016. Drivers and pressures—Untangling the terms commonly used in marine science and policy. *Journal of Environmental Management* 181:8-15. DOI: https://doi.org/10.1016/j.jenvman.2016.05.058.

O'Farrell, H., A. Grüss, S.R. Sagarese, E.A. Babcock, and K.A. Rose. 2017. Ecosystem modeling in the Gulf of Mexico: Current status and future needs to address ecosystem-based fisheries management and restoration activities. *Reviews in Fish Biology and Fisheries* 27(3):587-614. DOI: https://doi.org/10.1007/s11160-017-9482-1.

Ogden, J.C., J.D. Baldwin, O.L. Bass, J.A. Browder, M.I. Cook, P.C. Frederick, P.E. Frezza, R.A. Galvez, A.B. Hodgson, and K.D. Meyer. 2014. Waterbirds as indicators of ecosystem health in the coastal marine habitats of Southern Florida: 2. Conceptual ecological models. *Ecological Indicators* 44:128-147. DOI: https://doi.org/10.1016/j.ecolind.2014.03.008.

O'Gorman, E.J., J.E. Fitch, and T.P. Crowe. 2012. Multiple anthropogenic stressors and the structural properties of food webs. *Ecology* 93(3):441-448. DOI: https://doi.org/10.1890/11-0982.1.

Ojha, T., S. Misra, and N.S. Raghuwanshi. 2015. Wireless sensor networks for agriculture: The state-of-the-art in practice and future challenges. *Computers and Electronics in Agriculture* 118:66-84. DOI: https://doi.org/10.1016/j.compag.2015.08.011.

Olander, L., S. Mason, K. Warnell, and H. Tallis. 2018. *Building Ecosystem Services Conceptual Models.* National Ecosystem Services Partnership Conceptual Model Series No. 1. Durham, NC: Duke University, Nicholas Institute for Environmental Policy Solutions. https://nicholasinstitute.duke. edu/ conceptual-model-series.

Oliver, B., and J.J. Ramirez-Avila. 2019. Barrier Island restoration: A literature review. In *World Environmental and Water Resources Congress 2019.* Reston, VA: American Society of Civil Engineers. DOI: https://doi.org/10.1061/9780784482353.029.

Olds, A.D., I. Nagelkerken, C.M. Huijbers, B.L. Gilby, S.J. Pittman and T.A. Schlacher. 2017. Connectivity in coastal seascapes. In *Seascape Ecology*, edited by S. Pittman. Wiley-Blackwell. Pp. 261-292.

Olson, A.M., M. Hessing-Lewis, D. Haggarty, and F. Juanes. 2019. Nearshore seascape connectivity enhances seagrass meadow nursery function. *Ecological Applications* 29(5):e01897. DOI: https://doi. org/10.1002/eap.1897.

Omernik, J.M., and G.E. Griffith. 2014. Ecoregions of the conterminous United States: Evolution of a hierarchical spatial framework. *Environmental Management* 54:1249-1266. DOI: https://doi. org/10.1007/s00267-014-0364-1.

Orlando, S.P.J., L.P. Rozas, G.H. Ward, and C.J. Klein. 1993. *Salinity characteristics of Gulf of Mexico estuaries.* Silver Spring, MD: National Oceanic and Atmospheric Administration, National Ocean Service, Office of Ocean Resources Conservation and Assessment.

Ortega-Ortiz, A., A. Delgado-Estrella, and A. Ortega-Argueta. 2004. Marine mammals in the Gulf of Mexico: current knowledge and recommendations for their conservation. In *Environmental Analysis of the Gulf of Mexico: Special Publication Series No. 1.* Edited by M. Caso, I. Pisanty, E. Ezcurra, K. Withers, and M. Nipper. Galveston, TX: Harte Research Institute for Gulf of Mexico Studies.

Ortiz, J.-C., N.H. Wolff, K.R.N. Anthony, M. Devlin, S. Lewis, and P.J. Mumby. 2018. Impaired recovery of the Great Barrier Reef under cumulative stress. *Science Advances* 4(7):eaar6127. DOI: https://doi. org/10.1126/sciadv.aar6127.

Osland, M.J., A.C. Spivak, J.A. Nestlerode, J.M. Lessmann, A.E. Almario, P.T. Heitmuller, M.J. Russell, K.W. Krauss, F. Alvarez, D.D. Dantin, J.E. Harvey, A.S. From, N. Cormier, and C.L. Stagg. 2012. Ecosystem development after mangrove wetland creation: Plant–soil change across a 20-year chronosequence. *Ecosystems* 15(5):848-866.

Osland, M.J., N. Enwright, R.H. Day, and T.W. Doyle. 2013. Winter climate change and coastal wetland foundation species: Salt marshes vs. mangrove forests in the southeastern United States. *Global Change Biology* 19(5):1482-1494. DOI: https://doi.org/10.1111/gcb.12126.

Osland, M.J., R.H. Day, C.T. Hall, M.D. Brumfield, J.L. Dugas, and W.R. Jones. 2017a. Mangrove expansion and contraction at a poleward range limit: Climate extremes and land-ocean temperature gradients. *Ecology* 98(1):125-137. DOI: https://doi.org/10.1002/ecy.1625.

Osland, M.J., K.T. Griffith, J.C. Larriviere, L.C. Feher, D.R. Cahoon, N.M. Enwright, D.A. Oster, J.M. Tirpak, M.S. Woodrey, R.C. Collini, J.J. Baustian, J.L. Breithaupt, J.A. Cherry, J.R. Conrad, N. Cormier, C.A. Coronado-Molina, J.F. Donoghue, S.A. Graham, J.W. Harper, M.W. Hester, R.J. Howard, K.W. Krauss, D.E. Kroes, R.R. Lane, K.L. McKee, I.A. Mendelssohn, B.A. Middleton, J.A. Moon, S.C. Piazza, N.M. Rankin, F.H. Sklar, G.D. Steyer, K.M. Swanson, C.M. Swarzenski, W.C. Vervaeke, J.M. Willis, and K.V. Wilson. 2017b. Assessing coastal wetland vulnerability to sea-level rise along the northern Gulf of Mexico coast: Gaps and opportunities for developing a coordinated regional sampling network. *PLoS ONE* 12(9):e0183431. DOI: https://doi.org/10.1371/journal.pone.0183431.

Osland, M.J., R.H. Day, C.T. Hall, L.C. Feher, A.R. Armitage, J. Cebrian, K.H. Dunton, A.R. Hughes, D.A. Kaplan, A.K. Langston, A. Macy, C.A. Weaver, G.H. Anderson, K. Cummins, I.C. Feller, and C.M. Snyder. 2020. Temperature thresholds for black mangrove (Avicennia germinans) freeze damage, mortality and recovery in North America: Refining tipping points for range expansion in a warming climate. *Journal of Ecology* 108(2):654-665. DOI: https://doi.org/10.1111/1365-2745.13285.

Osland, M.J., P.W. Stevens, M.M. Lamont, R.C. Brusca, K.M. Hart, J.H. Waddle, C.A. Langtimm, C.M. Williams, B.D. Keim, A.J. Terando, E.A. Reyier, K.E. Marshall, M.E. Loik, R.E. Boucek, A.B. Lewis, and J.A. Seminoff. 2021. Tropicalization of temperate ecosystems in North America: The northward range expansion of tropical organisms in response to warming winter temperatures. *Global Change Biology* 27(13):3009-3034. DOI: https://doi.org/10.1111/gcb.15563.

Osland, M.J., A.R. Hughes, A.R. Armitage, S.B. Scyphers, J. Cebrian, S.H. Swinea, C.C. Shepard, M.S. Allen, L.C. Feher, J.A. Nelson, C.L. O'Brien, Colt R. Sanspree, D.L. Smee, C.M. Snyder, A.P. Stetter, P.W. Stevens, K.M. Swanson, L.H. Williams, J.M. Brush, J. Marchionno, and R. Bardou. 2022. The impacts of mangrove range expansion on wetland ecosystem services in the southeastern United States: Current understanding, knowledge gaps, and emerging research needs. *Global Change Biology* 28:3163-3187. DOI: https://doi.org/10.1111/gcb.16111.

Page, B.P., A. Kumar, and D.R. Mishra. 2018. A novel cross-satellite based assessment of the spatio-temporal development of a cyanobacterial harmful algal bloom. *International Journal of Applied Earth Observation and Geoinformation* 66:69-81. DOI: https://doi.org/10.1016/j.jag.2017.11.003.

Pahl-Wostl, C. 2009. A conceptual framework for analysing adaptive capacity and multi-level learning processes in resource governance regimes. *Global Environmental Change* 19(3):354-365. DOI: https://doi.org/10.1016/j.gloenvcha.2009.06.001.

Paine, R.T., M.J. Tegner, and E.A. Johnson. 1998. Compounded perturbations yield ecological surprises. *Ecosystems* 1(6):535-545. DOI: https://doi.org/10.1007/s100219900049.

Palmer, M.A., J.B. Zedler, and D.A. Falk. 2016. Ecological theory and restoration ecology. In *Foundations of restoration ecology,* edited by M.A. Palmer, J.B. Zedler and D.A. Falk. Washington, DC: Island Press/Center for Resource Economics.

Palomo, I., L. Willemen, E. Drakou, B. Burkhard, N. Crossman, C. Bellamy, K. Burkhard, C.S. Campagne, A. Dangol, J. Franke, S. Kulczyk, S. Le Clec'h, D. Abdul Malak, L. Muñoz, V. Narusevicius, S. Ottoy, J. Roelens, L. Sing, A. Thomas, K. Van Meerbeek, and P. Verweij. 2018. Practical solutions for bottlenecks in ecosystem services mapping. *One Ecosystem* 3:e20713.

Parashar, J., S.M. Bhandarkar, J. Simon, B.M. Hopkinson, and S.C. Pennings. 2021. *Estimation of abundance and distribution of salt marsh plants from images using deep learning.* Presented at 2020 25th International Conference on Pattern Recognition (ICPR), 10-15 Jan. 2021. DOI: https://doi.org/10.1109/ICPR48806.2021.9412264.

Passeri, D.L., S.C. Hagen, S.C. Medeiros, M.V. Bilskie, K. Alizad, and D. Wang. 2015. The dynamic effects of sea level rise on low-gradient coastal landscapes: A review. *Earth's Future* 3(6):159-181. DOI: https://doi.org/10.1002/2015EF000298.

Pastorok, R.A., A. MacDonald, J.R. Sampson, W. Pace, D.J. Yozzo, and J.P. Titre. 1997. An ecological decision framework for environmental restoration projects. *Ecological Engineering* 9(1):89-107. DOI: https://doi.org/10.1016/S0925-8574(97)00036-0.

Paull, L., S. Saeedi, M. Seto, and H. Li. 2013. Sensor-driven online coverage planning for autonomous underwater vehicles. *IEEE/ASME Transactions on Mechatronics* 18(6):1827-1838. DOI: https://doi.org/10.1109/tmech.2012.2213607.

Payne, J.R., T.J. Reilly, R.J. Martrano, G.P. Lindblom, M.C. Kennicutt Ii, and J.M. Brooks. 2005. Spill-of-opportunity testing of dispersant effectiveness at the Mega Borg oil spill. Presented at 2005 International Oil Spill Conference, IOSC 2005. May 15-19, Miami Beach, FL.

Pearlstine, L.G., J.M. Beerens, G. Reynolds, S.M. Haider, M. McKelvy, K. Suir, S.S. Romañach, and J.H. Nestler. 2020. Near-term spatial hydrologic forecasting in Everglades, USA for landscape planning and ecological forecasting. *Environmental Modelling & Software* 132:104783. DOI: https://doi.org/10.1016/j.envsoft.2020.104783.

Pecl, G.T., M.B. Araújo, J.D. Bell, J. Blanchard, T.C. Bonebrake, I.-C. Chen, T.D. Clark, R.K. Colwell, F. Danielsen, B. Evengård, L. Falconi, S. Ferrier, S. Frusher, R.A. Garcia, R.B. Griffis, A.J. Hobday, C. Janion-Scheepers, M.A. Jarzyna, S. Jennings, J. Lenoir, H.I. Linnetved, V.Y. Martin, P.C. McCormack, J. McDonald, N.J. Mitchell, T. Mustonen, J.M. Pandolfi ,N. Pettorelli, E. Popova, S.A. Robinson, B.R. Scheffers, J.D. Shaw, C.J.B. Sorte, J.M. Strugnell, J.M. Sunday, M.-N. Tuanmu, A. Vergés, C. Villanueva, T. Wernberg, E. Wapstra, and S.E. Williams. 2017. Biodiversity redistribution under climate change: Impacts on ecosystems and human well-being. *Science* 355(6332):eaai9214. DOI: https://doi.org/10.1126/science.aai9214.

Pendleton, E.A., E.R. Thieler, and S. Jeffress Williams. 2010. Importance of coastal change variables in determining vulnerability to sea- and lake-level change. *Journal of Coastal Research* 26(1(261)):176-183. DOI: https://doi.org/10.2112/08-1102.1.

Peralta, G., T. Bouma, J. Soelen, J. Pérez-Lloréns, and I. Hernández. 2003. On the use of sediment fertilization for seagrass restoration: A mesocosm study on Zostera marina L. *Aquatic Botany* 75:95-110. DOI: https://doi.org/10.1016/S0304-3770(02)00168-7.

Peters, D.P.C. 2010. Accessible ecology: Synthesis of the long, deep, and broad. *Trends in Ecology & Evolution* 25(10):592-601. DOI: https://doi.org/10.1016/j.tree.2010.07.005.

Peters, D.P.C., and G.S. Okin. 2017. A toolkit for ecosystem ecologists in the time of big science. *Ecosystems* 20(2):259-266. DOI: https://doi.org/10.1007/s10021-016-0072-1.

Peters, D.P.C., K.M. Havstad, J. Cushing, C. Tweedie, O. Fuentes, and N. Villanueva-Rosales. 2014. Harnessing the power of big data: Infusing the scientific method with machine learning to transform ecology. *Ecosphere* 5(6). DOI: https://doi.org/10.1890/es13-00359.1.

Petersen, J.K., J.W. Hansen, M.B. Laursen, P. Clausen, J. Carstensen, and D.J. Conley. 2008. Regime shift in a coastal marine ecosystem. *Ecological Applications* 18(2):497-510. DOI: https://doi.org/10.1890/07-0752.1.

Peters-Lidard, C.D., M. Clark, L. Samaniego, N.E.C. Verhoest, T. van Emmerik, R. Uijlenhoet, K. Achieng, T.E. Franz, and R. Woods. 2017. Scaling, similarity, and the fourth paradigm for hydrology. *Hydrology and Earth System Sciences* 21(7):3701-3713. DOI: https://doi.org/10.5194/hess-21-3701-2017.

Peterson, C.H., and R.N. Lipcius. 2003. Conceptual progress towards predicting quantitative ecosystem benefits of ecological restorations. *Marine Ecology Progress Series*. 264:297-307. DOI: https://doi.org/10.3354/meps264297.

Peterson, C.H., J.H. Grabowski and S.P. Powers. 2003. Estimated enhancement of fish production resulting from restoring oyster reef habitat: Quantitative valuation. *Marine Ecology Progress Series*. 264:249-264.

Peterson, C.H., K.P. Franklin, and E.E. Cordes. 2020. Connectivity corridor conservation: A conceptual model for the restoration of a changing Gulf of Mexico ecosystem. *Elementa: Science of the Anthropocene* 8(1). DOI: https://doi.org/10.1525/elementa.016.

Petkewich, M., K. Lackstrom, B.J. McCloskey, L.F. Rouen, and P.A. Conrads. 2019. *Coastal salinity index along the southeastern Atlantic Coast and the Gulf of Mexico, 1983 to 2018*. Reston, VA: U.S. Geological Survey Open-File Report 2019–1090.

Phillips, P.J., C.A. Hahn, P.C. Fontana, A.N. Yates, K. Greene, D.A. Broniatowski, and M.A. Przybocki. 2021. *Four Principles of Explainable Artificial Intelligence*. National Institute of Standards and Technology Interagency or Internal Report 8312. 43 pp. DOI: https://doi.org/10.6028/NIST.IR.8312

Piacenza, S.E., P.M. Richards, and S.S. Heppell. 2019. Fathoming sea turtles: Monitoring strategy evaluation to improve conservation status assessments. *Ecological Applications* 29(6). DOI: https://doi.org/10.1002/eap.1942.

Piggott, J.J., C.R. Townsend, and C.D. Matthaei. 2015. Reconceptualizing synergism and antagonism among multiple stressors. *Ecology and Evolution* 5(7):1538-1547. DOI: https://doi.org/10.1002/ece3.1465.

Pinckney, J.L., A.A. Quigg, and D.L. Roelke, 2017. Interannual and seasonal patterns of estuarine phytoplankton diversity in Galveston Bay, Texas, USA. *Estuaries and Coasts* 40(1): 310-316. DOI: https://doi.org/10.1007/s12237-016-0135-8

Platt, T., and S. Sathyendranath. 1993. Comment on "The remote sensing of ocean primary productivity: Use of a new data compilation to test satellite algorithms" by William Balch et al. *Journal of Geophysical Research: Oceans* 98(C9):16583-16584. DOI: https://doi.org/10.1029/93JC01314.

Poloczanska, E.S., C.J. Brown, W.J. Sydeman, W. Kiessling, D.S. Schoeman, P.J. Moore, K. Brander, J.F. Bruno, L.B. Buckley, M.T. Burrows, C.M. Duarte, B.S. Halpern, J. Holding, C.V. Kappel, M.I. O'Connor, J.M. Pandolfi, C. Parmesan, F. Schwing, S.A. Thompson, and A.J. Richardson. 2013. Global imprint of climate change on marine life. *Nature Climate Change* 3(10):919-925. DOI: https://doi.org/10.1038/nclimate1958.

Potter, I.C., J.R. Tweedley, M. Elliott, and A.K. Whitfield. 2015. The ways in which fish use estuaries: A refinement and expansion of the guild approach. *Fish and Fisheries* 16:230-239. DOI: https://doi.org/10.1111/faf.12050.

Powell, E.J., M.C. Tyrrell, A. Milliken, J.M. Tirpak, and M.D. Staudinger. 2017. A synthesis of thresholds for focal species along the U.S. Atlantic and Gulf Coasts: A review of research and applications. *Ocean & Coastal Management* 148:75-88. DOI: https://doi.org/10.1016/j.ocecoaman.2017.07.012.

Prasad, P. 2015. Recent trend in wireless sensor network and its applications: A survey. *Sensor Review* 35(2):229-236. DOI: https://doi.org/10.1108/SR-08-2014-683.

Pu, R., S. Bell, and C. Meyer. 2014. Mapping and assessing seagrass bed changes in Central Florida's west coast using multitemporal Landsat TM imagery. *Estuarine, Coastal and Shelf Science* 149:68-79. DOI: https://doi.org/10.1016/j.ecss.2014.07.014.

Pulster, E.L., A. Gracia, M. Armenteros, G. Toro-Farmer, S.M. Snyder, B.E. Carr, M.R. Schwaab, T.J. Nicholson, J. Mrowicki, and S.A. Murawski. 2020. A first comprehensive baseline of hydrocarbon pollution in Gulf of Mexico fishes. *Scientific Reports* 10(1). DOI: https://doi.org/10.1038/s41598-020-62944-6.

Purtlebaugh, C.H., C.W. Martin, and M.S. Allen. 2020. Poleward expansion of common snook Centropomus undecimalis in the northeastern Gulf of Mexico and future research needs. *PLoS ONE* 15(6 June). DOI: https://doi.org/10.1371/journal.pone.0234083.

Quinn, N.W.T., R. Ortega, P.J.A. Rahilly, and C.W. Royer. 2010. Use of environmental sensors and sensor networks to develop water and salinity budgets for seasonal wetland real-time water quality management. *Environmental Modelling & Software* 25(9):1045-1058. DOI: https://doi.org/10.1016/j.envsoft.2009.10.011.

Rabalais, N.N., and R.E. Turner. 2019. Gulf of Mexico hypoxia: Past, present, and future. *Limnology and Oceanography Bulletin* 28(4):117-124. DOI: https://doi.org/10.1002/lob.10351.

Rajpar, M.N., I. Ozdemir, M. Zakaria, S. Sheryar and A. Rab. 2018. Seabirds as Bioindicators of Marine Ecosystems. In *Seabirds*, edited by H. Mikkola. IntechOpen. https://doi.org/10.5772/intechopen.75458.

Rammer, W., and R. Seidl. 2019. Harnessing deep learning in ecology: An example predicting bark beetle outbreaks. *Frontiers in Plant Science* 10(1327). DOI: https://doi.org/10.3389/fpls.2019.01327.

Ramsay, H. 2017. The global climatology of tropical cyclones. In *Oxford research encyclopedia of natural hazard science*. Oxford, UK: Oxford University Press. DOI: https://doi.org/10.1093/acrefore/9780199389407.013.79.

Raposa, K.B., S. Lerberg, C. Cornu, J. Fear, N. Garfield, C. Peter, R.L.J. Weber, G. Moore, D. Burdick, and M. Dionne. 2018. Evaluating tidal wetland restoration performance using National Estuarine Research Reserve System reference sites and the Restoration Performance Index (RPI). *Estuaries and Coasts* 41(1):36-51. DOI: https://doi.org/10.1007/s12237-017-0220-7.

Raulerson, G.E., E.T. Sherwood, M.C. Burke, H.S. Greening, and A.J. Janicki. 2019. The Tampa Bay story: Fostering collaborative partnerships to restore and urban estuary. In *Working watersheds and coastal systems: Research and management for a changing future—Proceedings of the Sixth Interagency Conference on Research in the Watersheds*, edited by J.S. Latimer, C.C. Trettin, D.D. Bosch, and C.R. Lane. Asheville, NC: U.S. Department of Agriculture Forest Service, Southern Research Station. https://drive.google.com/file/d/1BnLYdaFE8ch3p-zJGS9cq-bb1ud_Y1jS/view.

Rebich, R.A., N.A. Houston, S.V. Mize, D.K. Pearson, P.B. Ging, and C.E. Hornig. 2011. Sources and delivery of nutrients to the northwestern Gulf of Mexico from streams in the South-Central United States. *Journal of the American Water Resources Association* 47(5):1061-1086. DOI: https://doi.org/10.1111/j.1752-1688.2011.00583.x.

Reja, M.Y., S.D. Brody, W.E. Highfield, and G.D. Newman. 2017. Hurricane recovery and ecological resilience: Measuring the impacts of wetland alteration post Hurricane Ike on the Upper TX Coast. *Environmental Management* 60(6):1116-1126. DOI: https://doi.org/10.1007/s00267-017-0943-z.

Renzi, J.J., Q. He, and B.R. Silliman. 2019. Harnessing positive species interactions to enhance coastal wetland restoration. *Frontiers in Ecology and Evolution* 7(131). DOI: https://doi.org/10.3389/fevo.2019.00131.

RESTORE Council (Gulf Coast Ecosystem Restoration Council). 2016. *Comprehensive Plan Update 2016: Restoring the Gulf Coast's Ecosystem and Economy*. New Orleans, LA: Gulf Coast Ecosystem Restoration Council. https://www.restorethegulf.gov/sites/default/files/CO-PL_20161208_CompPlanUpdate_English.pdf.

RESTORE Council. 2019. *Gulf Coast Ecosystem Restoration Council Planning Framework*. New Orleans, LA: Gulf Coast Ecosystem Restoration. https://restorethegulf.gov/sites/default/files/508_PlanningFramework_Final_201908.pdf.

RESTORE Council. 2021a. *FY 2020 Annual Report to Congress*. New Orleans, LA: Gulf Coast Ecosystem Restoration Council. https://www.restorethegulf.gov/sites/default/files/FY%202020%20Annual%20Report%20to%20Congress_508.pdf.

RESTORE Council. 2021b. *Gulf Coast Ecosystem Restoration Council Observational Data Plan (ODP) Guidelines Version 2.0*. New Orleans, LA: Gulf Coast Ecosystem Restoration Council. https://restorethegulf.gov/sites/default/files/20210520_Council_Observational_Data_Plan_Guidelines_Version%202.0_508.pdf.

Reusch, T.B.H., J. Dierking, H.C. Andersson, E. Bonsdorff, J. Carstensen, M. Casini, M. Czajkowski, B. Hasler, K. Hinsby, K. Hyytiäinen, K. Johannesson, S. Jomaa, V. Jormalainen, H. Kuosa, S. Kurland, L. Laikre, B.R. MacKenzie, P. Margonski, F. Melzner, D. Oesterwind, H. Ojaveer, J.C. Refsgaard, A. Sandström, G. Schwarz, K. Tonderski, M. Winder, and M. Zandersen. 2018. The Baltic Sea as a time machine for the future coastal ocean. *Science Advances* 4(5):eaar8195. DOI: https://doi.org/10.1126/sciadv.aar8195.

Rey, J.R., D.B. Carlson, and R.E. Brockmeyer, Jr. 2012. Coastal wetland management in Florida: Environmental concerns and human health. *Wetlands Ecology and Management* 20(3):197-211. DOI: https://doi.org/10.1007/s11273-011-9235-1.

Rezek, R.J., B.T. Furman, R.P. Jung, M.O. Hall, and S.S. Bell. 2019. Long-term performance of seagrass restoration projects in Florida, USA. *Scientific Reports* 9(1):15514. DOI: https://doi.org/10.1038/s41598-019-51856-9.

Richardson, A.D., K. Hufkens, T. Milliman, D.M. Aubrecht, M. Chen, J.M. Gray, M.R. Johnston, T.F. Keenan, S.T. Klosterman, M. Kosmala, E.K. Melaas, M.A. Friedl, and S. Frolking. 2018. Tracking vegetation phenology across diverse North American biomes using PhenoCam imagery. *Scientific Data* 5(1):180028. DOI: https://doi.org/10.1038/sdata.2018.28.

Ridge, J.T., P.C. Gray, A.E. Windle, and D.W. Johnston. 2020. Deep learning for coastal resource conservation: Automating detection of shellfish reefs. *Remote Sensing in Ecology and Conservation* 6(4):431-440. DOI: https://doi.org/10.1002/rse2.134.

Ridlon, A.D., A. Marks, C.J. Zabin, D. Zacherl, B. Allen, J. Crooks, G. Fleener, E. Grosholz, B. Peabody, J. Toft, and K. Wasson. 2021. Conservation of marine foundation species: Learning from native oyster restoration from California to British Columbia. *Estuaries and Coasts* 44(7):1723-1743. DOI: https://doi.org/10.1007/s12237-021-00920-7.

Rinaldo, T., K.A. Ramakrishnan, I. Rodriguez-Iturbe, and O. Durán Vinent. 2021. Probabilistic structure of events controlling the after-storm recovery of coastal dunes. *Proceedings of the National Academy of Sciences* 118(1):e2013254118. DOI: https://doi.org/10.1073/pnas.2013254118.

Robbins, L.L., and J.T. Lisle. 2017. Regional acidification trends in Florida shellfish estuaries: A 20+ year look at pH, oxygen, temperature, and salinity. *Estuaries and Coasts* 41(5):1268-1281. DOI: https://doi.org/10.1007/s12237-017-0353-8.

Robertson, D.M., and D.A. Saad. 2013. SPARROW models used to understand nutrient sources in the Mississippi/Atchafalaya river basin. *Journal of Environmental Quality* 42(5):1422-40. DOI: https://doi.org/10.2134/jeq2013.02.0066.

Robertson, D.M., and D.A. Saad. 2019. *Spatially referenced models of streamflow and nitrogen, phosphorus, and suspended-sediment loads in streams of the midwestern United States*. Reston, VA: U.S. Geological Survey. DOI: https://doi.org/10.3133/sir20195114.

Robertson, D.M., and D.A. Saad. 2021. Nitrogen and phosphorus sources and delivery from the Mississippi/Atchafalaya river basin: An update using 2012 SPARROW models. *Journal of the American Water Resources Association* 57(3):406-429. DOI: https://doi.org/10.1111/1752-1688.12905.

Robison, D., T. Ries, J. Saarinen, D. Tomasko, and C. Sciarrino. 2020. *Tampa Bay estuary program 2020 habitat master plan update*. Technical Report #07-20 of the Tampa Bay Estuary Program. http://tampabay.wateratlas.usf.edu/upload/documents/TBEP_07-20_Robison_2020HabitatMasterPlanUpdate.pdf.

Rodgers, K.D., and C.M. Swarzenski. 2019. *Compilation of estuarine salinity data for sites used in RESTORE Streamflow alteration assessments (ver. 2.0, June 2021)*. DOI: https://doi.org/10.5066/P9G2MFRX.

Rodgers, K.D., A.B. Hoos, V.L. Roland, and R.R. Knight. 2018. *Trend analysis results for sites used in RESTORE Streamflow alteration assessments*. Reston, VA: U.S. Geological Survey data release. DOI: https://doi.org/10.5066/P9YSE754.

Rodgers, K., V. Roland, A. Hoos, E. Crowley-Ornelas, and R. Knight. 2020. An analysis of streamflow trends in the Southern and Southeastern US from 1950–2015. *Water* 12(12):3345. DOI: https://doi.org/10.3390/w12123345.

Rodriguez, A.R., and K.L. Heck, Jr. 2020. Green turtle herbivory and its effects on the warm, temperate seagrass meadows of St. Joseph Bay, Florida (USA). *Marine Ecology Progress Series* 639:37-51. DOI: https://doi.org/10.3354/meps13285.

Roegner, G., A. Coleman, A. Borde, J. Tagestad, R. Erdt, J. Aga, S. Zimmerman, and C. Cole. 2019. *Quantifying restoration of juvenile salmon habitat with hyperspectral imaging from an unmanned aircraft system*. Richland, WA: Pacific Northwest National Laboratory. DOI: https://doi.org/10.13140/RG.2.2.23907.30246.

Rogers, K., N. Saintilan, and C. Copeland. 2012. Modelling wetland surface elevation dynamics and its application to forecasting the effects of sea-level rise on estuarine wetlands. *Ecological Modelling* 244:148-157. DOI: https://doi.org/10.1016/j.ecolmodel.2012.06.014.

Rohwer, Y., and E. Marris. 2021. Ecosystem integrity is neither real nor valuable. *Conservation Science and Practice* 3(4):e411. DOI: https://doi.org/10.1111/csp2.411.

Rollinson, C.R., A.O. Finley, M.R. Alexander, S. Banerjee, K.-A. Dixon Hamil, L.E. Koenig, D.H. Locke, M.L. DeMarche, M.W. Tingley, K. Wheeler, C. Youngflesh, and E.F. Zipkin. 2021. Working across space and time: Nonstationarity in ecological research and application. *Frontiers in Ecology and the Environment* 19(1):66-72. DOI: https://doi.org/10.1002/fee.2298.

Roman, C.T., and D.M. Burdick. 2012. *Tidal marsh restoration: A synthesis of science and management (the science and practice of ecological restoration series)*. Washington, DC: Island Press.

Roman, M.R., S.B. Brandt, E.D. Houde, and J.J. Pierson. 2019. Interactive effects of Hypoxia and temperature on coastal pelagic zooplankton and fish. *Frontiers in Marine Science* 6(MAR). DOI: https://doi.org/10.3389/fmars.2019.00139.

Roni, P., P.J. Anders, T.J. Beechie, and D.J. Kaplowe. 2018. Review of tools for identifying, planning, and implementing habitat restoration for Pacific salmon and steelhead. *North American Journal of Fisheries Management* 38(2):355-376. DOI: https://doi.org/10.1002/nafm.10035.

Rosati, J.D., and G.W. Stone. 2009. Geomorphologic evolution of barrier islands along the northern U.S. Gulf of Mexico and implications for engineering design in barrier restoration. *Journal of Coastal Research* 251:8-22. DOI: https://doi.org/10.2112/07-0934.1.

Ross, A.C., R.G. Najjar, M. Li, M.E. Mann, S.E. Ford, and B. Katz. 2015. Sea-level rise and other influences on decadal-scale salinity variability in a coastal plain estuary. *Estuarine, Coastal and Shelf Science* 157:79-92. DOI: https://doi.org/10.1016/j.ecss.2015.01.022.

Roy, E.D., E.A. Smith, S. Bargu, and J.R. White. 2016. Will Mississippi River diversions designed for coastal restoration cause harmful algal blooms? *Ecological Engineering* 91:350-364. DOI: https://doi.org/10.1016/j.ecoleng.2016.02.030.

Rundel, P.W., E.A. Graham, M.F. Allen, J.C. Fisher, and T.C. Harmon. 2009. Environmental sensor networks in ecological research. *New Phytologist* 182(3):589-607. DOI: https://doi.org/10.1111/j.1469-8137.2009.02811.x.

Russell, B.T., K.A. Cressman, J.P. Schmit, S. Shull, J.M. Rybczyk, and D.L. Frost. 2022. How should surface elevation table data be analyzed? A comparison of several commonly used analysis methods and one newly proposed approach. *Environmental and Ecological Statistics*. DOI: https://doi.org/10.1007/s10651-021-00524-1.

Rybovich, M., M.K. La Peyre, S.G. Hall, and J.F. La Peyre. 2016. Increased temperatures combined with lowered salinities differentially impact oyster size class growth and mortality. *Journal of Shellfish Research* 35(1):101-113. DOI: https://doi.org/10.2983/035.035.0112.

Ryo, M., C.A. Aguilar-Trigueros, L. Pinek, L.A.H. Muller, and M.C. Rillig. 2019. Basic principles of temporal dynamics. *Trends in Ecology & Evolution* 34(8):723-733. DOI: https://doi.org/10.1016/j.tree.2019.03.007.

Ryo, M., J.M. Jeschke, M.C. Rillig, and T. Heger. 2020. Machine learning with the hierarchy-of-hypotheses (HoH) approach discovers novel pattern in studies on biological invasions. *Research Synthesis Methods* 11(1):66-73. DOI: https://doi.org/10.1002/jrsm.1363.

Ryo, M., Angelov, B., Mammola, S., Kass, J.M., Benito, B.M. and F. Hartig. 2021. Explainable artificial intelligence enhances the ecological interpretability of black-box species distribution models. *Ecography*. 44:199-205. DOI: https://doi.org/10.1111/ecog.05360.

Sadinski, W., A.L. Gallant, M. Roth, J. Brown, G. Senay, W. Brininger, P.M. Jones, and J. Stoker. 2018. Multi-year data from satellite- and ground-based sensors show details and scale matter in assessing climate's effects on wetland surface water, amphibians, and landscape conditions. *PLoS ONE* 13(9):e0201951. DOI: https://doi.org/10.1371/journal.pone.0201951.

Sagi, O., and L. Rokach. 2018. Ensemble learning: A survey. *WIREs Data Mining and Knowledge Discovery* 8(4):e1249. DOI: https://doi.org/10.1002/widm.1249.

Sandifer, P.A., L.C. Knapp, T.K. Collier, A.L. Jones, R.-P. Juster, C.R. Kelble, R.K. Kwok, J.V. Miglarese, L.A. Palinkas, D.E. Porter, G.I. Scott, L.M. Smith, W.C. Sullivan, and A.E. Sutton-Grier. 2017. A conceptual model to assess stress-associated health effects of multiple ecosystem services degraded by disaster events in the Gulf of Mexico and elsewhere. *GeoHealth* 1(1):17-36. DOI: https://doi.org/10.1002/2016GH000038.

Sapkota, Y., and J.R. White. 2019. Marsh edge erosion and associated carbon dynamics in coastal Louisiana: A proxy for future wetland-dominated coastlines world-wide. *Estuarine, Coastal and Shelf Science* 226. DOI: https://doi.org/10.1016/j.ecss.2019.106289.

Sather, N.K., G.E. Johnson, D.J. Teel, A.J. Storch, J.R. Skalski, and V.I. Cullinan. 2016. Shallow tidal freshwater habitats of the Columbia River: spatial and temporal variability of fish communities and density, size, and genetic stock composition of juvenile Chinook salmon, *Transactions of the American Fisheries Society* 145(4):734-753. DOI: https://doi.org/10.1080/00028487.2016.1150878.

Sauer, J.R., J.E. Hines, J,E. Fallon, K.L. Pardieck, D.J. Ziolkowsk, Jr, and W.A. Link. 2011. *The North American breeding bird survey, results and analysis. 1966–2010. Version 12.07.2011.* Laurel, MD: USGS Patuxent Wildlife Research Center.

Scheffel, W.A., K.L. Heck, and M.W. Johnson. 2018. Tropicalization of the Northern Gulf of Mexico: Impacts of Salt Marsh Transition to Black Mangrove Dominance on Faunal Communities. *Estuaries and Coasts* 41(4):1193-1205. DOI: https://doi.org/10.1007/s12237-017-0334-y.

Schiff, K., P.R. Trowbridge, E.T. Sherwood, P. Tango, and R.A. Batiuk. 2016. Regional monitoring programs in the United States: Synthesis of four case studies from Pacific, Atlantic, and Gulf Coasts. *Regional Studies in Marine Science* 4:A1-A7. DOI: https://doi.org/10.1016/j.rsma.2015.11.007.

Schilling, K.E., K.-S. Chan, H. Liu, and Y.-K. Zhang. 2010. Quantifying the effect of land use land cover change on increasing discharge in the Upper Mississippi River. *Journal of Hydrology* 387(3):343-345. DOI: https://doi.org/10.1016/j.jhydrol.2010.04.019.

Schmale, D.G., III, A.P. Ault, W. Saad, D.T. Scott, and J.A. Westrick. 2019. Perspectives on harmful algal blooms (HABs) and the cyberbiosecurity of freshwater systems. *Frontiers in Bioengineering and Biotechnology* 7(JUN). DOI: https://doi.org/10.3389/fbioe.2019.00128.

Schmidt, B., B. Gemeinholzer, and A. Treloar. 2016. Open data in global environmental research: The Belmont Forum's Open Data survey. *PLoS ONE* 11(1):e0146695. DOI: https://doi.org/10.1371/journal.pone.0146695.

Schrandt, M.N., T.C. MacDonald, E.T. Sherwood, and M.W. Beck. 2021. A multimetric nekton index for monitoring, managing and communicating ecosystem health status in an urbanized Gulf of Mexico estuary. *Ecological Indicators* 123.

Scott, A.L., P.H. York, C. Duncan, P.I. Macreadie, R.M. Connolly, M.T. Ellis, J.C. Jarvis, K.I. Jinks, H. Marsh, and M.A. Rasheed. 2018. The role of herbivory in structuring tropical seagrass ecosystem service delivery. *Frontiers in Plant Science* 9. DOI: https://doi.org/10.3389/fpls.2018.00127.

Scott, M.D. L. Ramaswamy, and V. Lawson. 2016. CyanoTRACKER: A Citizen Science Project for Reporting Harmful Algal Blooms. *2016 IEEE 2nd International Conference on Collaboration and Internet Computing (CIC).* pp. 391-397. DOI: https://doi.org/10.1109/CIC.2016.058.

Seafarers, S.D., S. Lavender, G. Beaugrand, N. Outram, N. Barlow, D. Crotty, J. Evans, and R. Kirby. 2017. Seafarer citizen scientist ocean transparency data as a resource for phytoplankton and climate research. *PLoS ONE* 12(12):e0186092. DOI: https://doi.org/10.1371/journal.pone.0186092.

Seddon, N., A. Smith, P. Smith, I. Key, A. Chausson, C. Girardin, J. House, S. Srivastava, and B. Turner. 2021. Getting the message right on nature-based solutions to climate change. *Global Change Biology* 27(8):1518-1546. DOI: https://doi.org/10.1111/gcb.15513.

Sejnowski, T.J. 2020. The unreasonable effectiveness of deep learning in artificial intelligence. *Proceedings of the National Academy of Sciences* 117(48):30033-30038. DOI: https://doi.org/10.1073/pnas.1907373117.

Shafer, D.J., B. Herczeg, D.W. Moulton, A. Sipocz, K. Jaynes, L.P. Rozas, C.P. Onuf, and W. Miller. 2002. *Regional guidebook for applying the hydrogeomorphic approach to assessing wetland functions of northwest Gulf of Mexico tidal fringe wetlands, ERDC/EL TR-02-5.* Washington, DC: US Army Corps of Engineers, Engineer Research and Development Center.

Shafer, D., T. Roberts, M. Peterson, and K. Schmid. 2007. *A regional guidebook for applying the hydrogeomorphic approach to assessing the functions of tidal fringe wetlands along the Mississippi and Alabama gulf coast, ERDC/EL TR-07-2.* Vicksburg, VA: US Army Corps of Engineers, Engineer Research and Development Center.

Sharma, S., J. Goff, R.M. Moody, D. Byron, K.L. Heck, Jr., S.P. Powers, C. Ferraro, and J. Cebrian. 2016. Do restored oyster reefs benefit seagrasses? An experimental study in the northern Gulf of Mexico. *Restoration Ecology* 24(3):306-313. DOI: https://doi.org/10.1111/rec.12329.

Sharpley, A.N., P.J. Kleinman, P. Jordan, L. Bergström, and A.L. Allen. 2009. Evaluating the success of phosphorus management from field to watershed. *Journal of Environmental Quality*. 38:1981-1988. DOI: https://doi.org/10.2134/jeq2008.0056.

Shaver, D.J., K.M. Hart, I. Fujisaki, D. Bucklin, A.R. Iverson, C. Rubio, T.F. Backof, P.M. Burchfield, R.D.J.G.D. Miron, P.H. Dutton, A. Frey, J. Peña, D.G. Gamez, H.J. Martinez, and J. Ortiz. 2017. Inter-nesting movements and habitat-use of adult female Kemp's ridley turtles in the Gulf of Mexico. *PLoS ONE* 12(3). DOI: https://doi.org/10.1371/journal.pone.0174248.

Shenk, G.W., and L.C. Linker. 2013. Development and application of the 2010 Chesapeake Bay watershed total maximum daily load model. *Journal of the American Water Resources Association* 49(5):1042-1056. DOI: https://doi.org/10.1111/jawr.12109.

Shenk, G.W., J. Wu, and L.C. Linker. 2012. Enhanced HSPF model structure for Chesapeake Bay watershed simulation. *Journal of Environmental Engineering* 138(9):949-957.

Shepard, C., J. Brenner, K.L. Goodin, and K.W. Ames. 2018. Ecological resilience indicators for oyster ecosystems. In: *Ecological resilience indicators for five northern Gulf of Mexico ecosystems*. Arlington, VA: NatureServe. https://www.natureserve.org/sites/default/files/chapter_5_-_oyster_indicators.pdf

Shepard, C.C., C.M. Crain, and M.W. Beck. 2011. The protective role of coastal marshes: A systematic review and meta-analysis. *PLoS ONE* 6(11):e27374. DOI: https://doi.org/10.1371/journal.pone.0027374.

Sherwood, E.T., H.S. Greening, A.J. Janicki, and D.J. Karlen. 2016. Tampa Bay estuary: Monitoring long-term recovery through regional partnerships. *Regional Studies in Marine Science* 4:1-11. DOI: https://doi.org/10.1016/j.rsma.2015.05.005.

Shiu, Y., K.J. Palmer, M.A. Roch, E. Fleishman, X. Liu, E.-M. Nosal, T. Helble, D. Cholewiak, D. Gillespie, and H. Klinck. 2020. Deep neural networks for automated detection of marine mammal species. *Scientific Reports* 10(1):607. DOI: https://doi.org/10.1038/s41598-020-57549-y.

Sigman, D.M., and M.P. Hain. 2012. The biological productivity of the ocean. *Nature Education Knowledge* 3(10):21.

Silliman, B.R., E. Schrack, Q. He, R. Cope, A. Santoni, T. van der Heide, R. Jacobi, M. Jacobi, and J. van de Koppel. 2015. Facilitation shifts paradigms and can amplify coastal restoration efforts. *Proceedings of the National Academy of Sciences* 112(46):14295-14300. DOI: https://doi.org/10.1073/pnas.1515297112.

Silliman, B.R., P.M. Dixon, C. Wobus, Q. He, P. Daleo, B.B. Hughes, M. Rissing, J.M. Willis, and M.W. Hester. 2016. Thresholds in marsh resilience to the Deepwater Horizon oil spill. *Scientific Reports* 6(1):32520. DOI: https://doi.org/10.1038/srep32520.

Simenstad, C.A., and R.M. Thom. 1996. Functional equivalency trajectories of the restored Gog-Le-Hi-Te estuarine wetland. *Ecological Applications* 6(1):38-56. DOI: https://doi.org/10.2307/2269551.

Sinclair, J. 2011. *Oil and gas structures in Gulf of Mexico Data Atlas*. Stennis Space Center (MS): National Centers for Environmental Information. https://www.ncei.noaa.gov/maps/gulf-data-atlas/atlas.htm?plate=Offshore%20Structures.

Smith, R.D. 2003. *Assessment of riparian ecosystem integrity: San Jacinto and Upper Santa Margarita River watersheds, Riverside County, California*. Vicksburg, MS: Prepared for U.S. Army Corps of Engineers, Los Angeles District, Regulatory Branch.

Smith, R.D., A. Ammann, C. Bartoldus, and M.M. Brinson. 1995. *An approach for assessing wetland functions using hydrogeomorphic classification, reference wetlands, and functional indices.* Wetland Research Program Technical Report WRP-DE-9. Vicksburg, MS: US Army Corps of Engineers Waterways Experiment Station.

Smith, C.S., M.E. Rudd, R.K. Gittman, E.C. Melvin, V.S. Patterson, J.J. Renzi, E.H. Wellman, and B.R. Silliman. 2020. Coming to terms with living shorelines: A scoping review of novel restoration strategies for shoreline protection. *Frontiers in Marine Science* 7. DOI: https://doi.org/10.3389/fmars.2020.00434.

Smullen, J.T., J. Taft, and J. Macknis. 1982. Nutrient and sediment loads to the tidal Chesapeake Bay system. In *Chesapeake Bay program technical studies: A synthesis*, edited by E.G. Macalaster, D.A. Barker and M. Kasper. Washington, DC: Environmental Protection Agency.

Sobocinski, K.L., and R.J. Latour. 2015. Trophic transfer in seagrass systems: Estimating seasonal production of an abundant seagrass fish, Bairdiella chrysoura, in lower Chesapeake Bay. *Marine Ecology Progress Series* 523:157-174.

Solis, R., and G. Powell. 1999. Hydrography, mixing characteristics, and residence times of Gulf of Mexico estuaries. In *Biogeochemistry of Gulf of Mexico estuaries*, edited by T.S. Bianchi, J.R. Pennock and R.R. Twilley. New York: John Wiley.

Spaling, H., and B. Smit. 1993. Cumulative environmental change: Conceptual frameworks, evaluation approaches, and institutional perspectives. *Environmental Management* 17(5):587-600. DOI: https://doi.org/10.1007/BF02393721.

Spies, R.B., S. Senner, and C.S. Robbins. 2016. An overview of the northern Gulf of Mexico ecosystem. *Gulf of Mexico Science* 33(1). DOI: https://doi.org/10.18785/goms.3301.09.

Stephens, G., A. Freeman, E. Richard, P. Pilewskie, P. Larkin, C. Chew, S. Tanelli, S. Brown, D. Posselt, and E. Peral. 2020. The emerging technological revolution in earth observations. *Bulletin of the American Meteorological Society* 101(3):E274-E285. DOI: https://doi.org/10.1175/bams-d-19-0146.1.

Stevenson, J.C., and C.B. Piper. 1979. *The decline of submerged aquatic plants in Chesapeake Bay.* Maryland Department of Natural Resources and Region 3 Chesapeake Bay Program, Environmental Protection Agency. FWS/OBS-79/24.

Steyer, G.D., C.E. Sasser, J.M. Visser, E.M. Swenson, J.A. Nyman and R.C. Raynie. 2003. A proposed coast-wide reference monitoring system for evaluating wetland restoration trajectories in Louisiana. *Environmental Monitoring and Assessment* 81(1-3):107-117.

Stockbridge, J., A.R. Jones, and B.M. Gillanders. 2020. A meta-analysis of multiple stressors on seagrasses in the context of marine spatial cumulative impacts assessment. *Scientific Reports* 10(1):11934. DOI: https://doi.org/10.1038/s41598-020-68801-w.

Suir, G.M., C.E. Sasser, and J.M. Harris. 2020. Use of remote sensing and field data to quantify the performance and resilience of restored Louisiana wetlands. *Wetlands* 40(6):2643-2658. DOI: https://doi.org/10.1007/s13157-020-01344-y.

Suter, G. 1999. Developing conceptual models for complex ecological risk assessments. *Human and Ecological Risk Assessment* 5:375-396. DOI: https://doi.org/10.1080/10807039991289491.

Suter, G.W., II, S.B. Norton, and S.M. Cormier. 2002. A methodology for inferring the causes of observed impairments in aquatic ecosystems. *Environmental Toxicology and Chemistry* 21(6):1101-1111. DOI: https://doi.org/10.1002/etc.5620210602.

Suter, G.W., S.B. Norton, and S.M. Cormier. 2010. The science and philosophy of a method for assessing environmental causes. *Human and Ecological Risk Assessment: An International Journal* 16(1):19-34. DOI: https://doi.org/10.1080/10807030903459254.

Sweet, W., G. Dusek, J.T.B. Obeysekera, and J.J. Marra. 2018. Patterns and projections of high tide flooding along the U.S. coastline using a common impact threshold. DOI: http://doi.org/10.7289/V5/TR-NOS-COOPS-086.

Taddeo, S., I. Dronova, and N. Depsky. 2019. Spectral vegetation indices of wetland greenness: Responses to vegetation structure, composition, and spatial distribution. *Remote Sensing of Environment* 234:111467. DOI: https://doi.org/10.1016/j.rse.2019.111467.

Tao, B., H. Tian, W. Ren, J. Yang, Q. Yang, R. He, W. Cai, and S. Lohrenz. 2014. Increasing Mississippi River discharge throughout the 21st century influenced by changes in climate, land use, and atmospheric CO_2. *Geophysical Research Letters* 41(14):4978-4986. DOI: https://doi.org/10.1002/2014GL060361.

TBNMC (Tampa Bay Nitrogen Management Consortium). 2017. *Tampa Bay Nutrient Management Strategy 2017 reasonable assurance update document.* https://drive.google.com/file/d/1VdYE67_aZAhFq6JYQSKxx2WY2O5KwyUI/view.

Teichert, N., A. Borja, G. Chust, A. Uriarte, and M. Lepage. 2016. Restoring fish ecological quality in estuaries: Implication of interactive and cumulative effects among anthropogenic stressors. *Science of the Total Environment* 542:383-393. DOI: https://doi.org/10.1016/j.scitotenv.2015.10.068.

Testa, J.M., W.M. Kemp, L.A. Harris, R.J. Woodland, and W.R. Boynton. 2017. Challenges and directions for the advancement of estuarine ecosystem science. *Ecosystems* 20(1):14-22. DOI: https://doi.org/10.1007/s10021-016-0004-0.

Testa, J.M., R.R. Murphy, D.C. Brady, and W.M. Kemp. 2018. Nutrient- and climate-induced shifts in the phenology of linked biogeochemical cycles in a temperate estuary. *Frontiers in Marine Science* 5. DOI: https://doi.org/10.3389/fmars.2018.00114.

Texas Environment. 2011. *Galveston Bay Haunted by Toxic Chemicals, Industrial Waste.* https://environmenttexas.org/news/txe/galveston-bay-haunted-toxic-chemicals-industrial-waste.

Thessen, A.E. 2016. Adoption of machine learning techniques in ecology and earth science. *One Ecosystem* 1. DOI: https://doi.org/10.3897/oneeco.1.e8621.

Thom, R.M. 1997. System-development matrix for adaptive management of coastal ecosystem restoration projects. *Ecological Engineering* 8(3):219-232. DOI: https://doi.org/10.1016/S0925-8574(97)00012-8.

Thom, R.M., G.W. Williams, and H.L. Diefenderfer. 2005. Balancing the need to develop coastal areas with the desire for an ecologically functioning coastal environment: Is net ecosystem improvement possible? *Restoration Ecology* 13(1):193-203. DOI: https://doi.org/10.1111/j.1526-100X.2005.00024.x.

Thom, R., G. Williams, N. Evans, and E. Haas. 2011. Lower Columbia River and estuary habitat restoration prioritization framework. *Ecological Restoration* 29:94-110. DOI: https://doi.org/10.3368/er.29.1-2.94.

Tokoro, T., and T. Kuwae. 2017. A new procedure for processing eddy-covariance data to better quantify atmosphere-aquatic ecosystem CO2 exchanges. *Biogeosciences Discuss.* DOI: https://doi.org/10.5194/bg-2017-499.

Tomasko, D.A., C.A. Corbett, H.S. Greening, and G.E. Raulerson. 2005. Spatial and temporal variation in seagrass coverage in Southwest Florida: Assessing the relative effects of anthropogenic nutrient load reductions and rainfall in four contiguous estuaries. *Marine Pollution Bulletin* 50(8):797-805. DOI: https://doi.org/10.1016/j.marpolbul.2005.02.010.

Tomasko, D., M. Alderson, R. Burnes, J. Hecker, J. Leverone, G. Raulerson, and E. Sherwood. 2018. Widespread recovery of seagrass coverage in Southwest Florida (USA): Temporal and spatial trends and management actions responsible for success. *Marine Pollution Bulletin* 135:1128-1137. DOI: https://doi.org/10.1016/j.marpolbul.2018.08.049.

Tomasko, D., M. Alderson, R. Burnes, J. Hecker, N. Iadevaia, J. Leverone, G. Raulerson, and E. Sherwood. 2020. The effects of Hurricane Irma on seagrass meadows in previously eutrophic estuaries in Southwest Florida (USA). *Marine Pollution Bulletin* 156. DOI: https://doi.org/10.1016/j.marpolbul.2020.111247.

Tominack, S.A., K.Z. Coffey, D. Yoskowitz, G. Sutton, and M.S. Wetz. 2020. An assessment of trends in the frequency and duration of Karenia brevis red tide blooms on the South Texas coast (western Gulf of Mexico). *PLoS ONE* 15(9):e0239309. DOI: https://doi.org/10.1371/journal.pone.0239309.

Törnqvist, T.E., K.L. Jankowski, K.L. Jankowski, Y.X. Li, Y.X. Li, J.L. González, and J.L. González. 2020. Tipping points of Mississippi Delta marshes due to accelerated sea-level rise. *Science Advances* 6(21). DOI: https://doi.org/10.1126/sciadv.aaz5512.

Toth, L. 1995. *Principles and guidelines for restoration of river/floodplain ecosystems—Kissimmee River Florida* (2nd ed.). Boca Raton, FL: CRC Press. DOI: https://doi.org/10.1201/9780203741016-4.

Trifonova, N., M. Karnauskas, and C. Kelble. 2019. Predicting ecosystem components in the Gulf of Mexico and their responses to climate variability with a dynamic Bayesian network model. *PLoS ONE* 14(1). DOI: https://doi.org/10.1371/journal.pone.0209257.

Trujillo, E., A. Del Rio, G. Johnson, and R. Thom. 2021. *Puget Sound partnership's science panel to discuss cumulative effects evaluation and application to Puget Sound recovery.*

Tunnell, J.W., Jr. 2017. Shellfish of the Gulf of Mexico. In *Habitats and biota of the Gulf of Mexico: Before the Deepwater Horizon Oil Spill* (Vol. 1)., edited by C. Ward. New York: Springer. DOI: https://doi.org/10.1007/978-1-4939-3447-8_8.

Turner, R.E. 1997. Wetland loss in the northern Gulf of Mexico: Multiple working hypotheses. *Estuaries* 20(1):1-13. DOI: https://doi.org/10.2307/1352716.

Turner, R.E., and D.R. Cahoon. 1987. *Causes of wetland loss in the coastal central Gulf of Mexico* (Vol. 1). Executive Summary. Final Report. https://www.osti.gov/biblio/7192545-causes-wetland-loss-coastal-gulf-mexico-volume-executive-summary-final-report.

Turner, R.E., and D.R. Cahoon. 1988. *Causes of wetland loss in the coastal central Gulf of Mexico* (Vol. 2). Technical Narrative. Final Report. https://www.osti.gov/biblio/7027816-causes-wetland-loss-coastal-central-gulf-mexico-volume-technical-narrative-final-report.

Turner, R.E. and N.N. Rabalais. 1999. Suspended particulate and dissolved nutrient loadings to Gulf of Mexico estuaries. In *Biogeochemistry of Gulf of Mexico estuaries,* edited by T.S. Bianchi, J.R. Pennoch and R. R. Twilley. New York: John Wiley and Sons, Inc.

Turner, R.E., and N.N. Rabalais. 2019. The Gulf of Mexico. In *World seas: An environmental evaluation* (2nd ed.), edited by C. Sheppard. Cambridge, MA: Academic Press. DOI: https://doi.org/10.1016/B978-0-12-805068-2.00022-X.

Turner, R.E., J.J. Baustian, E.M. Swenson, and J.S. Spicer. 2006. Wetland sedimentation from hurricanes Katrina and Rita. *Science* 314(5798):449-452. DOI: https://doi.org/10.1126/science.1129116.

Turner, R.E., N.N. Rabalais, and D. Justić. 2012. Predicting summer hypoxia in the northern Gulf of Mexico: Redux. *Marine Pollution Bulletin* 64(2):319-324. DOI: https://doi.org/10.1016/j.marpolbul.2011.11.008.

Tweel, A.W., and R.E. Turner. 2012. Watershed land use and river engineering drive wetland formation and loss in the Mississippi River birdfoot delta. *Limnology and Oceanography* 57(1):18-28. DOI: https://doi.org/10.4319/lo.2012.57.1.0018.

Twilley, R.R., E.J. Barron, H.L. Gholz, M.A. Harwell, R.L. Miller, D.J. Reed, J.B. Rose, E.H. Siemann, R.G. Wetzel, and R.J. Zimmerman. 2001. *Confronting climate change in the Gulf Coast region: Prospects for sustaining our ecological heritage.* Cambridge, MA: Union of Concerned Scientists & Ecological Society of America, and Washington, DC: Ecological Society of America.

Twilley, R.R., J.W. Day, A.E. Bevington, E. Castañeda-Moya, A. Christensen, G. Holm, L.R. Heffner, R. Lane, A. McCall, A. Aarons, S. Li, A. Freeman, and A.S. Rovai. 2019. Ecogeomorphology of coastal deltaic floodplains and estuaries in an active delta: Insights from the Atchafalaya Coastal Basin. *Estuarine, Coastal and Shelf Science* 227:106341. DOI: https://doi.org/10.1016/j.ecss.2019.106341.

UNESCO-IOC. 2012. *Requirements for global implementation of the strategic plan for coastal GOOS.* GOOS Report 193. Paris: Intergovernmental Oceanographic Commission. https://cdn.ioos.noaa.gov/media/2019/08/requirements_global_implementation_strategic_plan_for_coastal_goos.pdf.

UNIDO (United Nations Industrial Development Organization). 2014. *Annual report.* https://www.unido.org/sites/default/files/2015-04/15-00722_Ebook_2.pdf.

U.S. Department of Health, Education, and Welfare. 1964. *Smoking and health: Report of the advisory committee to the surgeon general of the public health service.* Washington, DC: U.S. Government Printing Office (PHS Publication No. 1103).

USACE (U.S. Army Corps of Engineers). 2014. *Technical Letter 1100-2-1, Procedures to evaluate sea level change impacts, responses and adaptation.* Washington, DC: Department of the Army.

USACE. 2016. *Mississippi Coastal Improvements Program (MsCIP) Comprehensive Barrier Island Restoration: Hancock, Harrison, and Jackson Counties, Mississippi. Final Supplemental Environmental Impact Statement.* https://www.sam.usace.army.mil/Portals/46/docs/program_management/mscip/docs/MsCIP%20Final%20Sup%20SEIS/MsCIP%20SEIS_Final_JAN2016.pdf.

USACE. 2018. *Guidance for incorporating climate change impacts to inland hydrology in civil works studies, designs, and projects.* USACE ECB 2018-14. Washington, DC: Department of the Army.

USACE. 2000. Economic, social and regional considerations. In *Planning guidance notebook.* ER 1105-2-100. Washington, DC: Department of the Army.

USFWS (U.S. Fish and Wildlife Service). 2006. Coastal Wetlands Planning, Protection and Restoration Act Wetland Value Assessment Methodology Procedural Manual. 23 pp. https://lacoast.gov/reports/wva/WVA%20Procedural%20Manual.pdf.

USGCRP (U.S. Global Change Research Program). 2017. *Climate science special report: Fourth national climate assessment* (Vol. 1), edited by D.J. Wuebbles, D.W. Fahey, K.A. Hibbard, D.J. Dokken, B.C. Stewart, and T.K. Maycock. Washington, DC: U.S. Global Change Research Program. DOI: https://doi.org/10.7930/J0J964J6.

Valentine, D.L., G.B. Fisher, S.C. Bagby, R.K. Nelson, C.M. Reddy, S.P. Sylva, and M.A. Woo. 2014. Fallout plume of submerged oil from Deepwater Horizon. *Proceedings of the National Academy of Sciences* 111(45):15906-15911. DOI: https://doi.org/10.1073/pnas.1414873111.

Valverde, R.A., and K. Rouse Holzwart. 2017. Sea turtles of the Gulf of Mexico. In *Habitats and biota of the Gulf of Mexico: Before the Deepwater Horizon oil spill*, edited by C. Ward. New York: Springer. DOI: https://doi.org/10.1007/978-1-4939-3456-0_3.

Van de Broek, M., C. Vandendriessche, D. Poppelmonde, R. Merckx, S. Temmerman, and G. Govers. 2018. Long-term organic carbon sequestration in tidal marsh sediments is dominated by old-aged allochthonous inputs in a macrotidal estuary. *Global Change Biology* 24(6):2498-2512. DOI: https://doi.org/10.1111/gcb.14089.

van Katwijk, M.M., A. Thorhaug, N. Marbà, R.J. Orth, C.M. Duarte, G.A. Kendrick, I.H.J. Althuizen, E. Balestri, G. Bernard, M.L. Cambridge, A. Cunha, C. Durance, W. Giesen, Q. Han, S. Hosokawa, W. Kiswara, T. Komatsu, C. Lardicci, K.-S. Lee, A. Meinesz, M. Nakaoka, K.R. O'Brien, E.I. Paling, C. Pickerell, A.M.A. Ransijn, and J.J. Verduin. 2016. Global analysis of seagrass restoration: The importance of large-scale planting. *Journal of Applied Ecology* 53(2):567-578. DOI: https://doi.org/10.1111/1365-2664.12562.

Vazquez-Cuervo, J., S. Fournier, B. Dzwonkowski, and J. Reager. 2018. Intercomparison of in-situ and remote sensing salinity products in the Gulf of Mexico, a river-influenced system. *Remote Sensing* 10(10):1590.

Veatch, W.C. 2017. *2015 Updated Atlas of US Army Corps of Engineers historic daily tide data in coastal Louisiana.* Mississippi Valley Division: U.S. Army Corps of Engineers.

Vergés, A., P.D. Steinberg, M.E. Hay, A.G.B. Poore, A.H. Campbell, E. Ballesteros, K.L. Heck, D.J. Booth, M.A. Coleman, D.A. Feary, W. Figueira, T. Langlois, E.M. Marzinelli, T. Mizerek, P.J. Mumby, Y. Nakamura, M. Roughan, E. van Sebille, A.S. Gupta, D.A. Smale, F. Tomas, T. Wernberg, and S.K. Wilson. 2014. The tropicalization of temperate marine ecosystems: Climate-mediated changes in herbivory and community phase shifts. *Proceedings of the Royal Society B: Biological Sciences* 281(1789):20140846. DOI: https://doi.org/10.1098/rspb.2014.0846.

Vergés, A., C. Doropoulos, R. Czarnik, K. McMahon, N. Llonch, and A.G.B. Poore. 2018. Latitudinal variation in seagrass herbivory: Global patterns and explanatory mechanisms. *Global Ecology and Biogeography* 27(9):1068-1079. DOI: https://doi.org/10.1111/geb.12767.

Vinent, O.D., B.E. Schaffer, and I. Rodriguez-Iturbe. 2021. Stochastic dynamics of barrier island elevation. *Proceedings of the National Academy of Sciences* 118(1):e2013349118. DOI: https://doi.org/10.1073/pnas.2013349118.

Vinuesa, R., H. Azizpour, I. Leite, M. Balaam, V. Dignum, S. Domisch, A. Felländer, S.D. Langhans, M. Tegmark, and F. Fuso Nerini. 2020. The role of artificial intelligence in achieving the Sustainable Development Goals. *Nature Communications* 11(1):233. DOI: https://doi.org/10.1038/s41467-019-14108-y.

Vittor and Associates, Inc. 2014. *Calibrating a biological condition gradient model to the Mobile Bay estuary.* Prepared for Great Lake Environmental Center, Inc. and Environmental Protection Agency Region. https://www.mobilebaynep.com/images/uploads/library/MBNEP_BCG_final.pdf.

Vittor and Associates, Inc. 2019. *D'olive watershed monitoring study and development of a watershed condition framework.* Prepared for Mobile Bay National Estuary Program. https://www.mobilebaynep.com/assets/pdf/Watershed-Condition-Framework-Report-2019_12_31.pdf.

Vollenweider, R.A. 1976. Advances in defining critical loading levels for phosphorus in lake eutrophication. *Memorie dell'Istituto Italiano di Idrobiologia, Dott. Marco de Marchi Verbania Pallanza* 33:53-83.

Vollmer, N.L., and P.E. Rosel. 2013. A review of common bottlenose dolphins (*Tursiops truncatus truncatus*) in the northern Gulf of Mexico: Population biology, potential threats, and management. *Southeastern Naturalist* 12(m6):1-43.

Vos, K., K.D. Splinter, M.D. Harley, J.A. Simmons, and I.L. Turner. 2019. CoastSat: A Google Earth Engine-enabled Python toolkit to extract shorelines from publicly available satellite imagery. *Environmental Modelling & Software* 122. DOI: https://doi.org/10.1016/j.envsoft.2019.104528.

Walker, S., Dausman, A, and Lavoie, D. 2012. *Gulf of Mexico ecosystem science assessment and needs—A product of the Gulf coast ecosystem restoration task force science coordination team.*

Wallace, R.D., C.T. Bargeron, D.J. Moorhead, and J.H. LaForest. 2016. IveGot1: Reporting and tracking invasive species in Florida. *Southeastern Naturalist* 15(sp8):51-62. DOI: https://doi.org/10.1656/058.015.sp805.

Walters, C. 1997. Challenges in adaptive management of riparian and coastal ecosystems. *Conservation Ecology* 1(2):1. https://www.ecologyandsociety.org/vol1/iss2/art1/.

Waltham, N.J., M. Elliott, S.Y. Lee, C. Lovelock, C.M. Duarte, C. Buelow, C. Simenstad, I. Nagelkerken, L. Claassens, C.K.C. Wen, M. Barletta, R.M. Connolly, C. Gillies, W.J. Mitsch, M.B. Ogburn, J. Purandare, H. Possingham, and M. Sheaves. 2020. UN Decade on Ecosystem Restoration 2021–2030—What chance for success in restoring coastal ecosystems? *Frontiers in Marine Science* 7. DOI: https://doi.org/10.3389/fmars.2020.00071.

Waltham, N.J., C. Alcott, M.A. Barbeau, J. Cebrian, R.M. Connolly, L.A. Deegan, K. Dodds, L.A. Goodridge Gaines, B.L. Gilby, C.J. Henderson, C.M. McLuckie, T.J. Minello, G.S. Norris, J. Ollerhead, J. Pahl, J.F. Reinhardt, R.J. Rezek, C.A. Simenstad, J.A.M. Smith, E.L. Sparks, L.W. Staver, S.L. Ziegler, and M.P. Weinstein. 2021. Tidal marsh restoration optimism in a changing climate and urbanizing seascape. *Estuaries and Coasts* 44(6):1681-1690. DOI: https://doi.org/10.1007/s12237-020-00875-1.

Wamsley, T.V., M.A. Cialone, J.M. Smith, J.H. Atkinson, and J.D. Rosati. 2010. The potential of wetlands in reducing storm surge. *Ocean Engineering* 37(1):59-68. DOI: https://doi.org/10.1016/j.oceaneng.2009.07.018.

Wang, C., and R. Furrer. 2021. Combining heterogeneous spatial datasets with process-based spatial fusion models: A unifying framework. *Computational Statistics & Data Analysis* 161:107240. DOI: https://doi.org/10.1016/j.csda.2021.107240.

Wang, H., J. Lehrter, K. Maiti, K. Fennel, A. Laurent, N. Rabalais, N. Hussain, Q. Li, B. Chen, K.M. Scaboo, and W.J. Cai. 2020a. Benthic respiration in hypoxic waters enhances bottom water acidification in the northern Gulf of Mexico. *Journal of Geophysical Research: Oceans* 125(10). DOI: https://doi.org/10.1029/2020JC016152.

Wang, J., C. Li, F. Xu, and W. Huang. 2020b. Severe weather-induced exchange flows through a narrow tidal channel of Calcasieu Lake estuary. *Journal of Marine Science and Engineering* 8(2). DOI: https://doi.org/10.3390/jmse8020113.

Ward, N.D., J.P. Megonigal, B. Bond-Lamberty, V.L. Bailey, D. Butman, E.A. Canuel, H. Diefenderfer, N.K. Ganju, M.A. Goñi, E.B. Graham, C.S. Hopkinson, T. Khangaonkar, J.A. Langley, N.G. McDowell, A.N. Myers-Pigg, R.B. Neumann, C.L. Osburn, R.M. Price, J. Rowland, A. Sengupta, M. Simard, P.E. Thornton, M. Tzortziou, R. Vargas, P.B. Weisenhorn, and L. Gulfham-Myers. 2020. Representing the function and sensitivity of coastal interfaces in Earth system models. *Nature Communications* 11(1):2458. DOI: https://doi.org/10.1038/s41467-020-16236-2.

Wassmann, P., M. Reigstad, T. Haug ,B. Rudels, M.L. Carroll, H. Hop, G.W. Gabrielsen, S. Falk-Petersen, S.G. Denisenko, E. Arashkevich, D. Slagstad, and O. Pavlova. 2006. Food webs and carbon flux in the Barents Sea. *Progress in Oceanography* 71(2):232-287. https://doi.org/10.1016/j.pocean.2006.10.003.

Watson, S., D.A. Duecker, and K. Groves. 2020. Localisation of Unmanned Underwater Vehicles (UUVs) in Complex and Confined Environments: A Review. *Sensors (Basel)* 20:21. https://doi.org/10.3390%2Fs20216203

WCRP (World Climate Research Programme) Global Sea Level Budget Group. 2018. Global sea-level budget 1993–present. *Earth System Science Data* 10(3):1551-1590. DOI: https://doi.org/10.5194/essd-10-1551-2018.

Webb, J.A., S.J. Nichols, R.H. Norris, M.J. Stewardson, S.R. Wealands, and P. Lea. 2012. Ecological responses to flow alteration: Assessing causal relationships with eco evidence. *Wetlands* 32(2):203-213. DOI: https://doi.org/10.1007/s13157-011-0249-5.

Webb, J.A., K.A. Miller, M.J. Stewardson, S.C.d. Little, S.J. Nichols, and S.R. Wealands. 2015. An online database and desktop assessment software to simplify systematic reviews in environmental science. *Environmental Modelling & Software* 64(C):72-79. DOI: https://doi.org/10.1016/j.envsoft.2014.11.011.

Weinstein, B.G. 2018. Scene-specific convolutional neural networks for video-based biodiversity detection. *Methods in Ecology and Evolution* 9(6):1435-1441. DOI: https://doi.org/10.1111/2041-210X.13011.

Wells, R.S., H.L. Rhinehart, L.J. Hansen, J.C. Sweeney, F.I. Townsend, R. Stone, D.R. Casper, M.D. Scott, A.A. Hohn, and T.K. Rowles. 2004. Bottlenose dolphins as marine ecosystem sentinels: Developing a health monitoring system. *EcoHealth* 1(3):246-254. DOI: https://doi.org/10.1007/s10393-004-0094-6.

Wells, R.S., L.H. Schwacke, T.K. Rowles, B.C. Balmer, E. Zolman, T. Speakman, F.I. Townsend, M.C. Tumlin, A. Barleycorn, and K.A. Wilkinson. 2017. Ranging patterns of common bottlenose dolphins Tursiops truncatus in Barataria Bay, Louisiana, following the Deepwater Horizon oil spill. *Endangered Species Research* 33(1):159-180. DOI: https://doi.org/10.3354/esr00732.

Wernberg, T., D.A. Smale, F. Tuya, M.S. Thomsen, T.J. Langlois, T. de Bettignies, S. Bennett, and C.S. Rousseaux. 2013. An extreme climatic event alters marine ecosystem structure in a global biodiversity hotspot. *Nature Climate Change* 3(1):78-82. DOI: https://doi.org/10.1038/nclimate1627.

Wernberg, T., S. Bennett, R.C. Babcock, T. de Bettignies, K. Cure, M. Depczynski, F. Dufois, J. Fromont, C.J. Fulton, R.K. Hovey, E.S. Harvey, T.H. Holmes, G.A. Kendrick, B. Radford, J. Santana-Garcon, B.J. Saunders, D.A. Smale, M.S. Thomsen, C.A. Tuckett, F. Tuya, M.A. Vanderklift, and S. Wilson. 2016. Climate-driven regime shift of a temperate marine ecosystem. *Science* 353(6295):169-72. DOI: https://doi.org/10.1126/science.aad8745. PMID: 27387951.

Westgate, M.J., G.E. Likens, and D.B. Lindenmayer. 2013. Adaptive management of biological systems: A review. *Biological Conservation* 158:128-139. DOI: https://doi.org/10.1016/j.biocon.2012.08.016.

Whitall, D.R. 2008. Historical nitrogen and phosphorus loadings to the northern Gulf of Mexico NOS NCCOS 85. Silver Spring, MD: NOAA/NOS/Center for Coastal Monitoring and Assessment.

White, E.D., A.M.E. Meselhe, B. Couvillion, Z. Dong, S.M. Duke-Sylvester, and Y. Wang. 2017. *Coastal master plan: Attachment C2-22: Integrated Compartment Model (ICM) development.* Baton Rouge, LA: Coastal Protection and Restoration Authority.

White, J.R., R.D. DeLaune, D. Justic, J.W. Day, J. Pahl, R.R. Lane, W.R. Boynton, R.R. Twilley. 2019a. Consequences of Mississippi River diversions on nutrient dynamics of coastal wetland soils and estuarine sediments: A review. *Estuarine, Coastal and Shelf Science* 224:209-216. DOI: https://doi.org/10.1016/j.ecss.2019.04.027.

White, E.D., D.J. Reed, and E.A. Meselhe. 2019b. Modeled sediment availability, deposition, and decadal land change in coastal Louisiana marshes under future relative sea level rise scenarios. *Wetlands* 39(6):1233-1248. DOI: https://doi.org/10.1007/s13157-019-01151-0.

Whitney, L.D., and J.S. Hobgood. 1997. The relationship between sea surface temperatures and maximum intensities of tropical cyclones in the Eastern North Pacific Ocean. *Journal of Climate* 10(11):2921-2930. DOI: https://doi.org/10.1175/1520-0442(1997)010<2921:Trbsst>2.0.Co;2.

Whitt, C., J. Pearlman, B. Polagye, F. Caimi, F. Muller-Karger, A. Copping, H. Spence, S. Madhusudhana, W. Kirkwood, L. Grosjean, B.M. Fiaz, S. Singh, S. Singh, D. Manalang, A.S. Gupta, A. Maguer, J.J.H. Buck, A. Marouchos, M.A. Atmanand, R. Venkatesan, V. Narayanaswamy, P. Testor, E. Douglas, S. de Halleux, and S.J. Khalsa. 2020. Future vision for autonomous ocean observations. *Frontiers in Marine Science* 7. DOI: https://doi.org/10.3389/fmars.2020.00697.

Wickwire, T., and C.A. Menzie. 2010. The causal analysis framework: Refining approaches and expanding multidisciplinary applications. *Human and Ecological Risk Assessment: An International Journal* 16(1):10-18. DOI: https://doi.org/10.1080/10807030903459205.

Wieringa, S., E. Engebretsen, K. Heggen, and T. Greenhalgh. 2017. Has evidence-based medicine ever been modern? A Latour-inspired understanding of a changing EBM. *Journal of Evaluation in Clinical Practice* 23(5):964-970. DOI: https://doi.org/10.1111/jep.12752.

Wiesenburg, D.A., B. Shipp, F.J. Fodrie, S. Powers, J. Lartigue, K.M. Darnell, M.M. Baustian, C. Ngo, J.F. Valentine, and K. Wowk. 2021. Prospects for Gulf of Mexico environmental recovery and restoration. *Oceanography* 34(1):164-173.

Wilber, D.H., and R. Bass. 1998. Effect of the Colorado River Diversion on Matagorda Bay Epifauna. *Estuarine, Coastal and Shelf Science* 47(3):309-318. https://doi.org/10.1006/ecss.1998.0356.

Wilkinson, M.D., M. Dumontier, I.J. Aalbersberg, G. Appleton, M. Axton, A. Baak, N. Blomberg, J.-W. Boiten, L.B. da Silva Santos, P.E. Bourne, J. Bouwman, A.J. Brookes, T. Clark, M. Crosas, I. Dillo, O. Dumon, S. Edmunds, C.T. Evelo, R. Finkers, A. Gonzalez-Beltran, A.J.G. Gray, P. Groth, C. Goble, J.S. Grethe, J. Heringa, P.A.C. 't Hoen, R. Hooft, T. Kuhn, R. Kok, J. Kok, S.J. Lusher, M.E. Martone, A. Mons, A.L. Packer, B. Persson, P. Rocca-Serra, M. Roos, R. van Schaik, S.-A. Sansone, E. Schultes, T. Sengstag, T. Slater, G. Strawn, M.A. Swertz, M. Thompson, J. van der Lei, E. van Mulligen, J. Velterop, A. Waagmeester, P. Wittenburg, K. Wolstencroft, J. Zhao, and B. Mons. 2016. The FAIR Guiding Principles for scientific data management and stewardship. *Scientific Data* 3(1):160018. DOI: https://doi.org/10.1038/sdata.2016.18.

Willcock, S., J. Martínez-López, D.A.P. Hooftman, K.J. Bagstad, S. Balbi, A. Marzo, C. Prato, S. Sciandrello, G. Signorello, B. Voigt, F. Villa, J.M. Bullock, and I.N. Athanasiadis. 2018. Machine learning for ecosystem services. *Ecosystem Services* 33:165-174. DOI: https://doi.org/10.1016/j.ecoser.2018.04.004.

Williams, B.K. 2011. Adaptive management of natural resources—framework and issues. *Journal of Environmental Management* 92(5):1346-1353. DOI: https://doi.org/10.1016/j.jenvman.2010.10.041.

Williams, B.K., and E.D. Brown. 2014. Adaptive management: From more talk to real action. *Environmental Management* 53(2):465-479. DOI: https://doi.org/10.1007/s00267-013-0205-7.

Williams, I.D., C.S. Couch, O. Beijbom, T.A. Oliver, B. Vargas-Angel, B.D. Schumacher, and R.E. Brainard. 2019. Leveraging automated image analysis tools to transform our capacity to assess status and trends of coral reefs. *Frontiers in Marine Science* 6(222). DOI: https://doi.org/10.3389/fmars.2019.00222.

Windham-Myers, L., S. Crooks, and T.G. Troxler. 2018. *A blue carbon primer: The state of coastal wetland carbon science, practice, and policy*. Boca Raton: CRC Press. DOI: https://doi.org/10.1201/9780429435362

Wise, D.R., D.W. Anning, and O.L. Miller. 2019. *Spatially referenced models of streamflow and nitrogen, phosphorus, and suspended-sediment transport in streams of the southwestern United States* (ver. 1.1, June 2020): U.S. Geological Survey Scientific Investigations Report 2019-5106, 66 p., https://doi.org/10.3133/sir20195106.

Withers, P.J., P. Jordan, L. May, H.P. Jarvie, and N.E. Deal. 2014. Do septic tank systems pose a hidden threat to water quality? *Frontiers in Ecology and the Environment* 12(2):123-130. DOI: https://doi.org/10.1890/130131.

Würsig, B. 2017. Marine mammals of the Gulf of Mexico. In *Habitats and biota of the Gulf of Mexico: Before the Deepwater Horizon oil spill* (Vol. 2), edited by C.H. Ward. New York: Springer. DOI: https://doi.org/10.1007/978-1-4939-3456-0_5.

Wurtzebach, Z., and C. Schultz. 2016. Measuring ecological integrity: History, practical applications, and research opportunities. *BioScience* 66(6):446-457. DOI: https://doi.org/10.1093/biosci/biw037.

Wynn, R.B., V.A.I. Huvenne, T.P. Le Bas, B.J. Murton, D.P. Connelly, B.J. Bett, H.A. Ruhl, K.J. Morris, J. Peakall, D.R. Parsons, E.J. Sumner, S.E. Darby, R.M. Dorrell, and J.E. Hunt. 2014. Autonomous Underwater Vehicles (AUVs): Their past, present and future contributions to the advancement of marine geoscience. *Marine Geology* 352:451-468. DOI: https://doi.org/10.1016/j.margeo.2014.03.012.

Xu, J., Y. Wang, and Z.-M. Tan. 2016. The relationship between sea surface temperature and maximum intensification rate of tropical cyclones in the North Atlantic. *Journal of the Atmospheric Sciences* 73(12):4979-4988. DOI: https://doi.org/10.1175/JAS-D-16-0164.1.

Yan, Y., B. Zhao, J. Chen, H. Guo, Y. Gu, Q. Wu, and B. Li. 2008. Closing the carbon budget of estuarine wetlands with tower-based measurements and MODIS time series. *Global Change Biology* 14(7):1690-1702. DOI: https://doi.org/10.1111/j.1365-2486.2008.01589.x.

Yang, X., and Z. Liu. 2005. Quantifying landscape pattern and its change in an estuarine watershed using satellite imagery and landscape metrics. *International Journal of Remote Sensing* 26(23):5297-5323. DOI: https://doi.org/10.1080/01431160500219273.

Yang, Y.-Y., and G.S. Toor. 2018. Stormwater runoff driven phosphorus transport in an urban residential catchment: Implications for protecting water quality in urban watersheds. *Scientific Reports* 8(1):11681. DOI: https://doi.org/10.1038/s41598-018-29857-x.

Yang, L., C. Homer, J. Brock, and J. Fry. 2013. An efficient method for change detection of soil, vegetation and water in the northern Gulf of Mexico wetland ecosystem. *International Journal of Remote Sensing* 34(18):6321-6336. DOI: https://doi.org/10.1080/01431161.2013.800653.

Yarbro, L.A., and P.R. Carlson, Jr. 2016. *Seagrass integrated mapping and monitoring program: Mapping and monitoring report no. 2.* Tallahassee, FL: Florida Fish and Wildlife Conservation Commission, Fish and Wildlife Research Institute.

Yee, S., G. Cicchetti, T.H. DeWitt, M.C. Harwell, S.K. Jackson, M. Pryor, K. Rocha, D.L. Santavy, L. Sharpe, and E. Shumchenia. 2020. The ecosystem services gradient: A descriptive model for identifying levels of meaningful change. In *Ecosystem-based management, ecosystem services and aquatic biodiversity: Theory, tools and applications,* edited by T.G. O'Higgins, M. Lago and T.H. DeWitt. Cham: Springer International Publishing.

Yoe, C. 2012. *Guide to constructing the without project scenario (condition).* Institute for Water Resources 2012-R-03. https://www.iwr.usace.army.mil/Portals/70/docs/iwrreports/2012-R-03.pdf.

Yuill, B., D. Lavoie, and D.J. Reed. 2009. Understanding subsidence processes in coastal Louisiana. *Journal of Coastal Research*(10054):23-36. DOI: https://doi.org/10.2112/si54-012.1.

Zedler, J.B. 2017. What's new in adaptive management and restoration of coasts and estuaries? *Estuaries and Coasts: Journal of the Coastal and Estuarine Research Federation* 40(1):1-21. DOI: https://doi.org/10.1007/s12237-016-0162-5.

Zeng, Z., Y. Wang, and C.C. Wu. 2007. Environmental dynamical control of tropical cyclone intensity: An observational study. *Monthly Weather Review* 135(1):38-59. DOI: https://doi.org/10.1175/MWR3278.1.

Zhang, Y.K., and K.E. Schilling. 2006. Increasing streamflow and baseflow in Mississippi River since the 1940s: Effect of land use change. *Journal of Hydrology* 324(1):412-422. DOI: https://doi.org/10.1016/j.jhydrol.2005.09.033.

Zhang, R., B. Yan, H. Guo, Y. Zhang, B. Hu, H. Yang, L. Wang, and Y. Wang. 2019. A new environmental monitoring system based on WiFi technology. *Procedia CIRP* 83:394-397. DOI: https://doi.org/10.1016/j.procir.2019.04.088.

Zhong, S., K. Zhang, M. Bagheri, J.G. Burken, A. Gu, B. Li, X. Ma, B.L. Marrone, Z.J. Ren, J. Schrier, W. Shi, H. Tan, T. Wang, X. Wang, B.M. Wong, X. Xiao, X. Yu, J.-J. Zhu, and H. Zhang. 2021. Machine learning: New ideas and tools in environmental science and engineering. *Environmental Science & Technology*. DOI: https://doi.org/10.1021/acs.est.1c01339.

Ziegler, S.L., R. Baker, S.C. Crosby, D.D. Colombano, M.A. Barbeau, J. Cebrian, R.M. Connolly, L.A. Deegan, B.L. Gilby, D. Mallick, C.W. Martin, J.A. Nelson, J.F. Reinhardt, C.A. Simenstad, N.J. Waltham, T.A. Worthington, and L.P. Rozas. 2021. Geographic variation in salt marsh structure and function for nekton: A guide to finding commonality across multiple scales. *Estuaries and Coasts* 44(6):1497-1507. DOI: https://doi.org/10.1007/s12237-020-00894-y.

Zieman, J.C., and R.T. Zieman. 1989. The ecology of the seagrass meadows of the west coast of Florida: A community profile. *Biological Report—US Fish & Wildlife Service* 85(7.25):1-155.

Appendix A

Distribution and Status of Funds Derived from Deepwater Horizon–Related Settlements

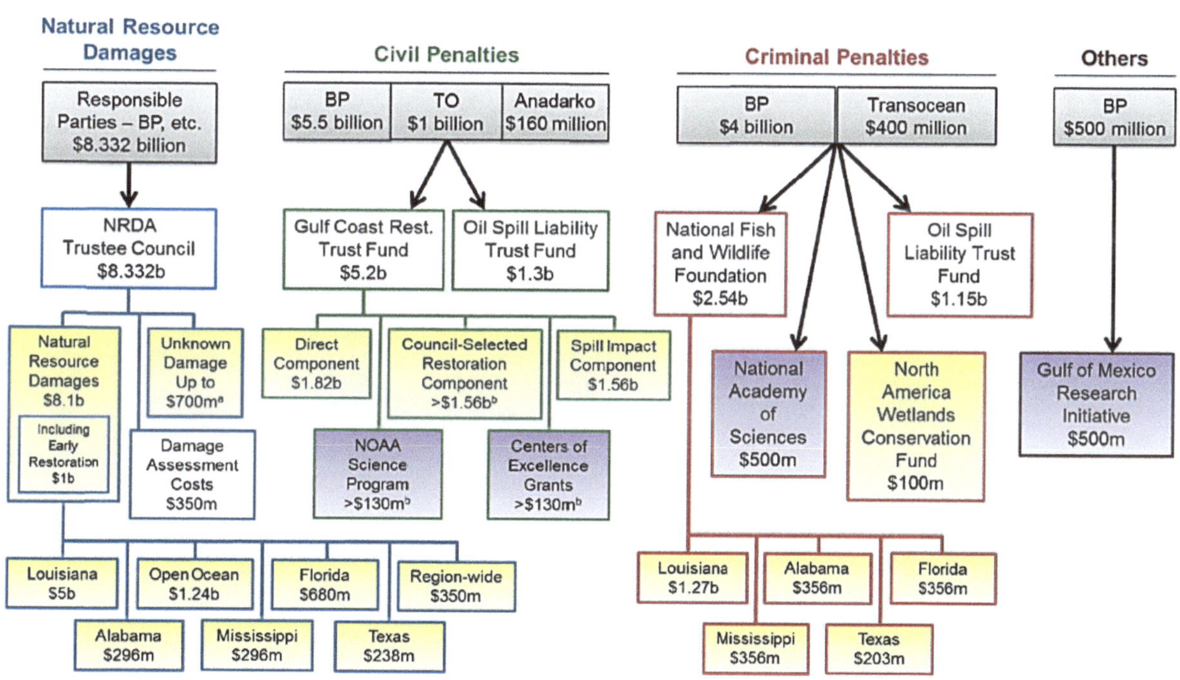

FIGURE A.1. The funding landscape for Deepwater Horizon settlements and penalties. Funding went to entities that fund restoration efforts (yellow shading) as well as to science programs (purple shading). SOURCE: Effective Monitoring to Evaluate Ecological Restoration in the Gulf of Mexico, NASEM (2017). Available at: https://www.nap.edu/catalog/23476/effective-monitoring-to-evaluate-ecological-restoration-in-the-gulf-of-mexico. See source for more information.

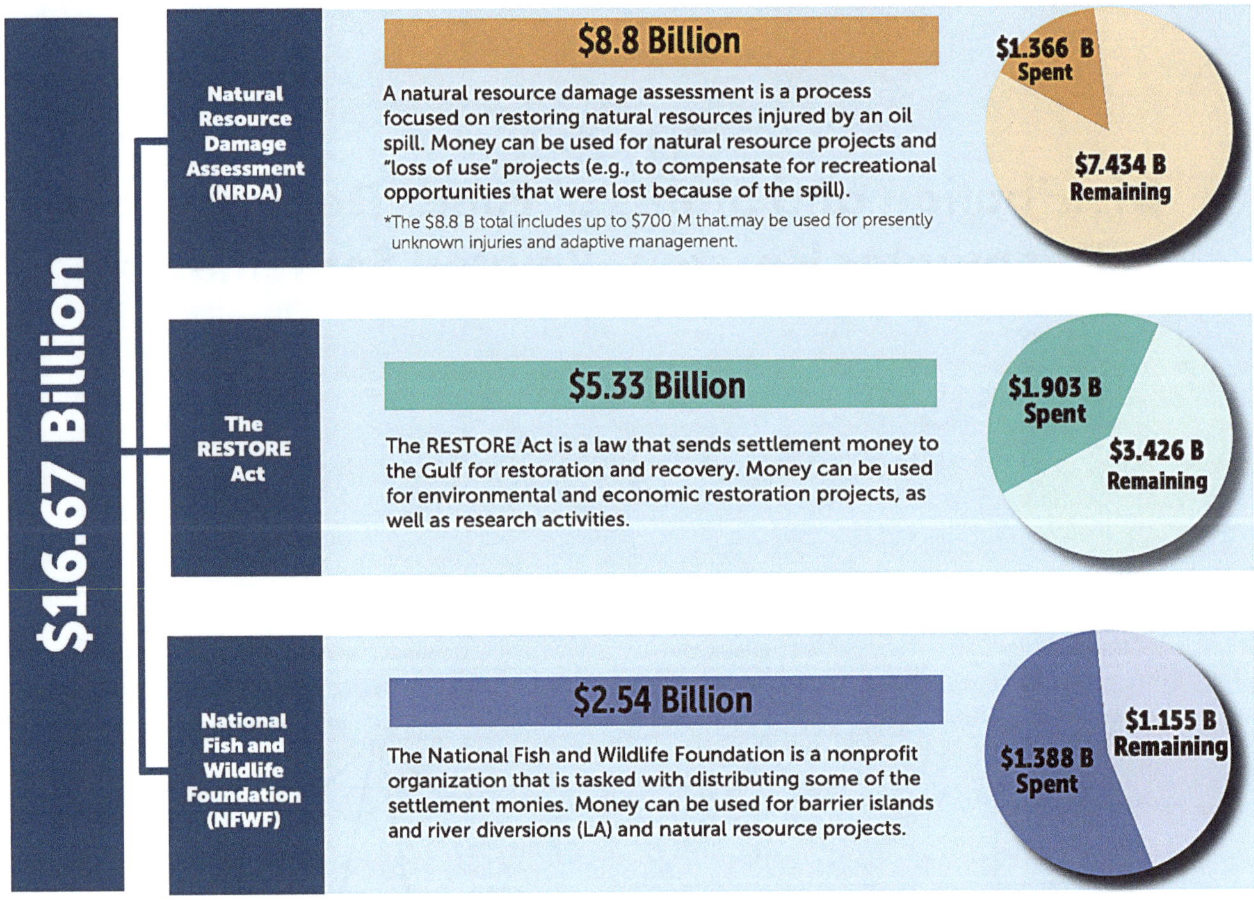

FIGURE A.2. Amount of Deepwater Horizon settlement and penalty funds spent and remaining as of April 2020. "Spent" means money that has already been spent on or designated for projects, programs, and planning. SOURCE: Environmental Law Institute Gulf Coast Recovery & Restoration: 10-Year Review. Available at http://eli-ocean.org/wp-content/blogs.dir/2/files/Gulf-Restoration-Recovery-10-Year-Review.pdf.

Appendix B
Committee Member and Staff Biographies

Committee Member Biographies

Holly Greening was both Executive Director and Senior Scientist of the Tampa Bay Estuary Program (TBEP), where she oversaw a unique federal, state, and local partnership dedicated to the preservation and restoration of Florida's largest open-water estuary. She managed TBEP's varied technical and public outreach efforts and served as the chief liaison between the program and the elected officials, scientists, regulators, and citizens that served on its various committees. Also, through her role as TBEP's Executive Director, she facilitated the development of Tampa Bay's successful nutrient management and seagrass recovery strategy. She has served on the Governing Board of the Estuarine Research Federation (now Coastal and Estuarine Research Federation) and four National Academies committees on coastal issues: the Committee on Causes and Management of Coastal Eutrophication, the Committee on Evaluation of NOAA's Sectoral Applications Research Program to Provide Climate Change Information to Resource Managers, the Committee on National Needs in Coastal Mapping and Charting, and the Committee on the Evaluation of Chesapeake Bay Program Implementation for Nutrient Reduction to Improve Water Quality. She also served as a member of the National Academies' Ocean Studies Board from 2005 to 2007. Upon retiring from TBEP after 27 years in 2018, Ms. Greening cofounded CoastWise Partners to provide volunteer assistance to coastal and watershed programs nationwide and internationally. In 1980, she received an M.S. in marine ecology from Florida State University.

Walter Boynton is a Professor Emeritus at the Chesapeake Biological Laboratory, University of Maryland Center for Environmental Science. Dr. Boynton is a coastal and estuarine ecosystem ecologist and his research spans sediment biogeochemistry, eutrophication, seagrass ecology, and coastal and estuarine restoration, among other topics. His approach has a strong synthesis and modeling emphasis, and he is especially interested in working with decision makers. Dr. Boynton has served on many local, regional, and national boards, including EPA Science Advisory Board panels that reviewed the hypoxic zone in the Gulf of Mexico and nutrient criteria in Florida. He has also worked with the U.S. Department of Justice on Gulf of Mexico oil spill issues. He is the recipient of several awards, including the Odum Award for Lifetime Achievement from the Coastal and Estuarine Research Federation and the Ruth Patrick Award from the Association for the Sciences of Limnology and Oceanography and the Mathias Medal from Sea Grant and the Chesapeake Research Consortium. He is also a past president of the Coastal and Estuarine Research Federation. He received a B.S. in biology from Springfield College, a M.S. in marine science from the University of North Carolina Chapel Hill, and a Ph.D. in environmental engineering from the University of Florida.

Bethany A. Carl Kraft is a Coastal Scientist and Senior Program Manager at Volkert, Inc. Ms. Kraft's expertise lies at the intersection of science and policy. She provides support for a wide range of ecosystem restoration projects across the Gulf Coast, including serving as ecology lead for restoration activities in Louisiana, Texas, Alabama, and Florida. She also supports clients' Deepwater Horizon restoration programs, working on projects for myriad resources including oysters, birds, wetlands, marine mammals, and sea turtles. Her current projects include marsh creation, watershed management planning, living shorelines, stream restoration and monitoring and adaptive management. Previously she was the Director of the Gulf Restoration Program at the Ocean Conservancy, where she led an interdisciplinary team to secure a science-based and community-supported restoration and ecosystem enhancement effort in the Gulf of Mexico. She also served a 1-year detail to the Gulf Coast Ecosystem Restoration Council (RESTORE Council) as the Director of External Affairs and previously served as the Executive Director at the Alabama Coastal Foundation, where she was active in Alabama's *Deepwater Horizon* oil spill response. She began her conservation career in Texas with the Texas Commission on Environmental Quality (TCEQ). Ms. Kraft holds an undergraduate degree in political science from Texas A&M University, a Graduate Certificate in coastal sciences from the University of New Orleans, and an M.S. in forest resources and conservation from the University of Florida.

Heida Diefenderfer is a Senior Research Scientist at the Pacific Northwest National Laboratory and a Faculty Fellow at the University of Washington. Dr. Diefenderfer is an ecologist with expertise in ecological restoration. Her research focus is on riverine and coastal ecosystems, including swamps, marshes, and submerged aquatic vegetation. She has a particular interest in multidisciplinary work that encompasses a range of temporal and spatial scales, including the watershed, riparian zone, and coastal nearshore. Dr. Diefenderfer's expertise in these systems includes aspects of ecohydrology, ecological engineering, geomorphology, landscape ecology, systems analysis and synthesis, cultural studies, and sustainability. In her present role, she leads applied research advancing ecological restoration methods for endangered species recovery, understanding of the cumulative effects of large-scale ecosystem restoration, spatial planning and modeling of wetland evolution associated with coastal resilience, blue carbon and climate-adaptation studies, and impacts-assessment for energy infrastructure. She has served on several technical advisory committees and project teams in the Pacific Northwest and elsewhere, including roles as an appointee of the Washington State Commissioner of Public Lands to the WA Natural Heritage Advisory Council, and as a system expert for the Louisiana Coastal Protection and Restoration Authority's adaptive management of large-scale ecosystem restoration. Dr. Diefenderfer has served as a reviewer for the Academies' Gulf Research Program and participated in the Academies' Keck Futures Initiative workshop on ecosystem services, a continuing research interest. Dr. Diefenderfer received a B.A. in biology with a thesis in landscape ecology from Reed College, a B.A. in liberal arts with emphasis on Native American studies from The Evergreen State College, an M.A. in English from Western Washington University, and a Ph.D. in ecosystem analysis from the College of Forest Resources.

Albert George was the first Director of Conservation for the South Carolina Aquarium in Charleston, South Carolina, and now serves as an advisor to the aquarium. His interests and expertise span marine science, climate change, public policy, social equity, and technology. As the Director of Conservation, he worked with policymakers and citizens to prepare for sea level rise and the impacts of climate change; his efforts included town halls, production of a documentary, and the development of a data visualization tool. Mr. George also participated in a working group at The Nature Conservancy on salt marshes and sea level rise, including along the coastal Gulf of Mexico. He is also a founder and has been the coordinator of the Georgia Green Economy Summit for the past decade. Previously, he worked on conservation issues in the Brazil Amazon Rainforest, served as the Director of Education at the Georgia Aquarium, consulted for Booz Allen Hamilton, worked as the Director of the Packard Scholar Program at Morehouse College, and was an education consultant for Stanford University; in addition, he was also an NIH Marc Fellow at Yale University, among other positions. Mr. George has a B.S. in marine biology from Savannah State University, a Master of Public Policy from the John F. Kennedy School of Government at Harvard University, and an M.S. in HTS/technology management from the Georgia Institute of Technology.

Kenneth L. Heck, Jr. is a Senior Scientist Emeritus at the Dauphin Island Sea Lab and Professor Emeritus at the University of South Alabama School of Marine Science. Dr. Heck is a marine ecologist whose research has focused on plant–animal interactions in coastal waters, with an emphasis on seagrass-dominated ecosystems. Recent efforts have focused on restoring northern Gulf of Mexico seagrass meadows and oyster reefs. He has more than four decades of experience on the U.S. Atlantic and Gulf coasts, and in Central America, Europe, and Australia. Dr. Heck has edited two volumes of scholarly works, coedited a special issue of the journal *Estuaries*, and published nearly 200 peer-reviewed articles. He has held editorial positions at the journals *Systematic Zoology*, *Estuaries and Coasts*, *Gulf of Mexico Science*, and *Marine Ecology Progress Series*. In addition, he regularly serves on advisory and review panels for state and federal agencies, including the Scientific Advisory Committees for the Mobile Bay and Pensacola/Perdido Bays National Estuary Programs, and EPA and NOAA panels concerning nutrient criteria in Florida waters and the effects of sea level rise on the northern Gulf of Mexico. Dr. Heck also served as a consultant to NOAA on the effects of the *Deepwater Horizon* disaster on Gulf of Mexico seagrass meadows, and on a National Research Council panel charged with developing monitoring and evaluations protocols for on-going restoration activities in the Gulf of Mexico. He is a past president of the Coastal and Estuarine Research Federation (CERF), a scientific organization of some 1,100 marine and estuarine scientists and is a corecipient of CERF's Odum Award for Lifetime Achievement. Dr. Heck received a B.S. from the University of West Florida and M.S. and Ph.D. degrees from Florida State University.

Barbara A. Kleiss is a Research Professor and Coordinator of the River Science and Engineering Certificate Program at Tulane University's School of Science and Engineering. Her research is focused on the rivers and wetlands in the lower Mississippi River valley. Dr. Kleiss's current interests include developing an improved understanding of the functions of the residual Mississippi River floodplains, further understanding sediment dynamics in the river and their measurement, and developing programs by which principles of river science and engineering can be more readily conveyed to river management professionals across the country. Her research has included work on sediment deposition and nitrogen dynamics in bottomland hardwood wetlands in Mississippi River tributaries; water chemistry, ecology, and groundwater assessment of over 40 rivers systems in the Mississippi Embayment; participated in the development of Level IV ecoregions for the lower Mississippi Valley; and assessment of the efficacy of river diversions. She has also been involved in creating, developing, and directing large interdisciplinary research programs associated with the Mississippi River and its delta as the Chief of the U.S. Geological Survey's National Water Quality Assessment Program's Mississippi Embayment project and the Director of both the Louisiana Coastal Area program and the U.S. Army Corps of Engineer's Mississippi River Science and Technology program. She earned a B.S. in biology from Spring Hill College, an M.S. in biology from the University of Southern Mississippi, and a Ph.D. in oceanography and coastal sciences from Louisiana State University.

Catherine L. Kling is the Tisch University Professor of Environmental, Energy, and Resource Economics in the Dyson School of Applied Economics at Cornell University and the Faculty Director of the Atkinson Center for a Sustainable Future. She specializes in the economic valuation of ecosystem services and the integrated assessment modeling for water quality modeling. Dr. Kling currently chairs the National Academies' Water Science and Technology Board and is a member of the *PNAS* editorial board. She has been a member of nine Academies study committees, including several focused on water resources and agricultural issues. She served as president of the Association of Environmental and Resource Economists, held editorial positions at 10 economics journals, and has published over 100 journal articles and book chapters. She is currently the editor of the *Review of Environmental Economics and Policy*. She is an elected Fellow of the Association of Environmental and Resources Economists, the Agricultural & Applied Economics Association, and the American Association for the Advancement of Science. She is also a University Fellow at Resources for the Future, a member of the National Academy of Sciences, and served for 10 years on EPA's Science Advisory Board. She received her B.A. in business and economics from the University of Iowa and Ph.D. in economics from the University of Maryland, College Park.

Larry McKinney is the Chair for Gulf Strategies at Harte Research Institute for Gulf of Mexico Studies at Texas A&M University, Corpus Christi. In his current role, Dr. McKinney works with an interdisciplinary team that integrates science, policy, and socioeconomic expertise for an economically and environmentally sustainable Gulf of Mexico. He has over 50 years of experience working in the Gulf of Mexico region as both a researcher and resource manager. His research interests include fisheries, benthic ecology, marsh restoration, and assessing ecosystem change. Dr. McKinney has led or chaired numerous science and policy efforts in the Gulf of Mexico region, including the State of the Gulf Summits, the Texas OneGulf consortium, the Gulf of Mexico University Research Collaborative, the Texas Sea Grant Advisory Committee, and the EPA Science Advisory Committee for the Gulf of Mexico. He has also served on several boards, including those of the Gulf of Mexico University Research Consortium, Texas State Audubon Society, and Texas State Aquarium. He is a member of the Texas Academy of Science, where he is a past president. In 2019, he was invited to present testimony to the President's Ocean Policy Taskforce. Dr. McKinney has a B.S. in zoology and a Ph.D. in biology from Texas A&M University.

Deepak R. Mishra is the Merle C. Prunty, Jr. Professor in the Department of Geography at the University of Georgia, Athens (UGA). He is also the Director of UGA's Small Satellite Research Lab (SSRL), which is focused on designing and building CubeSats for environmental remote sensing applications. He is a faculty fellow of UGA's AI Institute, faculty member of UGA Marine Institute, and affiliate director of GA Space Grant Consortium. Dr. Mishra's research focuses on studying inland waters and coastal ecosystems using a wide array of geospatial technologies, from ground-based sensor networks to satellite-based modeling and mapping. He combines field-based remote sensing with satellite remote sensing to model and map water quality and vegetation biophysical characteristics of coastal systems. His current projects include developing a satellite-based decision-support tool for coastal salt marsh conservation and restoration in the southeastern United States; modeling and mapping cyanobacterial harmful algal blooms dynamics in inland waters; combining Eddy Flux tower and satellite-based data to model and predict gross primary productivity and carbon sequestration potential of wetlands; and developing the cyberinfrastructure for monitoring coastal marshes and inland water quality. Dr. Mishra has served in editorial positions at several journals, including ISPRS Journal of Photogrammetry and Remote Sensing, Remote Sensing, and GIScience and Remote Sensing. He has been a member of many advisory committees, technical and review panels, and research groups. He has served as a Fulbright Specialist and investigator on international research projects in countries such as India, Brazil, and Uruguay. He received his M.Sc. in earth sciences from Pondicherry University, M. Tech. in civil engineering from the Indian Institute of Technology, and Ph.D. in natural resources from the University of Nebraska, Lincoln.

Staff Biographies

Thelma Cox is a program coordinator with the Gulf Research Program. She provides administrative support for three consensus studies aimed to document progress toward ensuring an outcome of a safe, healthy, and resilient Gulf of Mexico over time in partial fulfillment of the charge of the Gulf Research Program (GRP). Prior to joining GRP, she was a program coordinator for the Board on Behavioral, Cognitive, and Sensory Sciences in the Division of Behavioral and Social Sciences and Education. She also served as an administrative assistant for various boards within the formerly known Institute of Medicine (IOM) and the program unit of Health and Medicine Division. Since joining the National Academy of Sciences in 1986, she has worked on a diversity of activities and provided administrative support on numerous committees, roundtables, and forums. She is a recipient of the National Research Council Recognition Award and three

IOM Staff Achievement Awards. She received an Associate of Science in Business Administration (ASBA) degree from Averett University.

Deborah Glickson is the Director of the Board on Earth Sciences and Resources and the Water Science and Technology Board at the National Academies of Sciences, Engineering, and Medicine. Dr. Glickson also directs the standing Committee on Solid Earth Geophysics. Since joining the National Academies staff, she has worked on 17 consensus studies and workshops, including understanding the long-term evolution of the coupled natural–human system on the Gulf Coast for the GRP; decadal surveys on Earth and ocean science for the National Science Foundation, as well as next-generation Earth systems sciences; future water priorities and assuring laboratory data quality for the U.S. Geological Survey, and studies on marine hydrokinetic energy, methane hydrates, coal mining and human health, and geoscience education. Dr. Glickson received an M.S. in geology from Vanderbilt University and a Ph.D. in oceanography from the University of Washington. Her doctoral research focused on magmatic and tectonic contributions to mid-ocean ridge evolution and hydrothermal activity at the Endeavour Segment of the Juan de Fuca Ridge. After finishing her Ph.D., Dr. Glickson was a NOAA John A. Knauss Marine Policy Fellow and worked on coastal and ocean policy and legislation in the U.S. Senate. She was also the Associate Director of the Cooperative Institute for Ocean Exploration, Research, and Technology at Florida Atlantic University-Harbor Branch Oceanographic Institution.

Megan May was an associate program officer with the Ocean Studies Board from April 2020 to January 2022. During her time with the Ocean Studies Board she worked on consensus studies for environmental trends in the Gulf of Mexico and ocean plastics. Previous projects Megan has worked on include a workshop to identify community-driven science for NSF's P2C2 program, a consensus study for oil in the sea, and activities related to the U.S. National Committee for the Decade of Ocean Sciences. Prior to working for the National Academies, Megan was a NOAA John A. Knauss Marine Policy Fellow in the United States Senate in the Office of Tammy Baldwin, where she worked on the Agriculture, Environment, and Natural Resources portfolio. Megan also taught at Bard College for their intensive Citizen Science course. Megan has a B.A in biology from DePauw University, a certificate in science, technology, and policy from MIT, and a Ph.D. in marine environmental microbiology from MIT and Woods Hole Oceanographic Institution Joint Program in Oceanography.

Laura Windecker is a program officer with the Environmental Protection and Stewardship Board at the Gulf Research Program. Since joining the National Academies of Sciences, Engineering, and Medicine in 2016, she has worked on all aspects of grants management for the Gulf Research Program, including developing funding opportunities and managing the program's environmental grants portfolio. Prior to joining the National Academies, she was a NOAA John A. Knauss Marine Policy Fellow working as Advisor to the Director of the U.S. Committee on the Marine Transportation System, a federal interagency committee, where she served as project manager of a Report to Congress. Laura has a Ph.D. in marine science from the University of California, Santa Barbara, where her research focused marine phytoplankton and their role in global carbon dynamics. Laura also has an M.S. in oceanography from the Graduate School of Oceanography at the University of Rhode Island, and an A.B. in physics and marine science from Bowdoin College. Her first oceanographic research cruise was as an undergraduate aboard the SSV Westward with Sea Education Association.

Appendix C

People Who Provided Input to the Committee

Meeting 1 (August 12, 2020)

- Laura Bowie, Gulf of Mexico Alliance
- Greg Grandy, Coastal Protection and Restoration Authority of Louisiana
- David Green, Texas General Land Office
- Jessica Henkel, Gulf Coast Ecosystem Restoration Council
- Amy Hunter, Alabama Department of Conservation and Natural Resources
- Gareth Leonard, Florida Fish and Wildlife Conservation Commission
- Paul Mickle, Mississippi Department of Marine Resources
- Jonathan Porthouse, National Fish and Wildlife Foundation
- Lisa Robertson, Florida Department of Environmental Protection
- Buck Sutter, Gulf Coast Ecosystem Restoration Council

Meeting 2 (September 29, 2020)

- Barry Keim, Louisiana State University
- Kirk Rogers, U.S. Geological Survey
- Torbjörn Törnqvist, Tulane University

Meeting 3 (November 9-10 & 16, 2020)

- Mandy Karnauskas, National Oceanic and Atmospheric Administration
- Michael Osland, U.S. Geological Survey
- Matt Posner, Pensacola and Perdido Bays Estuary Program
- Nancy Rabalais, Louisiana State University
- Whitney Scheffel, Pensacola and Perdido Bays Estuary Program
- Gregory Steyer, U.S. Geological Survey
- Bob Stokes, Galveston Bay Foundation
- Roberta Swann, Mobile Bay Estuary Program

Meeting 4 (December 14, 2020)

- Robert Spies, Exxon Valdez Oil Spill Trustee Council
- Christopher Swarzenski, U.S. Geological Survey